T0211576

Messier's Nebulae and Star Clusters

The Practical Astronomy Handbooks are a new concept in publishing for amateur and leisure astronomy. These books are for active amateurs who want to get the very best out of their telescopes and who want to make productive observations and new discoveries. The emphasis is strongly practical: what equipment is needed, how to use it, what to observe, and how to record observations in a way that will be useful to others. Each title in the series will be devoted either to the techniques used for a particular class of object, for example observing the Moon or variable stars, or to the application of a technique, for example the use of a new detectors, to amateur astronomy in general. The series will build into an indispensable library of practical information for all active observers.

Harold Hill *A Portfolio of Lunar Drawings*

Astronomical drawing still has an important place alongside photography in the same way that photography has not supplanted artist in the fields of botany and ornithology, for example. Indeed, since astronomical images tend to shimmer because of movements in the Earth's atmosphere, even under the best possible conditions, drawings constructed by an artist who takes advantage of the fleeting moments of perfect vision are often more detailed than photographs.

   No one can fail to be impressed by the beauty and artistry of this work and, to the initiated, the accuracy and attention to detail is remarkable. This is a book for astronomers amateur and professional alike and for those who would simply like to get to know the moon a little better.

Messier's Nebulae and Star Clusters

KENNETH GLYN JONES

Second edition

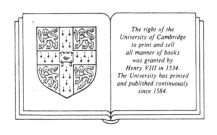

*The right of the
University of Cambridge
to print and sell
all manner of books
was granted by
Henry VIII in 1534.
The University has printed
and published continuously
since 1584.*

CAMBRIDGE UNIVERSITY PRESS

Cambridge
New York   Port Chester
Melbourne   Sydney

CAMBRIDGE UNIVERSITY PRESS
Cambridge, New York, Melbourne, Madrid, Cape Town, Singapore, São Paulo

Cambridge University Press
The Edinburgh Building, Cambridge CB2 8RU, UK

Published in the United States of America by Cambridge University Press, New York

www.cambridge.org
Information on this title: www.cambridge.org/9780521370790

First published by Faber & Faber Ltd 1968
Second edition published by Cambridge University Press 1991
This digitally printed version 2008

A catalogue record for this publication is available from the British Library

Library of Congress Cataloguing in Publication data

Jones, Kenneth Glyn.
Messier's nebulae and star clusters/Kenneth Glyn Jones.–2nd ed.
    p.   cm.
Includes bibliographical references.
1. Nebulae – Catalogs.   2. Stars – Clusters – Catalogs.   I. Title.
QB851.06   1990   89–22199
523.1′125–dc19   CIP

ISBN 978-0-521-37079-0 hardback
ISBN 978-0-521-05849-0 paperback

For Brenda

Contents

List of plates

List of tables and figures

Tables

Figures

Preface to the second edition

The original edition of this book was written more than twenty years ago. During that interval new technology has provided astronomers, both amateur and professional, with at least a triple increase in telescope power and efficiency. This by itself makes the observation of the objects in Messier's catalogue a much easier exercise, and, if one can afford such luxuries as digital setting circles or computer-controlled direction, the task of 'bagging' them all may be reduced to a relatively facile exploit. Indeed, more than a few *virtuosi* have demonstrated that it is possible to log the complete list during a single night's observing session – a feat which involves both good planning and good luck in finding clear skies on the optimal date.

This book, however, is addressed to the observer who is not yet tempted by record-breaking exploits but who wishes to make a first acquaintance with the magnificent treasures of Messier's collection, and then to learn how to know them more intimately. For this purpose a distinctly leisurely approach is likely to prove the more satisfying, so that new skills of observing technique and a wider grasp of perception may gradually be acquired. Eventually, a more preferable objective than the one-night Messier's 'marathon' may be the demonstration of one's ability to pick up all the above-horizon objects in Messier's list at any one time – without the aid of this present guide or any other – to show that one knows them all by heart as it were. The feat may not impress the sensationalists but it might receive an approving nod or two from the discerning.

Additional material in this edition includes an important historical amendment to take account of the previously unknown contribution of G. B. Hodierna as a pioneer observer of nebulae a century before Messier's time. In twenty years, too, a great deal of new information has been gathered which concerns the classification and physical nature of many star clusters and nebulae, and those of Messier's catalogue (being relatively bright objects) have benefited greatly from this process. In the case of objects such as M 1, M 31 and M 82, research papers are voluminous and there are very few of the remainder which have escaped a more modern scrutiny. I have endeavoured to incorporate as much of this additional data as time and space have allowed, but there is little doubt that new and important information will continue to be added to this rich storehouse of astronomical splendours which first engaged Charles Messier's attention more than two hundred years ago.

Acknowledgements to the second edition

I would like to take this opportunity to express my gratitude to all those friends and colleagues who have rendered invaluable help in the process of revision. To Rosemary Naylor and Dr Simon Mitton I offer my appreciation of the vital initial stimulus to undertake the work. To Robert Argyle of the RGO my thanks are due for a bountiful supply of data on double stars and supernovae, and I am deeply grateful to Peter Hingley of the RAS and Webb Society colleagues Edmund Barker, Steve Hynes, John Isles, Brian McInnerny, Don Miles, Ronald Morales, Colin Pither, Mike Swan and Martin Yates, who have so generously put their various expert skills at my disposal. Finally, to my wife I admit the debt which any married writer owes, and add to this my thanks, not only for assisting in the task of proof-reading, but even professing to enjoy it.

Preface to the first edition

Sooner or later every amateur astronomer turns his newly-acquired telescope on to Messier 31 (The Great Nebula in Andromeda) and, comparing what he sees with the illustration published in almost any popular book on astronomy, experiences a considerable disappointment. The sight of the Orion Nebula (M 42) may revive his spirits a little but after searching the atlas positions of other photographic spectaculars such as the 'Crab' Nebula (M 1) or the Spiral Nebula in Triangulum (M 33) and, being rewarded often with nothing but blank sky or a few faint, uninteresting stars, he gives up looking for Messier Objects as a bad job.

This is a pity, for almost all these objects can be found and seen satisfactorily in a 3-inch refractor or a 4-inch reflector. The amateur of today, in fact, is likely to have much better equipment than the best of that available to Messier himself and although one cannot hope to see all the detail that appears in the beautiful photographs taken with such instruments as the 120-inch reflector of the Lick Observatory, nevertheless many of these objects are really splendid sights in a 6-inch telescope. At 8 inches aperture (a common size for amateurs today), detail can be made out even in some of the distant galaxies of the Virgo group where Messier was able to see no more than a faint 'nebula without a star'.

It must, however, be admitted at the start that there is little scope for 'serious' work by the amateur among the Messier Objects: the study of remote star clusters and galaxies requires complex and expensive equipment and the co-operative work of many expert observatory staff. Fortunately one does not *have* to devote one's hobby to serious ends and in this case, a systematic search for, and careful observation of all the objects in Messier's famous catalogue can be a very pleasing and instructive task. By this means the amateur is introduced to a wide variety of important and historical objects and is given valuable practice in finding his way around the heavens: he will also be rewarded with a view of some of the finest sights to be seen in the telescope.

The fact that Messier's list is complete at 102 objects also provides the observer with a limited objective – even if the limit is arbitrary – and, having observed them all, the experience gained will serve him well when he decides to turn to more serious and constructive study in fields where the amateur has important scope, such as that of the variable stars or the nearer planets.

Messier observed all of the objects in his catalogue from the latitude of Paris which is close to 49° N: observers in the U.S.A (the whole of which, except Alaska, is south of this parallel) will thus have a view at least as good as Messier's. British observers are less favourably placed but all except about eight of the objects can readily be seen from latitudes as high as 55° N.

The object of this book is to make the search for the Messier Objects as easy and as systematic as possible and to present as much information, both historical and modern, as the amateur may need in an easily accessible form.

Many amateur astronomical societies, especially in the United States and Canada, have formed 'Messier Clubs' in which younger members are encouraged and assisted to find and observe all the objects in Messier's list. In Great Britain, too, The Webb Society, founded in 1967 and named after Rev T. W. Webb (1807–85), unites amateurs whose interests lie in the observation of nebulae, star clusters and double stars. It is hoped that this book will help to supplement these activities.

It is assumed that the observer has a telescope of at least 3 inches aperture which he can set reasonably accurately in the meridian (i.e. north–south) and that he can identify the brighter stars and major constellations.

No special equipment is required, but suggestions for choice of telescope, observing site, 'optional extras', etc., are discussed in Chapter 3.

Adequate star maps and identification diagrams to assist in finding and identifying each object are provided, but the possession of a star atlas such as *Norton's* may well prove a useful addition for general viewing.

Telescope sketches are included for each object with the field and orientation indicated: these are not meant to be accurate astronomical drawings but are for identification purposes only. All were made by the author at the telescope, using a crude 4-inch reflector of 3 feet focal length to obtain an approximation to Messier's own view and an 8-inch Newtonian of entirely 'home' construction to obtain greater detail.

The information given for each object has been gathered from many sources, most of which are acknowledged in the Bibliography. In every case, Messier's own description has been quoted, the source being an original volume of the *Connaissance des Temps* for 1784.

Acknowledgements to the first edition

Most of the sources from which information has been gathered are listed in the Bibliography in Appendix 6 and further references are given in Notes on Sources in Chapter 5.

The historical notes published by Mrs Helen Sawyer Hogg in the *Journal* of the R.A.S. of Canada in 1947 (Vol. XLI, Nos. 6, 7, 8), although I did not encounter them until I had covered much the same ground myself, provided much valuable information regarding catalogues of nebulous objects of the 18th century. I am also grateful to Mrs Hogg for the data taken from the valuable catalogue of clusters in the *Handbuch der Physik*.

The several articles by Owen Gingerich in *Sky and Telescope* concerning Messier and his catalogue – and especially on the 'missing' Messier Objects – furnished much material and many useful clues for further research.

Many other volumes of *Sky and Telescope* were consulted, particularly the *Deep Sky Wonders* column of Walter Scott Houston, and it could be said that almost every issue of that deservedly popular periodical provided something of value to the author.

For personal assistance I would like to express my thanks to the Librarian of the British Astronomical Association, Mr J. L. White, F.R.A.S., and also to Mrs H. E. Cox and Miss D. M. Farmer for their most helpful co-operation. Similar valuable assistance, too, was generously given by Mrs D. Harrington and Dr Madison of the Library of the Royal Astronomical Society at Burlington House, London.

My special gratitude is due to Mr D. Overbeek of Germiston, Transvaal, South Africa, who kindly allowed me the use of the excellent observing facilities at his observatory in order to view the more southerly objects. For generous assistance in providing photographs of many of the Messier Objects, I would like to extend my thanks to Mr Christos Papadopoulos of Johannesburg and to Dr Hans Vehrenberg of Dusseldorf, and for friendly encouragement, guidance and help, I am deeply grateful to Miss Olwen Hedley, Mr John Anthony Cope, Mr E. Arthur Cunningham, Mr F. H. Smith and Mr H. N. D. Wright, F.R.A.S.

Finally, to my wife are due not only an acknowledgement of thanks for sharing the labours of proofreading, but also a declaration of my appreciation of a long forbearance.

Introduction

Charles Messier's famous catalogue of *Nebulae and Star-Clusters*, completed in 1781, was far from being the first of its kind. The compiling of lists seems to have been a compulsive human occupation from the beginning of history and although the visible stars may have, at first, appeared uncountable and the scarce-visible nebulae hardly worth counting, this would have been more of a challenge than a deterrent.

The first star catalogue known to have been compiled was that of Hipparchus who observed at Rhodes between 146 and 127 BC. According to Pliny, this work was prompted by the appearance of a 'nova' or new star in the constellation of Scorpius in 134 BC. Hipparchus was the first to position the stars by means of celestial co-ordinates and he gave latitudes and longitudes – based on the ecliptic – for some 850 stars, arranged into six magnitudes. He also mentioned two nebulous objects: they were the 'Beehive' or 'Praesepe' cluster in Cancer (M 44) which Hipparchus referred to as 'a little cloud' and the double star cluster in the hilt of Perseus' sword (*h* and χ *Persei* – not in Messier's list) which he called 'a cloudy spot'.

The next star catalogue – and the first of which we have a complete record – was that of Ptolemy who observed at Alexandria during the years AD 127 to 151. Ptolemy's most famous contribution to astronomy was his *Great Syntaxas* which, translated into Arabic under the title of the *Almagest* became a work of the greatest authority throughout the whole mediaeval world. Ptolemy's star catalogue was probably based upon Hipparchus' earlier list but was enlarged to more than 1000 entries and included seven nebulous objects. Two of these (M 44 and *h* and χ *Persei*) were taken from Hipparchus but the third was a new addition, 'a nebula behind the sting of Scorpius', a bright galactic cluster which Messier was to include in his list as M 7. A fourth object was the constellation of Coma Berenices, composed of many faint stars which form a recognised cluster but not included by Messier. The three remaining objects are neither true nebulae nor clusters but merely accidental groupings of faint stars which looked 'nebulous' to the naked eye: they may be referred to as 'asterisms'.

The first true 'nebula', as distinct from a star cluster, was included by the Persian astronomer, Al Sufi, in his *Book of the Fixed Stars*, compiled for the epoch AD 964. This was the Great Nebula in Andromeda (M 31) which Al Sufi mentioned as a familiar object in his time: it is not known who was the first to observe this famous nebula. Al Sufi also included in his catalogue 'a nebulous star' one cubit (2° 20') North of the star δ *Navis*. This object is almost certainly the star o *Velorum* which is surrounded by several fainter stars and borders on the bright galactic cluster, IC 2391. Al Sufi's list contains another seven nebulous objects, six of them being identical with those in Ptolemy's *Almagest* with the addition of an 'asterism' in Vulpecula, sometimes referred to as 'Briocchi's Cluster'.

In the 15th century, Ulugh Begh, a grandson of Tamerlane, founded an observatory at Samarkand and also compiled a star catalogue: this list, however, contained only the same nebulous objects as in Ptolemy's catalogue.

Tycho Brahe's catalogue of 777 stars, completed in 1590, was justly famed for its observational accuracy, but the seven nebulous objects it contained were, with the single exception of the 'Praesepe' cluster (M 44) all asterisms and have not found any place in modern catalogues of nebulae and star clusters. Strangely, Tycho did not mention the Andromeda Nebula (M 31) although he catalogued a star quite close to it.

The *Prodromus Astronomiae* of Hevelius, published posthumously in 1690 was the last star catalogue to be compiled without the aid of the telescope and although, among the 1500 star positions, he included 16 nebulous objects, only two of them, M 31 and M 44, are genuine nebulae or clusters, the remainder being no more than asterisms of one kind or another. Even so, one of these, in Ursa Major, was to be included in Messier's list as 'two faint stars near each other' (M 40).

The invention of the telescope in the early years of the seventeenth century brought about a revolution in astronomical observation. In the hands of Galileo the instrument was exploited to examine many nebulae more closely, and he soon revealed that most of them were capable of being resolved into clusters of faint stars. He discovered no new objects of this kind.

Apart from the discovery in 1610, by Pieresc, that the star θ *Orionis* was surrounded by faint stars and nebulosity, it seemed that no new nebulae gave themselves up to telescope scrutiny until 1665 when A. Ihle picked up a bright 'nebulous star' in Sagittarius – the first 'globular cluster', M 22. However, in 1984, the examination of a remarkable document, printed in Palermo, Sicily in 1654, has revealed that the hitherto obscure astronomer, G. B. Hodierna, using a small refractor with a fixed magnification of ×20, had observed more than forty objects, nineteen of which have been identified as true nebulae or star clusters. In addition to his re-discovery of the Andromeda Nebula (M 31) he was the first to observe six of the star clusters of Messier's list; M 6, M 36, M 37, M 38, M 41 and M 48. Another four objects: NGC 2362, NGC 6231, NGC 6530 and possibly NGC 2451 (not listed by Messier) were first discovered by him.

The first Astronomer Royal, John Flamsteed, published his great star catalogue, *Historia Coelestis Britannica* in 1712 and followed this with a revised and enlarged edition containing 2866 star positions in 1725. Flamsteed made several references to 'nebulae' or 'nebulous stars' and while some of these were well-known objects, i.e. the Coma Cluster, *h* and χ *Persei*, M 31, M 44 and the Orion Nebula (M 42), three others were newly discovered by Flamsteed himself. These are: the bright cluster with gaseous nebulosity in Sagittarius (M 8), a galactic cluster in Canis Major (M 41) and the cluster surrounding the star *12 Monocerotis* (not observed by Messier). The few remaining 'nebulous' objects in Flamsteed's catalogue are mere asterisms.

In the *Philosophical Transactions* of the Royal Society for 1715, Halley listed six 'luminous spots or patches': these were all true nebulae or clusters and comprised the globular cluster «ω» *Centauri*, discovered by Halley at St Helena in 1677, another globular, M 13, in Hercules, also discovered by Halley in 1714 and four objects known previously: M 11, M 22, M 31 and M 42.

A little later, in the *Philosophical Transactions* for 1733, William Derham (1657–1735), a Winchester clergyman and Fellow of the Royal Society, made a list of 16 'nebulous stars', 14 of which were extracted from Hevelius' star catalogue and the last two from Halley's catalogue of southern stars. Except for the first and last objects, M 31 and M 7, the remainder were none of them true nebulae nor clusters but either patches in the Milky Way or just groups of faint stars which looked 'nebulous' only to the naked eye. Derham's list, however, received a wide publication, being reprinted in the *Mémoirs* of the French Academy of Sciences in 1734 and was also included by the French scientist, de Maupertuis in his book *Discours sur la Figure des Astres* in 1742. Messier and many other later observers must have spent many fruitless hours searching for these nugatory or non-existent objects in Derham's list.

In about 1746, Philippe de Chéseaux (1718–51) gave the positions and descriptions of some true clusters and 'truly nebulous stars' which he had observed. About eight of these were original discoveries but his list, included in a letter to Réamur, remained unpublished and was largely unknown to most other observers.

The Abbé Nicholas Louis de la Caille (generally referred to as Lacaille) was a very important observer who, in a short life (1713–62) catalogued the positions of over 10 000 stars down to the seventh magnitude and made many other important contributions to astronomy. Working at the Cape of Good Hope from 1751 to 1752 he observed and described 42 nebulae and made an attempt at a classification of such objects in a list published in 1755.

This brings us to Charles Messier (1730–1817). Messier's interest in making a catalogue of nebulous objects was, at first, anyway, something of a negative one for he was primarily a comet observer. Between 1760 and 1798 he discovered 13 comets and his keenness for the scent prompted Louis XV to call him 'The ferret of comets'.

The occasion of Messier's first concern with nebulous objects was in Aug. 1758 when he was looking for a comet he had found some weeks before. He was observing near the star ζ *Tauri* when he came upon a hazy object looking very much like a faint comet. It soon became apparent, however, that this object was stationary among the stars and therefore was not a comet and Messier decided to make a catalogue of all such confusing objects within reach of his telescope. The object in Taurus was marked on his comet chart and later listed as M 1 but it was not in fact an original discovery by Messier for it had been seen before in 1731 by Bevis. The record of this observation however had subsequently been lost and was not known to Messier. The name 'Crab' Nebula by which it is now generally known was not given to it until 1846 when it was seen in Lord Rosse's great telescope and rather imaginatively described as having 'claw-like appendages' (see M 1).

Soon after this Messier became involved in the search for the return of the comet which had been predicted by Edmund Halley to reappear towards the end of 1758. Halley's comet was first detected by Palitzch in Germany on Dec.

25th–26th, 1758, but news of this did not reach Paris until much later and Messier made an independent discovery on Jan. 21st, 1759. For some reason, however, Delisle, Messier's superior, would not allow his assistant's observations to be published.

It was on the chart of Halley's comet that Messier in 1760 marked his second object, a globular cluster in Aquarius, describing it as 'resembling the nucleus of a comet'. This, too, had been seen before – by Maraldi in 1746 – also when looking for a comet.

In 1764 Messier began to make a more systematic search for objects for his catalogue and the work went on intermittently, together with comet discovery and other astronomical observations, throughout the years 1765–70 when he checked and reobserved many of the objects reported earlier by Hevelius, Halley and Derham, weeding out those he found to be missing or mere 'asterisms', i.e. small, line-of-sight groupings of faint stars not in actual clusters.

The first list was published in the *Mémoirs* of the French Royal Academy of Sciences for 1771 under the title of: *A Catalogue of Nebulae and Star Clusters discovered among the Fixed Stars above the horizon of Paris. Observed at the Observatory of the Navy with different instruments by M. Messier.*

The catalogue contained positions and descriptions of the objects numbered 1 to 45, but No. 40, an object from Hevelius and Derham – the position of which had been noted by de Maupertuis in his *Discours sur la Figure des Astres* in 1742 – was described by Messier to be only 'two faint stars near one another . . .' and no nebula was seen (M 40).

Immediately after his paper to the Academy, Messier started work on an additional list, observing four more clusters, Nos. 46, 47, 48 and 49 on Feb. 19th, 1771. Nos. 47 and 48 were not positioned with Messier's usual care and were to provide something of a puzzle for subsequent observers. Further objects were added during the course of his comet observations and by April 1780 he made up the number of objects to a total of sixty-eight and this list was published in the Almanac *Connaissance des Temps* for 1783. Two additional objects, M 69 and M 70, were discovered on Aug. 31st, 1780, and these were inserted, separately from the list, at the end of the same volume.

During the course of the following year many more objects were observed until the total stood at exactly 100 and a revised and extended catalogue was published in the *Connaissance des Temps* for 1784. However, a last minute inclusion was made of three additional objects discovered by a younger colleague of Messier, Pierre Méchain, bringing the total up to 103. Again, as will be discussed later, two further objects, Nos. 91 and 102, subsequently proved to be 'missing' and they are often omitted from Messier's list.

Méchain's contribution to the catalogue, now generally attributed to Messier alone, is often underestimated and sometimes ignored. He, too, was interested in comet work, discovering eight comets between 1781 and 1799, and during his observations came upon many new nebulous objects. These were generally passed on to Messier who checked and recorded them in his lists. It was Méchain too, who pointed out to Messier the large number of nebulae in the constellations Virgo and Coma Berenices where Messier had observed previously only three objects: Nos. 58, 59 and 60. More than twenty of the 103 objects in the final catalogue of 1784 were acknowledged by Messier as having been found by Méchain.

However, it is Messier's name which is generally remembered, although it is an odd turn of fate which has led to his fame resting upon a catalogue of objects he was really trying to avoid while the successful 'ferret of comets' has no comet which bears his name.

Several other lists of nebulae and clusters were compiled during Messier's time, notably by J. E. Bode at Berlin and J. G. Koehler at Dresden, both being published in the *Astronomisches Jahrbuch*; Bode's in 1777 and Koehler's in 1780. Koehler's list contained 20 objects with another three added later; most of them had been included in Messier's first catalogue but Koehler found several of them independently and also made the first discovery of M 59, M 60 and M 67 before Messier.

Bode's list was more ambitious: it contained 75 objects, to which two further objects were added later. He claimed it to be a *Complete Catalogue of Nebulae and Star-Clusters*, embracing those which had been found by him and other observers such as Flamsteed, Halley, Hevelius, Lacaille, Messier and others.

In comparison with Messier's catalogue, however, Bode's has many unsatisfactory features. In the first place he did not observe all the objects himself – in fact, he seems to have examined only about 24 of them; secondly, he included many of the worthless and misleading asterisms of Derham's list, adding one or two more of his own for good measure and, finally, the list contains more errors than should have been perpetrated by one who was soon to become the director of a major observatory. Bode did, however, discover five new nebulae, all of which were later included in Messier's final catalogue. Bode's list has been almost forgotten and if this was due to the superior and more enduring qualities of Messier's work, it cannot be said to have involved any injustice.

As might be expected, Messier's list was not the last catalogue of nebulous objects: with the increasing power and resolution of the new telescopes constructed by Herschel and later by the Earl of Rosse, more and more was to be revealed in the hitherto unfathomed depths of the heavens.

William Herschel (1738–1822) was one of the greatest astronomical observers of all time: he discovered the planet Uranus in 1781 and between 1786 and 1802 compiled three lists of nebulae and clusters containing more than 2500 entries. John Herschel (1792–1871) extended his father's work, discovering 525 new objects by 1825 and then, from Table Mountain near Capetown, he began a further survey of the southern hemisphere which produced another 1700. He returned to England in 1838 and then combined all the previous lists into one *General Catalogue* containing some 5000 entries. This was published in 1864 in the *Philosophical Transactions* of the Royal Society.

This catalogue was brought up to date in 1888 by J. L. E. Dreyer (1852–1926) and renamed the *New General Catalogue* (NGC) which, augmented by the first (1895) and second (1908) *Index Catalogues* (IC), gives a combined total of more than 13 000 entries and is the basic catalogue for nebulae and clusters still in use today.

All the Messier Objects are included in it and have NGC numbers except M 40 which, as mentioned earlier is an insignificant asterism, and M 45 (the Pleiades). Although M 73 too, is only an asterism (of four stars) it was included in Sir John Herschel's catalogue of 1864 under the number 4617 and later embraced in the NGC as No. 6994. M 25, a galactic cluster in Sagittarius, discovered by de Chéseaux in 1746, has a number in the *Index Catalogue* (IC 4725).

Messier's catalogue, as covered in this book, embraces all the objects, M 1 to M 103, excluding only M 102 and including seven additional objects. (The reasons for including M 47, M 48 and M 91, and adding M 104 to M 110 are given in Chapter 1.)

With the exception of the two unimportant asterisms of M 40 and M 73, the remaining objects have all proved to be very important subjects for astronomical study and there is hardly a work on stellar evolution, on the size and form of the Galaxy or of wider cosmology theory that does not mention some of the Messier Objects. (No doubt the title of, say, M 67 is brief and more easily remembered than the equivalent NGC 2682 and this may be one of the reasons why Messier's name has lasted, but the fact remains that his catalogue was the first accurate and systematic list of nebulae with any size and scope.)

The 109 objects may be categorised as follows:

Asterisms	2
Planetary Nebulae	4
Galactic Clusters	27
Extra-Galactic Systems	39
Supernova Remnant	1
Gaseous Nebulae	7
Globular Clusters	29

Each of these classes is described in more detail in Chapter 2.

The brightest and also the nearest of the Messier objects is M 45 – the Pleiades – magnitude 1.6; distance about 400 light-years. Messier included this object in his list probably as a 'make-weight' and because it had appeared in earlier catalogues. It is not likely that comet observers, who are notoriously eagle-eyed, would have confused this with a comet. It is not included in the NGC although it is a true cluster of associated stars.

The faintest object is M 76, a planetary nebula in Perseus, sometimes known as the 'Little Dumb-bell'. Its magnitude is generally given as 12.2 but several observers consider it brighter than this and it is not very difficult to pick up with a 6-inch in good conditions.

The most distant Messier Object is the face-on spiral galaxy in Cetus, M 77. This is at least 16 Megaparsecs (Mpc) – 52 million light-years – distant and may be as far as 20 Mpc – 65 million light-years. Its magnitude is 8.9 and, though small, it is not difficult to see, even in a 3-inch refractor or 6-inch reflector.

The best known Messier Object is almost certainly M 31 – the Great Nebula in Andromeda – its magnitude is generally given as 4.8 but some observers have put it as high as 3.5. It is easily visible to the naked eye, providing that the sky is clear. Although the eye sees only the central portion, even in a telescope, it actually has the largest apparent size of any heavenly body, covering a total area of nearly two square degrees or more than four times as large as the full Moon. As it is an external galaxy, distance about 680 kiloparsecs (kpc) – about 2.2 million light-years – it is usually considered to be the most distant object detectable by unaided vision. However, numerous observers, especially in the United States, have seen M 33, a galaxy in Triangulum, without any optical aid and this is slightly more distant at 720 kpc, about 2.35 million light-years.

However, the observer with keen eyesight and a well-developed competitive sense should try for M 64, the 'Black-Eye Nebula', a spiral galaxy in Coma Berenices, which is much further away, being about 13.5 Mpc or nearly 44 million light-years distant. The visual magnitude of M 64 has been quoted as high as 6.6, which makes it fractionally brighter than M 33 at 6.7 but this is almost certainly too optimistic a figure for no one has yet claimed it as a naked-eye object.

As to which are the finest objects to be seen in the telescope, this, of course is largely a subjective matter but certainly M 31 (with its companion M 32), and M 42, the Great Orion (gaseous) Nebula are superb objects. The finest planetary nebula is undoubtedly M 27 in Vulpecula: M 13 in Hercules is the largest and brightest of the globular clusters but M 5 in Serpens and M 22 in Sagittarius are close rivals. M 11 in Scutum and M 35 in Gemini are both fine galactic clusters but probably each observer has his own favourite among these objects. Of the spiral galaxies, probably M 51 in Canes Venatici and M 104 in Virgo are the most impressive.

As can be seen from Maps 1–6 the Messier Objects are not evenly distributed in the sky: the galaxies are most frequent near the *north galactic polar regions* with the densest concentration in the area 10 to 12 hours of R.A. and 5° to 15° N Dec. This region is shown to a larger scale in Map 7.

The galactic clusters appear to be more widely scattered but they are mostly distributed around the *galactic equator* and so are nearly all found in or close to the Milky Way.

The globular clusters also show a very distinct concentration, being very numerous in the constellations of Scorpius, Ophiuchus and Sagittarius, that is, towards the galactic centre which is situated in the latter constellation at about R.A. 17 h 45 min and 29° S Dec. This area also is shown in large scale in Map 8.

The nearest Messier Object to the galactic centre is the globular cluster M 62 which is a little less than 1 kpc – about 3000 light-years – from the galactic nucleus: the nucleus itself being about 8.2 kpc – nearly 27 000 light-years distant from the Sun.

Some of the objects in this part of the sky are the most difficult to observe from the British Isles, seven of them having declinations lower than 30° S and so not rising higher than 8° above the horizon for an observer at 52° N latitude.

The most southerly of all the objects is the galactic cluster M 7 which at 34°

48' S Dec. would have a maximum altitude of only 5° 15' even if viewed from the most southerly part of the mainland of Great Britain: The Lizard (lat. 49° 57' N). Given exceptional conditions this object might be visible from South Cornwall as it is a large cluster about 50' in diameter and has a total magnitude of about 3.5 but M 70 (Dec. 32° 21' S, mag. 9.6) and M 83 (Dec. 29° 37' S, mag. 10.1) present a real problem for British observers, especially as all these objects are visible only in the early summer when nights are at their shortest.

The latitude of Paris where Messier made all his observations is just below 49° N and the whole of the U.S.A. except, of course, Alaska, is south of this parallel so that U.S. observers, especially those in the southern states, have a distinct geographical advantage. However, those observers who have to live in higher latitudes may find some advantage in the fact that all the objects below 30° S Dec. are best placed for viewing during the months May, June and July. A small, portable telescope or a good pair of binoculars included in the summer vacation luggage may well enable the enthusiast to complete his Messier programme at leisure.

Many of the Messier Objects are also radio sources: M 1, the 'Crab' Nebula in Taurus, and M 82, an unusual irregular galaxy in Ursa Major, are both very active in the radio spectrum and their emissions have been studied extensively. This phenomenon is discussed in more detail in Chapter 2.

Table 1. *The Messier Objects in constellation*

ANDROMEDA M 31, M 32, M 110	OPHIUCHUS M 9, M 10, M 12, M 14,
AQUARIUS M 2, M 72, M 73	M 19, M 62, M 107
AURIGA M 36, M 37, M 38	ORION M 42, M 43, M 78
CANCER M 44, M 67	PEGASUS M 15
CANES VENATICI M 3, M 51,	PERSEUS M 34, M 76
M 63, M 94, M 106	PISCES M 74
CANIS MAJOR M 41	PUPPIS M 46, M 47, M 93
CAPRICORNUS M 30	SAGITTA M 71
CASSIOPEIA M 52, M 103	SAGITTARIUS M 8, M 17, M 18, M 20,
CETUS M 77	M 21, M 22, M 23, M 24, M 25, M 28,
COMA BERENICES M 53, M 64,	M 54, M 55, M 69, M 70, M 75
M 85, M 88, M 91, M 98, M 99,	SCORPIUS M 4, M 6, M 7, M 80
M 100	SCUTUM M 11, M 26
CYGNUS M 29, M 39	SERPENS M 5, M 16
GEMINI M 35	TAURUS M 1, M 45
HERCULES M 13, M 92	TRIANGULUM M 33
HYDRA M 48, M 68, M 83	URSA MAJOR M 40, M 81, M 82,
LEO M 65, M 66, M 95, M 96,	M 97, M 101, M 108, M 109
M 105	VIRGO M 49, M 58, M 59, M 60, M 61,
LEPUS M 79	M 84, M 86, M 87, M 89, M 90, M 104
LYRA M 56, M 57	VULPECULA M 27
MONOCEROS M 50	

1 Astronomical background to the Messier Objects: the 'missing' and 'additional' Objects

Although this book is not intended to cover much in the way of general descriptive astronomy, it may be useful at this point to give a brief account of the Universe as a whole so that the different types of objects that Messier observed can be fitted into the overall picture. In fact, many of the Messier Objects, being on the whole, easily visible, have played a large part in revealing much of our present knowledge of stellar evolution and general cosmology.

In Messier's day, it must be remembered, there was very little idea of the shape of the Galaxy, almost no realistic estimation of its size and no one had any conception of the existence of extra-galactic systems although – as we now know – 33 of the objects in his catalogue were in fact external galaxies.*

The Galaxy Our Sun is merely one of the 100 000 million stars of which the Galaxy is composed. If we could see our Galaxy from outside it would probably look very much like M 31, the Great Spiral in Andromeda, or the model galaxy illustrated in Figure 1, consisting of a bright, dense, central nucleus of stars

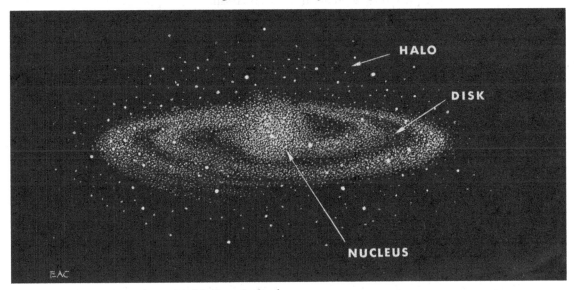

Fig. 1. Model of a spiral galaxy

* There had been some prophetic voices: one of the most remarkable being that of Christopher Wren who, in his inaugural address as professor of astronomy at Gresham College, London in 1657, thought that future astronomers might find 'the Galaxy to be myriads of stars; and every nebulous star appearing as if it were the firmament of some other world, at an incomprehensible distance, buried in the vast abyss of intermundious vacuum. . . .'

and globular clusters surrounded by a wide, relatively thin disc of stars, clusters of stars, gas and dust, some of which is arranged in spiral arms. Our own Sun lies in one of these spiral arms about two-thirds out from the centre to the edge and a little to the north of the central plane. It is the concentration of stars into this central plane or disc which appears to us as the Milky Way.

Around the nucleus, in what might be termed a kind of spherical 'halo' are more globular clusters moving in orbits at various angles to the plane of the Milky Way. These globular clusters can be seen in a similar pattern in M 31 and also in M 104 – the 'Sombrero' Nebula in Virgo. Also in the halo are many separate stars not associated with the galactic disc and similar to those in the globular clusters. Typical of these halo stars is the variable star *RR Lyrae*. There is also a gaseous halo which has been detected by radio astronomers.

The stars in the disc are generally much like the Sun, being very luminous and of a high surface temperature. They are comparatively young stars and there is little doubt that star-formation is still taking place in some parts of the gaseous nebulae of the disc. M 42, the Orion Nebula, shows distinct evidence of this. The galactic clusters are composed of similar stars and sometimes contain gas as well. All these younger members of the Galaxy have been classed together by Walter Baade into what he has termed Population I.

On the other hand, the globular clusters and the *RR Lyrae* stars of the halo are, on the average, cooler and less luminous than the Sun and have a lower metal content. They are much older than the stars of the Milky Way and, with the planetary nebulae, have been classified as Population II.

The nucleus of the Galaxy is not directly visible to us as it is obscured by dense clouds of inter-stellar gas and dust that lie between us and the centre. Outside the visible spectrum however, the intense radio source known as Sagittarius 'A' is thought to emanate from the nucleus. The globular clusters and the planetary nebulae appear to be more concentrated towards the galactic centre and there is little doubt that this nucleus, as it appears in so many other galaxies, is the 'hub' of the whole system.

The essential dimensions of our Galaxy are:

Diameter of the disc	about 80 000 light-years
Distance of the Sun from nucleus	about 26 500 light-years
Diameter of nucleus	about 5000 light-years
Diameter of halo of globular clusters	about 130 000 light-years

These dimensions are far from being final and many astronomers consider that they may all have to be increased, perhaps by as much as 25 per cent, when techniques of distance measurement are further refined.

Extra-galactic systems When we move outside our own Galaxy the scale of things changes enormously. The nearest star to the Sun is *Proxima Centauri*, a little more than 4 light-years away: the nearest external galaxy is 160 000 light-years distant. This is the Large Magellanic Cloud, an irregular galaxy, visible as a bright, diffuse object, mag. 0.9, in declination 69° S.

There are at least 1000 million observable galaxies, most of them being merely faint specks on photographic plates and at immense distances. These external galaxies are not randomly distributed but are often grouped in clusters, some containing hundreds of members. Our own Galaxy is a member of what is often called the Local Group which consists of about 30 members including M 31, M 32, M 33 and both the Large and Small Magellanic Clouds.

The nearest spiral galaxies outside our Local group are M 81, M 82 and M 83 – all about 8½ million light-years away: the furthest of the Messier Objects is M 77, about 52 million light-years distant.

Galaxies are divided into two main groups, the *elliptical* and the *spiral*. The elliptical galaxies are rather like huge globular clusters having little or no gaseous material and no spiral arms: examples of this class among the Messier Objects are M 32, M 60 and M 87. The spiral galaxies are sub-divided into two types; the *normal spirals* in which the spiral arms extend from a central nucleus and the *barred spirals* in which the spiral arms are seen to be connected to the extremities of a central 'bar'. The only barred spirals in Messier's list are M 58, M 91 and M 95: most of the galaxies which he observed are normal spirals but he described them merely as 'nebulous stars' and had no idea of their extra-galactic nature.

THE 'MISSING' MESSIER OBJECTS

Before examining Messier's catalogue in more detail it would be as well to deal with some of the anomalies which the list contains and to attempt to clear up a little of the 'mystery' which surrounds the objects mentioned earlier: M 47, M 48, M 91 and M 102.

The final catalogue published in 1781 contained a total of 103 objects: of these, M 40, as an asterism is generally omitted from present-day lists while M 73, also an asterism, is somewhat inconsistently, generally included.

In the positions given by Messier for the four 'mystery' objects nothing resembling their descriptions was found by later observers and many explanations have been offered as to what these objects might have been. We can now examine these in detail.

The position which Messier gave for M 47 falls in the Milky Way but in an area which reveals no obvious grouping that could be said to resemble a cluster. In the *New General Catalogue* the position was given an entry by Dreyer under the number 2478 and described tersely as a 'cluster'; but Auwers, who had surveyed the Messier Objects thoroughly and had supplied Dreyer with their up-to-date positions considered that the cluster, if it existed, was probably a very loose and poor one and that no one had observed it since Messier's day. The French astronomer Flammarion who was greatly interested in Messier's work also concluded in 1919 that M 47 did not exist as a cluster.

However, in 1934 it was suggested by Oswald Thomas in his book *Astronomie* that M 47 could be identified as the nearby cluster NGC 2422 and Dr T. F. Morris of the R.A.S. of Canada has given an explanation of Messier's probable error in computing his position of M 47 from his comparison star *2 Navis* (now *2 Puppis*). By simply changing the sign of Messier's right ascension and declination differences from this star the original position can now be made to coincide with NGC 2422.

This is a quite convincing explanation especially as NGC 2422 lies within two degrees of another Messier cluster, M 46, which is fainter and less easily distinguished than the former. It is inconceivable that Messier could have missed NGC 2422 and the observer may see for himself that in anything but the best seeing conditions M 46 may be very difficult to pick up while NGC 2422, being composed of fewer but brighter stars will stand out clearly.

There can be little doubt that NGC 2422 (which was also classified by Wm Herschel as 38 H. VIII) is really the object that Messier listed as 'cluster of stars a short distance from M 46 . . .' and numbered M 47 (see M 47).

No cluster has ever been observed either, in Messier's position for M 48* but Dr Morris has pointed out that, at exactly the same right ascension as Messier gave but in a declination almost 5° S of his there is a reasonably conspicuous cluster, NGC 2548, which answers the original description which was 'cluster of very faint stars without nebulosity; a short distance from the three stars that form the beginning of the Unicorn's tail'.

The identification here is less probable than that of M 47 and NGC 2422 but it is a reasonable one if only for the fact that there is no other cluster near which could in any way be considered as a candidate. The suggestion also has the support of several other observers including Owen Gingerich who, when at Harvard College Observatory, made an extensive study of Messier's catalogue and methods of working.

The next 'missing' object is M 91 and this is the most difficult of all to solve: the position given by Messier lies in the Coma–Virgo Group of galaxies and although there is no object answering to his description in the position he gave, the difficulty lies not in a lack of alternatives but in the number of them.

The most 'popular' solution has a touch of irony and is somewhat unkind to the 'comet-ferret', for several researchers have suggested that the object really was a comet which has, of course, long since moved on. There is, however, no strong evidence of this and it must be considered improbable that Messier, who was virtually obsessed with comet discovery and who was so successful in this field should have been so paradoxically misled.

Several observers have accepted the nearest object to Messier's position as M 91: this is the spiral galaxy NGC 4571 which is about 26' N and about 8' W of Messier's location. This attribution is also made in the *Anagalactic Nebulae* section of the 1964 edition of the *Atlas Coeli Katalog*.

Gingerich objects that this galaxy is too faint for Messier to have detected with his telescopes and it must be admitted that, at a visual magnitude of nearly 12 it is a good deal fainter than any of the other galaxies that Messier observed in this region. However, Messier did observe the planetary nebulae M 76 and M 97 both of the 12th magnitude and although these may have

---

* In J. L. E. Dreyer's introductory notes to the first part of the Index Catalogue (1895), he stated that M 48 was omitted from the NGC. He added a position which corresponds with Messier's erroneous place and the description, 'Cluster of small stars'.

been available higher in the sky for him than NGC 4571, the possibility that this galaxy was the object that Messier intended as M 91 cannot be ruled out.

In an article in *Sky & Telescope* Gingerich puts forward his own suggestion that M 91 was, in fact, a duplicate observation of M 58 which Messier had discovered two years earlier, the positions being of similar right ascension and differing in declination by about $2\frac{3}{4}°$. This difference Gingerich explains was a simple mistake by Messier in the reduction of the observation.

The objection to this is that M 58 is a fairly bright object being a little brighter than mag. 9 and has, moreover, a very clear and distinct outline which makes it an unmistakable object among the galaxies of the Coma–Virgo group. Messier's description of M 91 shows that he considered it to be even fainter than M 89 of which he says 'Its light is extremely weak and faint and it is not possible to detect it without difficulty'.

The object which Messier observed as M 91 must have been at least as faint as mag. 10 and as such cannot have been a mistaken duplicate observation of M 58 as Gingerich maintains. (As M 58 lies only about 1° to the S of M 89 the observer may easily compare the relative brightness of these two objects for himself when observing in this area.)

In 1969, however, a Texas amateur astronomer, W. C. Williams investigated the problem anew and convincingly demonstrated that Messier had almost certainly reduced the position of M 91 by laying off differences in R.A. and Dec. from M 89, but in plotting, had applied them to M 58 – the bright galaxy he had used as reference for some seven other objects in the area. Applying this correction, Williams was led to a position within one minute of arc of the barred spiral galaxy, NGC 4548, which at mag. 10.2, would have been within Messier's light-grasp. This identity for M 91 is now generally accepted. (See M 91.)

The last of the 'missing objects' is M 102, one of the three objects discovered by Méchain and added to Messier's final catalogue published in 1781 in the French almanac, *Connaissance des Temps* (printed three years in advance) for 1784. No positions were given for M 102 and M 103 as Messier had no time to check them before publication: M 102 was described as 'a very faint nebula situated between omicron *Bootis* and iota *Draconis*: near to it is a 6 mag. star.'

Now o *Bootis* is more than 40° to the S of ι *Draconis* and Admiral Smyth, in his *'Bedford' Catalogue* of 1844 took it that omicron (o) *Bootis* was an obvious misprint for theta (θ) *Bootis* which is in the right area. Between these two stars lies a somewhat scattered group of five nebulae. Smyth observed one of these which he describes as 'small but bright with 4 small stars spreading across N field of it' and, with reservations, 'as probably the object seen by Méchain'. He identifies it as No. 1910 in Sir J. Herschel's catalogue which is now known as NGC 5879. However, he also describes it as the brightest of the group, a distinction which belongs not to NGC 5879 but to NGC 5866, nearly $1\frac{1}{2}°$ away to the SW.

In 1917 Shapley and Davis, in a review of the Messier Objects in Vol. 29 of the *Proceedings* of the Astronomical Society of the Pacific, identified M 102 with the brighter nebula NGC 5866 and Becvar's present edition of the *Atlas Coeli* does the same, listing it as an elliptical galaxy mag. 10.8 visual. In the 4th edition of *Norton's Atlas*, M 102 is not mentioned but on Map 11, NGC 5866 appears in Herschel's classification as 215[1] and NGC 5879 as 757[2].

NGC 5866 has a group of six faintish stars to the north of it, one of them about mag. 8 and within 2' of the nebula, and being the brightest galaxy in the area, is the one which fits Smyth's description most closely.

NGC 5879 is a good deal fainter but, having a 7th magnitude star within 7' to the north, is nearer to Messier's original description and this object was identified with M 102 in early editions of *Norton's Atlas*.

In addition to these conjectures, J. L. E. Dreyer, in *Notes and Corrections to the NGC* p. 283, assumed that ι *Draconis* was an error for ι *Serpentis* and thought that M 102 might possibly be the same as NGC 5928.

In spite of all this, however, it appears that M 102 has existed from the beginning only as an all-too-human 'clerical error'. In the *Journal* of the R.A.S. of Canada for 1947 Dr Helen S. Hogg revealed that Méchain had written to J. Bernoulli in Berlin explaining the mistake, the letter being later published in the *Berliner Astronomiches Jahrbuch* for 1786. A translation of the relevant paragraph of the letter is as follows: 'On page 267 of the "*Connaissance des Temps* for 1784" M. Messier lists under No. 102 a nebula which I have discovered between omicron *Bootis* and iota *Draconis*: this is nothing but an error. This nebula is the same as the preceding No. 101. In the list of my nebulous stars communicated to him M. Messier was confused due to an error in the sky-chart.'

This would appear to dispose of the whole affair: although it is not quite clear whether the mistake was really Méchain's or Messier's, the acknowledgement by the discoverer of M 102 that it was in fact identical with M 101 cannot easily be set aside.

Nevertheless a lingering doubt remains: there *is* a group of nebulae near ι *Draconis* and Méchain could have seen at least the brightest of them, NGC 5866. Oddly enough, this nebula is almost exactly one whole hour of R.A. in advance of M 101 and only a little over 1° N in dec. Did Méchain observe it and then, registering or plotting its position one hour of R.A. in error, mistakenly *think* it a duplicate observation of M 101?

This tempting conjecture must be resisted however for M 101, although a faint and difficult object visually, is one of the largest of the Messier nebulae being at least 10'×8' at first seeing while NGC 5866 is small: about 3'×1' and if Méchain made any notes at all this fact would have been immediately apparent.

As a final 'red herring' there appears on Messier's own copy of the 1781 catalogue hand-written positions for *both* M 101 and M 102. As, however, the position given for M 102 still reveals no nebulous object the original uncertainty repeats itself and in the face of so much doubtful and contradictory evidence, Méchain's explanation to Bernoulli must be accepted at its face value and M 102 omitted from any genuine list of the Messier Objects.

THE ADDITIONAL OBJECTS

When Méchain wrote to Bernoulli in 1783 explaining the error concerning M 102, he also gave descriptions of six new objects he had found and gave positions for three of them: these particulars were printed with the rest of the letter in the 1786 *Astronomisches Jahrbuch*.

All of these objects have since been identified and at different times proposals have been made for adding them to Messier's original list of 103.

The first perpetrator – if that is the right word – of an apocryphal addition to the Messier list was Camille Flammarion who discovered on Messier's own copy of the 1781 catalogue a hand-written position for 'a very faint nebula in Virgo' – No. 4594 in the NGC. Flammarion wrote in the *Bulletin* of the Astronomical Society of France for 1921 'I have added it to Messier's Catalogue under the title of Messier 104'. This addition is now pretty generally accepted, although Smyth listed it under Wm Herschel's designation of 43 H. I and in the 14th edition of *Norton's Atlas* on Map 9 the object is similarly shown as 43[1].

The next three additional objects are also among those which Méchain included in his letter and these were soon identified as Nos. 3379, 4258 and 6171 in the NGC. Their addition to Messier's list as M 105, M 106 and M 107 was suggested by Dr H. S. Hogg who discovered the existence of Méchain's forgotten letter in 1947.

The remaining two of Méchain's six additional objects had in fact been mentioned in another manuscript addition to Messier's own copy of his catalogue: one, added to the entry for M 97 states 'near-by this nebula is seen another which has not been measured and also a third which is near Gamma Ursae Majoris' and a position for NGC 3992 follows the latter.

Owen Gingerich has identified the first of these as NGC 3556 and has proposed that this be added as M 108 and NGC 3992 as M 109. All these additions have received the sanction of the *Atlas Coeli* and its *Katalog* but, apart from M 104 they are seldom included in lists of Messier Objects.

The trouble with this sort of thing is to know where to stop. Messier himself published no further lists after the one of 1781 which ended with No. 103: all the additions are found in correspondence or in manuscript notes and although Messier wrote in the *Connaissance des Temps* for 1801 'Since the publication of my catalogue I have observed still others: I will publish them in the future according to the order of R.A. . . . ' he did not in fact ever do this.

One further object, however, did get into print when an engraving of Messier's detailed drawings of the Great Nebula in Andromeda (M 31) was published in the *Mémoirs* of the French National Institute for 1807. In this illustration he notes that M 31 itself was seen by Marius in 1611, the companion nebula to the south (M 32) by Le Gentil in 1749 and also that the other companion nebula (NGC 205) had been discovered by himself in 1773.

It is odd that Messier did not include this nebula in either the first or second supplements to his catalogue and it is not mentioned in any of the additional notes to his own copy of the final catalogue. Nevertheless there is as much

justification for including NGC 205 in the list with the title M 110 as there is for any of the others which were discovered not by Messier but by Méchain.*
(For Messier's drawing of M 31 see p. 128.)

This would appear to be the limit of possible addition: the catalogue strictly should end with M 103, omitting M 102 as a 'mistake'. Popular usage may perhaps justify the inclusion of M 104: the remaining objects are more generally known by their NGC numbers and giving them Messier numbers as well provides only a duplicate identity which, in other circles might be termed as 'alias' and just as likely to lead to confusion.

A full and detailed list of the additional objects is given separately (see Table 2).

---

* The discovery of NGC 205 is generally attributed to Caroline Herschel who saw it in 1783. This date is ten years after that claimed by Messier but long before that of the publication of the illustration of 1807. Messier's claim was supported by D'Arrest in his *Siderum Nebulosorum* of 1867 where he describes NGC 205 as 'Primum inventa a Messiero anno 1772; denuo a Carol. Herschelia, anno 1784'.

Table 2. *The additional Messier Objects*

Mess. No.	NGC	Wm H.	R.A. (1950)	Dec.	Type	Const.	Mag.	Discov.	Incl. by
104	4594	43 H. I	12 h 37.3 m	S 11° 21'	Sb Gal.	Vir.	8.7	Méchain	Flamm. 1921
105	3379	17 H. I	10 h 45.2 m	N 12° 51'	E1 Gal.	Leo	9.2	Méchain	Hogg 1947
106	4258	43 H. V	12 h 16.5 m	N 47° 35'	Sb Gal.	U. Ma.	8.6	Méchain	Hogg 1947
107	6171	40 H. VI	16 h 29.7 m	S 12° 57'	Glob.	Oph.	9.2	Méchain	Hogg 1947
108	3556	46 H. V	11 h 08.7 m	N 55° 57'	Sc Gal.	U. Ma.	10.7	Méchain	Gingerich 1960
109	3992	61 H. IV	11 h 55.0 m	N 53° 39'	Sb Gal.	U. Ma.	10.8	Méchain	Gingerich 1960
110	205	18 H. V	00 h 37.6 m	N 41° 25'	E5 Gal.	And.	9.4	Messier?	Glyn Jones 1966

2 The Messier Objects classified

Table 3. *The Messier Objects classified according to type*

ASTERISMS	M 40, M 73
GASEOUS NEBULAE	M 8, M 16, M 17, M 20, M 42, M 43, M 78
GALACTIC CLUSTERS	M6, M7, M11, M 18, M21, M23, M 24*, M 25, M 26, M 29, M 34, M 35, M 36, M 37, M 38, M 39, M 41, M 44, M 45, M 46, M 47, M 48, M 50, M 52, M 67, M 93, M 103
PLANETARY NEBULAE	M 27, M 57, M 76, M 97
GLOBULAR CLUSTERS	M 2, M 3, M 4, M 5, M 9, M 10, M 12, M 13, M 14, M 15, M 19, M 22, M 28, M 30, M 53, M 54, M 55, M 56, M 62, M 68, M 69, M 70, M 71, M 72, M 75, M 79, M 80, M 92, M 107
ELLIPTICAL GALAXIES	M 32, M 49, M 59, M 60, M 86, M 87, M 89, M 105, M 110
SPIRAL GALAXIES (SO)	M 84, M 85
SPIRAL GALAXIES (Normal)	M 31, M 33, M 51, M 61, M 63, M 64, M 65, M 66, M 74, M 77, M 81, M 83, M 88, M 90, M 94, M 96, M 98, M 99, M 100, M 101, M 104, M 106, M 108, M 109
SPIRAL GALAXIES (Barred)	M 58, M 91, M 95
IRREGULAR GALAXY	M 82
SUPERNOVA REMNANT	M 1

THE GASEOUS OR DIFFUSE NEBULAE

Messier's list contains seven objects which are generally classed as *gaseous nebulae* although *bright diffuse nebulae* would perhaps be a better description. They are: M 8 (the Lagoon Nebula), M 16, M 17 (the Omega Nebula), M 20 (the Trifid Nebula), M 42 and M 43 (the Great Orion Nebula) and M 78 — another bright nebula in Orion.

M 45 (the Pleiades) also contains bright nebulous matter but this was not visible in Messier's telescopes and M 45 is always classed as a galactic cluster.

The disc of the Galaxy contains not only stars and star-clusters but also a considerable amount of interstellar gas and dust. The amount of interstellar material is very variable: in the vicinity of the Sun there is probably as much material in the gas between the stars as there is in the stars themselves. Much of this is dark, obscuring matter and a good deal of our Galaxy, including the nucleus, is hidden from us by such clouds.

---

* See footnote to page 19 and M 24.

However, some of the gas is luminous and it is this which is visible in the gaseous nebulae but in many cases bright and dark material are mixed together in the same cloud. M 8 and M 16 are examples where stars, bright nebulae and dark obscuring matter are involved although the dark areas are mostly in the form of small 'globules' only visible in fairly large telescopes and with the aid of photography (see M 8 and M 16).

Some diffuse bright nebulae are variable in brightness; Hind's Nebula, NGC 1554–5 in Taurus and Hubble's Nebula, NGC 2261 in Monoceros are two examples, but there are none of these among the Messier Objects and little is known about the cause of their variability.

The *dark nebulae* are known to be composed of very small dust particles of a size that makes them opaque in the visible part of the spectrum but they can be penetrated to some degree by the longer wavelengths used by radio astronomers. Messier did not include any of these in his catalogue either, but the 'Horsehead' Nebula in Orion and the 'Coal Sack' in the Southern Cross are well known examples and there is a small dark nebula very close to the west of the galactic cluster M 24 (see M 24).

The bright diffuse nebulae are divided into two main types according to their appearance when viewed spectroscopically. When William Huggins, the English amateur astronomer who pioneered stellar spectroscopy, first examined the stars and nebulae with this instrument he found that the spectra of the diffuse nebulae were similar to that given off by a luminous gas and it was thought for nearly 50 years that all bright diffuse nebulae were of this kind.

However in 1912 V. M. Slipher of Lowell Observatory found that the very faint nebulosity of M 45, the Pleiades, gave a continuous spectrum like that of the stars and that this light was reflected and scattered radiation from the stars in the cluster. The name *reflection nebulae* is given to this type.

Those nebulae whose spectra showed them to be gaseous as Huggins described were characterised by having bright 'emission' lines in the green and ultra-violet and at first the green lines were attributed to radiation from a hypothetical element called *'Nebulium'*. It was soon realised, however, that there was no place in the periodic table for such an element and in 1927 the American astro-physicist I. S. Bowen showed that such 'forbidden' lines could be produced by ionised oxygen and nitrogen in conditions of very low density.

These are the *emission nebulae* and M 8, M 16, M 17, M 20 and M 42 are all of this type. M 43, near the NE corner of M 42, emits both emission and reflection spectra while M 78 and M 45 are entirely reflection nebulae.

After Bowen's discovery, Edwin Hubble showed that the determining factor as to whether a diffuse nebula gives an emission or reflection spectrum depends upon the temperature of the stars involved in it. Although some cases overlap, temperatures over about 20 000 °C excite emission in the gas while cooler stars illuminate the nebula by reflection. He also noted that the brighter the star, the more extended was the area of luminosity.

The gaseous nebulae are all of irregular outline and generally emit a greenish or bluish light. They are all inhabitants of the disc and are seen, therefore, in or close to, the Milky Way. The gas of which they are composed is very diffuse and contains, on the average, about 9 atoms of hydrogen to every 1 of helium.

M 42 is the finest and brightest of the gaseous nebulae in the heavens and it is visible to the naked eye as the middle star of Orion's sword (θ *Orionis*, mag. 4.0). It is a splendid and fascinating sight even in the smallest of telescopes. V. M. Fessenkov has calculated that the critical density of a gas to enable stars to form is of the order of 10^{-22} g/cm^3: in M 42 the density is above this figure in parts and there is little doubt that in this nebula star formation is occurring at the present time (see M 42).

Most of the constellation of Orion can be seen in long exposure photographs to be enveloped in nebulosity and in Becvar's *Katalog* more than 40 separate bright nebulae are listed in this constellation. In small telescopes the nebulosity of M 8, M 16 and M 20 may be less conspicuous than the stars involved but parts of it can easily be seen in the best conditions. In M 17 and M 78 however, the glowing gas shines strongly and the stars are fainter: M 17 is also a strong radio source.

The diffuse bright nebulae in Messier's list are all quite easily picked up but the numerous others of this kind, such as the nebulosity in M 45 are often very faint or too tenuous to be seen except in telescopes of fair aperture and short focal length like the so-called 'rich-field' telescopes that comet observers prefer.

GALACTIC CLUSTERS

The galactic clusters are very numerous: more than 500 of them are known and catalogued, and this must represent only a fraction of the total for many more must be hidden by obscuring gas and dust or lost among the great star-clouds of the Milky Way. However, while they may be common, they are certainly not common-place for many of them are not only strikingly beautiful but also have helped to shed considerable light upon present theories of stellar evolution.

Messier included 27 galactic clusters in his list: M 6, M 7, M 11, M 18, M 21, M 23, M 24*, M 25, M 26, M 29, M 34, M 35, M 36, M 37, M 38, M 39, M 41, M 44, M 45, M 46, M 47, M 48, M 50, M 52, M 67, M 93 and M 103.

This section represents the brightest and nearest of these clusters but the superb double-cluster in the sword-handle of Perseus which was known even in the time of Ptolemy, is a notable and inexplicable omission.

Like the gaseous nebulae, the galactic clusters are Population I objects, are members of the disc system and are to be found in or near the Milky Way. Moreover, these two classes of object are not entirely separate, for some

---

*   Strictly speaking, the object which Messier observed and catalogued as M 24 is not a true cluster but a Milky Way star-cloud and thus is really one of the 'non-objects' like M 40 and M 73. It does, however, contain within its borders a true cluster, NGC 6603 (not observed by Messier), and M 24 is included as a galactic cluster for this reason.

gaseous nebulae such as M 8, M 16 and M 20 contain clusters while some clusters, like M 45, contain gas.

The simplest classification of galactic clusters is that of Shapley of which the categories are: (c) very loose and irregular; (d) loose and poor; (e) intermediate rich; (f) fairly rich and (g) considerably rich and concentrated.

This classification for each of the Messier clusters is given on the appropriate pages in Chapter 4.

Another important observer, R. J. Trumpler, used a much more detailed classification based on photographs which uses four main degrees of concentration, each subdivided into one of three groups according to the numbers of stars present. A third category describes the range of brightness in the cluster stars.

In this classification, M 44, for example, is of type I,2,r: 'I' standing for 'strongly concentrated', '2' for 'stars uniformly spread over a large range of apparent magnitude' and 'r' for 'rich: containing more than 100 stars'.

Galactic clusters, of course, are not just chance collections of stars but are groups of associated members, fairly close to one another and travelling together as a group through space around the disc of the Galaxy. We can decide whether a particular star is or is not an actual member of the cluster by measuring the radial (line-of-sight) velocity relative to us. Those stars which have a common radial velocity are members of the cluster, while those whose radial velocities differ in any degree merely happen to lie, at the moment, in the same direction as the cluster and because of this different velocity will eventually move away.

In addition to this it may be assumed that clusters of stars which move together must have had a common origin and that they all condensed as stars at about the same time. Their present differences in size, surface temperature and spectral type can give us some insight into the physics of stellar evolution. A great deal has been discovered in this way by Sandage, Otto Struve and others.

Of the Messier clusters, M 11, M 41, M 44, M 45 and M 67 have all been studied along these lines. The results show that M 67 is one of the oldest galactic clusters known, having existed for perhaps 10 000 million years and having a possible life expectancy of 5000 million years more. M 11, M 41 and M 44 are younger clusters and have a shorter life span while M 45 (the Pleiades) is younger still and has a projected future of only 250 million years or so.

The future life expectancy of a cluster can be related to the degree of concentration of the stars composing it for the general gravitational field of the Galaxy tends to disrupt it and the looser the cluster, the greater this effect is likely to be. In addition, encounters with interstellar gas clouds may also pick off the more weakly bound members of the cluster and again, the more densely knit clusters are most likely to survive.

The younger clusters generally contain some gas which has not yet condensed into stars: M 45 is one of these and in photographs the whole of this cluster can be seen to be enmeshed in faintly glowing gas. It can also be seen visually especially around the star Merope but this is best brought out in a 'rich-field' telescope of large aperture and short focal length.

The older clusters like M 67 have generally lost all of their dust and gas and further star formation is unlikely to take place.

Apart from their value to astronomical theory the galactic clusters may also satisfy the aesthetic sense, for many of them form striking and often very pleasing patterns. The unmistakable 'cruciform' shape of M 38 in Auriga and the fan-like pattern of M 11 in Scutum which Admiral Smyth compared to 'a flight of wild ducks' are well known examples. The rich curves and festoons of bright stars in M 35 in Gemini so impressed Lassell that he declared that no one could see it for the first time without an exclamation!

There are many more of nearly equal beauty and no doubt each observer will have his own favourites as he discovers them.

The galactic and globular clusters compared

At this point it might be helpful to point out in a little more detail the differences between the galactic and globular clusters. In most cases the differences are quite evident to the eye: the galactic clusters are generally fairly loose in structure; they display a more or less irregular outline and can be distinctly seen, even in small telescopes, to contain stars of a wide range of magnitude.

Globular clusters, on the other hand, usually appear to be much more compact; they are often clearly circular or slightly elliptical in appearance and the range of brightness of the component stars is frequently small.

There are cases where the visual differences are not so distinct: the small galactic cluster, NGC 6603, which lies within M 24, appears quite dense and circular in outline while M 55, although a genuine globular cluster, is quite irregular in appearance and loose in structure.

There are, however, many other differences between the two types of cluster which are revealed by closer examination. Galactic clusters seldom contain variable stars while globular clusters are often rich in *RR Lyrae*, short-period variables and also contain longer-period variables of the *W Virginis* and *RV Tauri* type.

Differences of distribution are also noticeable: the galactic clusters all lie within or very close to the Milky Way, showing that they are associated with the disc of the Galaxy. Globular clusters are more concentrated towards the nucleus and are also found at great distances above and below the galactic plane, forming part of the halo population.

The most significant distinctions between the two kinds of cluster, however, are those revealed by the spectroscope.

In the galactic clusters the brightest stars are the hot, blue or white stars of spectral types O or B and in many cases these stars are found to be in rapid rotation. The bright stars of the Pleiades, M 45, are a good example of this.

A quite different composition is seen in the globular clusters: here the brightest stars are red giants of a much lower surface temperature and show no evidence of rapid rotation. Red giant stars are sometimes found within the borders of galactic clusters but often they turn out to be non-members when their radial velocities are measured. The hot blue stars are quite unknown in the globulars.

The brighter stars in galactic clusters are also found to have significantly different chemical composition from those in globular clusters. In the former, the stars have a comparatively high content of the heavier metallic elements and this shows up clearly in their spectra. In the globulars, metallic lines are often weak and their metallic content is low.

As will be described in the section on stellar evolution later in this chapter, these distinctions reveal very different ages and evolutionary history in the two types of cluster. The globulars are the 'old men' of the Galaxy: although they sometimes show rare bursts of activity (novae have been observed, for example, in M 14 and M 80) their fires burn low and they have attained an enduring stability in which change is imperceptibly slow.

The stars of the galactic clusters are mostly in their hot youth and squander their energies in reckless abandon: for many of them their lives may be glittering but brief.

GLOBULAR CLUSTERS

Messier observed 29 globular clusters; they are: M 2, M 3, M 4, M 5, M 9, M 10, M 12, M 13, M 14, M 15, M 19, M 22, M 28, M 30, M 53, M 54, M 55, M 56, M 62, M 68, M 69, M 70, M 71, M 72, M 75, M 79, M 80, M 92, M 107.

M 22, however, had been seen by Abraham Ihle as early as 1665 while Halley discovered M 13 in about 1714, describing it as 'but a little patch; but it shows itself clearly to the naked eye when the sky is serene and the Moon absent'.

The two brightest globulars in the heavens were not seen by Messier as they have too southerly a declination to be visible from northern Europe. These are (ω) *Centauri* (mag. 4) which was charted in Bayer's star atlas *Uranometria* before the invention of the telescope and recognised as a cluster by Halley in 1677, and *47 Tucanae*, also mag. 4, but being of declination 72° 21' S is not readily visible from anywhere in the northern hemisphere.

Of the Messier globulars M 13 is the brightest at mag. 5.7 followed by M 22 (mag. 5.9), M 92 (mag. 6.1), M 5 (mag. 6.2) and M 3 and M 4 (both mag. 6.4). All of these are fine objects even in small telescopes and can be at least partly resolved into stars in a 4-inch or 6-inch telescope. At first glance it may be thought that most of the others look so much alike that they are not really worth searching for. This is certainly not the case for a closer examination and comparison will soon show that each one has its own individual character, from the gem-like symmetry of M 53 to the straggling arrangement of M 12 and the angularity of M 71.

Harlow Shapley has classified the globular clusters according to their degree of concentration; class I being the most highly condensed and class XII the least concentrated. Shapley's class is given for each cluster on its individual page. This classification was made from detailed star-counts on large-scale photographic plates and the visual appearance in a small telescope may not always seem to fit this description precisely. However, those of class VIII and above, like M 4, M 12 and M 71, are more easily resolved, especially in the central regions, than the others.

In 1950 only about 100 globular clusters were known, the discoverers of these – as globular clusters – being: Wm Herschel (33), James Dunlop (21), Messier (14), John Herschel (5) and Lacaille (4), no other observer being credited with more than two. The total has now risen to 138 (in 1987) but the Galaxy probably contains at least as many again, hidden from us by the dust and gas in our spiral arm and the obscurities toward and beyond the galactic nucleus.

The globular clusters appear to be very unevenly distributed in the sky, being confined almost entirely to one hemisphere, and a glance at Maps 1 to 6 will show, even among the small sample of globulars in Messier's list, that out of 28, 27 lie on Maps 5 and 6 in an area of 90° of dec. and less than 9 hours of R.A. Only one, M 79 in Lepus, appears in the remaining four-fifths of the sky visible from Paris.

This, however, does not represent their real distribution but is due to the position of our Sun in the galactic disc: the globulars are really spread around the galactic nucleus in a huge spheroidal halo whose major axis is probably about 130 000 light-years.

Although many of the globulars appear irregular when seen in the telescope – M 13, for example, seems often to have an almost spiral arrangement of its outlying stars – photographs show than when all the fainter stars are included, the total outline is very nearly circular. On closer measurement, however, a slight flattening can be detected in many of them, M 13 having compression of about 1 : 20 with the P.A. of the major axis about 125°.

Many of the globular clusters contain pulsating variable stars of the *RR Lyrae* type, M 3 being the richest with 189, ω *Centauri* having 165 and M 15, 100. M 13 contains only about 15 of these short-period variables but Shapley measured four of them and found their apparent magnitude to be nearly 15. As RR Lyrae stars have an *absolute magnitude* (the brightness they would have if placed at 10 parsecs from the Earth) of very near to zero, the difference of 15 magnitudes gives a distance of M 13 of about 10 kpc. The latest estimate is a little less than this; about 6.9 kpc – about 22 500 light-years.

By measuring the total integrated magnitude of a cluster an estimate can be made of the number of stars contained in it, and the average number for a typical globular cluster is not less than 100 000. Shapley and Pease actually counted, on photographic plates, a total of 70 000 stars in M 22. M 13 is a very rich cluster and may contain as many as half a million.

Shapley employed globular clusters to make the important measurement of the distance of the nucleus of the Galaxy from the Sun. He assumed that the nucleus was at the centre of the halo of globulars as it appears to be in M 31 and M 104 and so, knowing the distances of a number of these clusters, he was able to arrive at a distance of about 8.2 kpc for the galactic nucleus.

Weizaecker's theory for the formation of the Galaxy explains the globular clusters as condensing out of the original spherical gas-cloud in masses large enough to enable them to survive as clusters. The smaller, less dense clouds were unstable and were soon disrupted to form a halo of separate Population II stars including the *RR Lyrae* and *W Virginis* variables and planetary nebulae. All these objects would continue to move in orbits around the nucleus which would be inclined at various angles to the plane of the Milky Way and this, in fact, is what we observe. These orbits, too, are thought to be much more elliptical than those of the stars in the disc which have roughly circular motion.

The globular clusters thus have ages which are probably almost as great as the Galaxy itself; about 6000 million years, and during this time they would all have passed many times through the galactic disc and close to the nucleus. This would have the effect of sweeping away any uncondensed material out of the cluster, and observation has shown that the globulars are remarkably free from dust and gas. M 13 has an orbit which has a period of perhaps 100 million years and so must have gone through this process 20 or 30 times.

The disruptive effects of the general gravitational field of the Galaxy have a greater effect on the looser clusters, and these are found to have orbits which keep them for most of the time well away from the galactic centre. Many globular clusters are found very near the nucleus and have orbits of small major axis. These must have spent all their existence in the region of the nucleus and must be very compact and dense to continue to exist in these circumstances. M 62 is the nearest of the Messier globulars to the nucleus and this is a very compact cluster, its distance from the galactic centre being a little less than 1 kpc.

The globular clusters have been investigated in a similar manner to the galactic clusters to obtain a pattern of their age and evolution. M 2, M 3, M 5, M 10, M 13, M 15 and M 92 have all played an important part in these studies and, although one of the oldest of the galactic clusters, M 67 has been found to be about the same age as the globular cluster, M 3; the evolutionary stages in M 67 are thought to have been very different from M 3 by reason of the much higher metal content of the Population I stars of the former.

It is interesting to speculate on the appearance the night sky would have if our Sun were a member of an average globular cluster containing, say, 100 000 stars as bright as the Sun within a radius of perhaps 40 parsecs. Many of the brightest stars would be so near that they would have an apparent magnitude of minus 11 or 12 and would be nearly as bright as the full moon, while even the faintest of the 100 000 would be visible to the naked eye compared with the 6000 or so visible in our heavens.

The skies would be brilliant and no doubt beautiful but they would be very frustrating to astronomers, and the Messier Objects would be all completely lost in the glittering spectacle of the luminous night.

Globular clusters have been detected in several other galaxies including M 31, M 32, M 87 and M 104, and one very faint object on the borders of Ursa Major and Leo was found to be a globular cluster at the enormous distance of about 475 000 light-years and seemingly isolated in space outside our Galaxy. This globular is called Abell No. 4 after its discoverer who altogether found 13 new, faint globular clusters in 1956 using a 48-inch Schmidt telescope.

The ellipticity of some globular clusters has often been suspected to be due to their rotation (just as the earth is distended at the equator due to its rotation), but until recently this has not been detected despite many efforts to do so. However, in 1964, G. A. Harding, using the 74-inch reflector of Radcliffe Observatory, found a difference in the radial velocity of stars on opposing sides of the 'equator' of the globular cluster ω *Centauri*. By spectroscopic examination of 13 faint stars he found a rotational velocity of 0.72 km/s per parsec of distance from the axis of the cluster. This gives a rotational period of about 10 million years. Now that this effect has been established for ω *Centauri*, it may well be determined for other globulars which exhibit noticeable ellipticity.

PLANETARY NEBULAE

Planetary nebulae are among the most interesting objects in the heavens: they are relatively few in number, exhibit a wide variety of form (some have an almost bizarre appearance) and their true origin and evolution are still very uncertain.

Messier's list contains only four of these objects: M 27 ('The Dumb-bell'), M 57 ('The Ring'), M 76 ('The Little Dumb-bell') and M 97 ('The Owl').

One thing that is quite certain about the planetary nebulae is that they have no connection at all with the planets: they are generally supposed to have been given this name in 1785 by William Herschel who noticed that many of them appeared to have small, hazy greenish discs not unlike the planet Uranus which he had discovered earlier in 1781. However, Darquier, in 1779, had already described M 57 as 'looking like a fading planet' so it seems that Herschel might merely have adopted this description for the whole class.

We know that the planetary nebulae are all members of our own Galaxy and that each is associated with a very small, hot central star of a surface temperature of anything between 25 000 °C and 120 000 °C. These blue or white dwarf stars are very massive and dense and are surrounded by a shell of very tenuous gas which is excited to glow by ultra-violet radiation from the star. This volume of glowing gas is very large, having a diameter that may be measured in light-years. The actual size of the object may indeed be very much larger than it appears, because what it is seen, even on the photographic plate, is only the gas which is excited. There may be larger areas of material surrounding the nebula which are not visible to us. The density even of the visible shell is extremely low; perhaps containing only a few atoms per cubic inch while, in contrast, the density of the central star is at the other extreme, being of the order of tons per cubic inch.

Many planetaries are annular or ring-shaped like M 57: this is the appearance a partly transparent luminous shell would have shown from a distance, but there are very many differences of visual form to be seen. Most of them are of a distinct elliptical shape or have marked polar concentrations on a major axis which often results in a 'dumb-bell' appearance as in M 27 and M 76. Others have wing-like appendages, often very faint, called 'ansae' while in a few a filamentary structure can be detected especially near the outer edge. There is no doubt that the processes which occur in the formation of planetary nebulae are far from being simple.

The outer material of the shell has been observed in many cases to be moving away from the centre and it is probable that it has been ejected at some time from the central star. Some have been discovered with two concentric shells of glowing gas as if two separate ejections had occurred a million years or so apart.

Recent observations, too, have shown that most planetaries are visibly growing in size; the mean rate of growth being about $1\frac{1}{2}$ per cent per century. The most rapid growth has appeared in NGC 246, an oval planetary about 460 parsecs distant in Cetus which appears to be increasing in size by about 7 seconds of arc per 100 years. M 57 is one of these expanding objects but has a slower rate of growth of about 1 second per 100 years.

Planetary nebulae are often distinguished under the Vorontzov–Veljaminov classification according to their structure as follows:

Type I   Stellar
Type IIa   Oval, evenly bright, concentrated
Type IIb   Oval, evenly bright, without concentration
Type IIIa   Oval, unevenly bright
Type IIIb   Oval, unevenly bright, with brighter edges
Type IV   Annular
Type V   Irregular
Type VI   Anomalous

The brightest planetary is NGC 7009, the remarkable 'Saturn' Nebula in Aquarius having a visual magnitude of about 7, but this object is quite small; with dimensions of about $35'' \times 25''$ it is less than half the size of M 57. The largest of these objects is NGC 7293, also in Aquarius, which is almost 15' in diameter but although its total magnitude is about 6.5 its diffuseness makes it a relatively difficult object to see.

Altogether some 1600 planetary nebulae are now known, more than 200 of them having been discovered in recent years by R. Minkowski.

It is possible that planetary nebulae are formed as nova or supernova transitions in dwarf stars but a great deal more needs to be known about them before any definite conclusion can be reached. They are not randomly distributed throughout the Galaxy but show a distinct concentration towards the nucleus and in a plane more or less perpendicular to the Milky Way. They can be detected at very great distances in the Galaxy, some being more than 9000 parsecs, nearly 30 000 light-years, away. Most of them are of great age and they are all classed as Population II objects. Baade discovered five planetaries in M 31 and others have been found in both Magellanic Clouds.

M 1 (The 'Crab' nebula)   M 1 is given a detailed description under its own heading in Chapter 4. Once thought to be an anomalous type of planetary nebula, its recognition as the remnant of a supernova is now beyond dispute. Supernovae occur in massive stars which have progressed rapidly through their hydrogen, helium and carbon burning stages of their evolution, and running out of fusion energy, collapse with catastrophic results. The rapid gravitational implosion is followed by an explosion in which the outer portions of the star are blown off as a huge, expanding shell of plasma, leaving an extremely energetic neutron star as the core. In the case of M 1 it is the outer shell which is visible in small telescopes.

EXTRA-GALACTIC NEBULAE (GALAXIES)

Messier, although he included 33 galaxies in his catalogue, had no idea of their true nature and in his time, no doubt, the growing knowledge of our own Galaxy was itself staggering enough to inhibit any thought of more remote and vaster systems outside it. The objects of Messier's list which were not obviously star clusters were generally described either as 'nebulae' or 'nebulous stars' and little was conjectured as to their structure, size or distance. When William Herschel turned his superior telescopes and his wider intellectual vision to these nebulae he soon realised that they were to be numbered in thousands and that they were very remote objects indeed. In 1811 he considered that some of them which were not resolvable by his instruments must be outside the Galaxy, and although he conceded in 1817 that his telescope had not yet 'fathomed the profundity of the Milky Way' he still held the view that extra-galactic systems existed.

In 1845 Lord Rosse, with his ponderous but effective 6-foot aperture telescope, had detected the spiral pattern of M 51, and later others such as M 63, M 99 and M 101 were seen to have a similar form and the term *spiral nebulae* came into use. However, they were still thought to be members of the Galaxy and when, a little later, Huggins showed with his spectroscope that many nebulae were composed of glowing gas it was thought that all unresolved nebulae were gaseous and within the confines of our system.

Conclusive proof that other systems existed beyond our own Galaxy naturally depended on the measurement of the very great distances involved and for a very long time this was one of the most intractable problems of astronomy.

The final demonstration that the spiral nebulae are, in fact, external to the Galaxy was not achieved until as late as 1924 when Edwin Hubble succeeded in identifying Cepheid variables in the Great Nebula in Andromeda (M 31). By many devious means he was thus able to measure the distance of this galaxy and the value he found – about 1 million light-years – although we now know it to be too small, nevertheless placed it unmistakably outside our system.

Hubble also began in 1926 to classify the galaxies in a more systematic manner by examining a group of 167 of them in the constellation Virgo, all known to be about the same (very great) distance from us. The visible differences in shape and structure among members of this group must be due to actual differences and not merely a consequence of unequal distances. Hubble's first division was into the three classes of spiral, elliptical and irregular: the spirals were then subdivided into the normal type with arms radiating from a central nucleus and the barred spirals in which the arms emerged from the ends of a central bar, while the ellipticals were classified according to the amount of ellipticity.

Anything unusual tended to be thrown into the category of 'irregular', although of course it was realised that many of the spindle-shaped and streaky nebulae were merely spiral galaxies seen edge-on or nearly so. In fact there are very few among the brighter galaxies which are now so misshapen as to be labelled as irregular although minor differences from the norm tend to get the even more ominous epithet of 'peculiar'.

Later, Hubble discovered a class of galaxies which appeared to lie between the ellipticals and the spirals and this category of 'stripped spirals' was given the designation SO by Baade and Spitzer. The final scheme is illustrated in Figure 2.

The elliptical galaxies range from E0 to E6, the number representing the amount of apparent compression, E0 being spherical and E6 having a minor axis about 40 per cent of the major.

The normal spirals range from Sa in which the spiral arms are tightly coiled, through Sb, to Sc where the arms are at their most open configuration. A similar range applies to the barred spirals which are designated SBa, SBb and SBc.

It was first thought that there might be some evolutionary sequence from E0 to E6 and then branching out via SO to Sa, Sb, Sc or SBa, SBb, SBc but today the general hypothesis is that the evolutionary sequence is probably in the other direction – the Sc and the SBc galaxies being the younger forms and the ellipticals the older.

Elliptical galaxies appear to be more numerous than spirals: according to Baade, of the 18 nearest galaxies to us, 10 are elliptical, while in the original

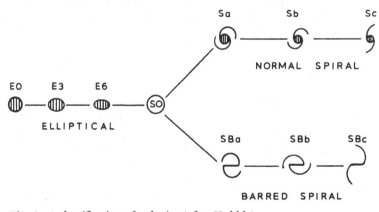

Fig. 2. A classification of galaxies (after Hubble)

group of 167 in Virgo which Hubble first examined, only two were spiral.

The elliptical galaxies and the nuclei of the spirals are composed mainly of red, cooler stars of Population II while the stars of the spiral arms are largely blue, super-giant stars of Population I together with some Population II.

It follows from this that it is mainly in the spiral arms of galaxies that star formation is recent: the elliptical galaxies are old systems, free from gas and dust and where star genesis has virtually ceased.

With masterly skill and patience Baade, in 1944, was the first to resolve the nucleus of M 31 and two of the nearest elliptical galaxies into individual stars. It was during these observations that he realised that the reason these objects were difficult to resolve was that they were composed of comparatively faint red stars and this led him to the idea of two largely separate populations of stars.

A later system of classification was introduced by Hubble and Sandage in 1940 which included a very open type of spiral, Sd. In 1956, de Vaucouleurs evolved a system which included more than 40 types, based upon their photographic appearance.

Elliptical galaxies Of the Messier Objects, nine are elliptical galaxies: M 32 (E2), M 49 (E1), M 59 (E5), M 60 (E2), M 86 (E3), M 87 (E0), M 89 (E0), M 105 (E1), M 110 (E5).

As explained previously these galaxies are composed mainly of Population II stars and have ages of 5000–6000 million years. They are generally free from gas and obscuring dust and it is thought that this state may be brought about by comparatively frequent collisions with other galaxies. This process is similar to that which is thought to occur with the globular clusters and, in fact, large globular clusters and small elliptical galaxies are very similar in many respects. It is quite likely that M 32, the elliptical companion of M 31, may have passed more than once through the disc of the Great Spiral in its long history, just as the globular clusters must have done.

The shapes of elliptical galaxies range from a perfectly circular outline (E0) down to elliptical or lenticular forms with a limit at about E6. The visual or photographic image, however, merely represents the appearance from our viewpoint and the circular, E0 types may really be elliptical when seen from a different angle.

Some of the ellipticals are really small for galaxies – dwarf systems with diameters less than 2 kpc are known – and there may well be others smaller still which we have not yet been able to detect. The nearest elliptical galaxy to us is the dwarf elliptical in Ursa Minor which is about 70 kpc away and has a diameter of only 3000 light-years or so, and there are at least six such dwarf systems in our Local Group within a radius of 1 million light-years.

Other ellipticals are giant systems, such as M 87 in the Virgo Group of galaxies: this object has received a considerable amount of attention for it has been seen to be surrounded by many globular clusters and it also displays a curious 'jet'.

All the elliptical galaxies in Messier's list can be easily seen as they are compact objects, and, because they contain no dark material, they generally appear to have a characteristic 'pearly' appearance. Two ellipticals can be seen together in the same low-power field in Virgo: these are M 59 and M 60.

SPIRAL GALAXIES

SO galaxies (M 84, M 85)

Although Hubble gave these a separate class, they may be looked upon as spiral galaxies without spiral arms (if this is not too Irish). They often have the flattened disc-like form of spirals, but the actual spiral arms are not present and they are generally free from clouds of dust and gas and newly evolving stars.

They are found most frequently in the dense clusters of galaxies where there is the risk of most frequent collision, and this may well be the reason why the dust and gas are absent.

A new subclassification of these 'stripped spirals' is proposed: SOa, SOb and SOc which is intended to indicate the extent of freedom from gas and dust, SOa being the oldest and 'cleanest' form.

Normal spiral galaxies

Sa M 65, M 96

Sb M 31, M 63, M 64, M 66, M 77, M 81, M 90, M 94, M 98, M 104, M 106, M 109

Sc M 33, M 51, M 61, M 74, M 83, M 88, M 99, M 100, M 101, M 108

The normal spirals generally consist of two or more spiral arms emanating from a dense, spheroidal nucleus. They range from those with a bright, central core with closely coiled arms – the older, Sa types such as M 65 – to those with an almost insignificant nucleus and wide, loose arms of which M 33 is a fine example.

Our own Galaxy is generally considered to be an Sb type of intermediate development but from our position within one of the arms it is difficult to tell. The rotation of spiral galaxies is now known to be generally in the sense that the spiral arms are convex to the direction of rotation as if they were 'winding up'. This has been observed in M 31 and NGC 253, an Sc galaxy in Sculptor.

It is thought that the spiral arms must be relatively short lived, having a total life span – unless they could be rejuvenated in some way – of only 200 million years or so.

Normal spirals appear to outnumber barred spirals in the ratio of about two to one. They are seen in many different configurations depending on the angle which the plane of the disc makes with our line of sight. M 104 is nearly edge-on to us while M 31 is inclined at about 15°, M 81 at about 60° and M 51 is practically face-on.

Barred spirals (M 58, M 91, M 95)

These galaxies generally have only two spiral arms which extend from opposite ends of a conspicuous central 'bar'. The earlier forms are open like the letter S but the later, SBa forms are more like the Greek theta (θ). The composition of stars and gas follows a similar distribution to that in the normal spirals but it is thought that these galaxies may evolve from the irregular types, since the Large Magellanic Cloud, which is classed as irregular, shows some appearance of developing a 'bar' and possibly one spiral arm.

The 'bar' in M 58 can be seen aligned approximately E–W but it requires at least an 8-inch or 10-inch aperture to see well.

Irregular galaxies (only M 82) Those resolved seem to contain mainly Population I stars but they may be accompanied by many globular clusters which contain mostly Population II stars. The best known irregular galaxies are the two Magellanic Clouds, our nearest neighbours in the Local Group. De Vaucouleurs suspects that the Smaller Cloud may be developing into a barred spiral also but this is difficult to tell as it appears almost edge-on to us. Other observers consider that the 'organised' appearance is merely due to tidal action.

Although M 82 is generally classed as irregular it may in fact be a 'peculiar' type of edge-on spiral.

Clusters of galaxies Of the 102 objects in Messier's catalogue 33 are galaxies: W. Herschel classified the objects which he observed and in the first three categories of nebulae in his lists well over 1000 are known to be external galactic systems. With the introduction of astronomical photography and use of larger apertures it soon became apparent that these objects were to be numbered in millions. In a modern instrument like the 48-inch Schmidt camera the exposure of a single plate may reveal more galaxies than there are entries in the *New General Catalogue* which lists some 13 000 nebulae and clusters.

The galaxies are not distributed evenly in space but show a distinct tendency to congregate in clouds or clusters and even into super-aggregations of groups of clusters containing hundreds and sometimes thousands of members. In Coma Berenices some 800 galaxies appear to be concentrated into a region only 4 or 5 million light-years in diameter. This, by astronomical standards, represents a state of considerable traffic congestion and it has been considered that in these circumstances, collisions between galaxies would be comparatively frequent – perhaps two or three encounters for each galaxy every 1000 million years. The chance of individual stars colliding would be extremely remote as they are well separated in the galaxies but the gas and dust would be heated by collision and swept away. The fact that all the galaxies in this group appear to be elliptical and SO types seems to support this hypothesis.

The gas in collision would also produce large amounts of energy in the radio spectrum. Several known radio sources such as the one known as Cygnus 'A' are suspected as being caused by such interpenetrating galaxies, but other theories exist and it may be a long time before we have an adequate explanation of these remote and tremendous events.

In the constellation Cetus there is a group of galaxies which appears to exhibit a certain amount of organisation among the 800-odd members of which it is composed. The alignment of the axis of so many of them in a P.A. of 130° is well above any chance arrangement. In addition, an unusually large number of them is seen in an edge-on configuration, but this may be due to the fact that face-on galaxies appear more diffuse and their lower surface brightness makes them less likely to be observed.

Besides the rich cluster of bright galaxies on the borders of Virgo and Coma Berenices there is another group within reach of the amateur's telescope in the extreme SE part of Virgo, quite close to the globular cluster M 5. This group contains 16 galaxies brighter than 13th magnitude and most of these could be picked up in an 8-inch or 10-inch reflector.

Supernovae Apart from the fact that M 1 is itself a supernova remnant, many of the galaxies in Messier's list have produced visible supernovae, the first of which, *S Andromedae* in M 31 (1885 A), was the brightest and most famous. The following table lists all S/Ns which have been found up to 1989.

Table 4. *Supernovae discovered in Messier galaxies*

M 31	1885 A			
M 49	1969 Q?			
M 58	1988 A	1989 M		
M 59	1939 B			
M 61	1926 A	1961 I	1964 F	
M 63	1971 I			
M 66	1973 R	1989 B		
M 82	1986 D			
M 83	1923 A	1950 B	1957 D	1968 L	1983 N
M 84	1957 B			
M 85	1960 R			
M 87	1919 A			
M 99	1967 H	1972 Q	1986 I	
M 100	1901 B	1914 A	1959 E	1979 C
M 101	1909 A	1951 H	1970 G	
M 106	1931 K			
M 108	1969 B			
M 109	1956 A			

It can be seen that M 83 is the most prolific, having exhibited five supernovae in 50 years. This total is surpassed only by one other system, NGC 6946, an Sc galaxy, mag. 8.9 in Cepheus, which produced six supernovae between 1917 and 1980.

3 Observing: equipment, conditions and method

The factors which contribute to successful viewing of the Messier Objects can be reviewed under three main headings:

1. The telescope with its mounting and accessories
2. Seeing conditions, astronomical and meteorological
3. Observing method

Each of these will be discussed separately in some detail and all that needs to be said in introduction is that although the best equipment will yield the most satisfactory results, the word 'best' is not necessarily equated with 'most expensive'. Successful observations will depend on how well all three main requirements are met, of which the last is by no means the least important.

THE TELESCOPE

Messier's own telescopes were, by any modern standards, very poor indeed, especially with regard to resolution, for he was unable, for example, to distinguish any separate stars in the globular cluster M 13. The amateur of today, even with limited means and simple equipment, can expect to see much more than Messier himself was able to distinguish in most of the objects in his catalogue.

However, even with the best equipment, one must not expect to see too much: the superb photographs of the spiral galaxies and other objects taken with the 120-inch Lick telescope are due not only to the excellent optical instrument but also to the fact that the photographic plate is able to store the faint images until they are bright enough to register in the emulsion. The eye cannot do anything of this sort but it, and the brain behind it, can be taught to 'see' in several special ways and this will be covered in *Observing Method*.

The telescope: aperture The accompanying graph (Figure 3) shows the faintest magnitude star which can be seen for a given diameter of object glass or mirror, and Table 4 shows the Messier Objects in decreasing order of magnitude.

It would seem that all the objects except M 76, M 97, M 98 and M 100 could be seen in a 2-inch refractor or with binoculars of 50 mm clear aperture. However, none of the objects are single stars and where they are diffuse objects, as many of them are, they may be more difficult to see than the figure for magnitude may suggest. The graph also assumes that seeing conditions are good and that the telescope is optically sound.

Nevertheless, allowing a little latitude in all these conditions, a good 3-inch refractor or 4-inch reflector should reveal all the objects with the four exceptions above while a decent 6-inch reflector would certainly show them all in a dark sky. It must be remembered, though, that besides the greater light-gathering properties of a large aperture there is also a corresponding increase in the resolution or definition of the instrument, and for the observation of fine detail, the larger the aperture the better.

It should be noted that the magnitudes quoted in the table below are integrated magnitudes, i.e. the total intensity of the light observed, regardless of the area over which it is spread. The values of the magnitude limit in the graph are for a *star*, the apparent area of which is effectively zero. The light coming from extended objects such as nebulae and galaxies is often spread over a considerable area and the surface brightness is correspondingly reduced. The factor which most concerns the observer is contrast, or brightness gradient relative to the background sky. So many variables are involved that it becomes extremely difficult to draw up a list of values which would apply to all objects under all conditions. Experience, however, is the best guide, and when the observer has worked through the Messier catalogue once or twice he will have the best of all assessments of their visibility – his own.

The telescope: field of view For an objective of given focal length the area of sky visible will vary with the eyepiece in use and will always be inversely proportional to magnification. For a particular objective/eyepiece combination, the magnification can be cal-

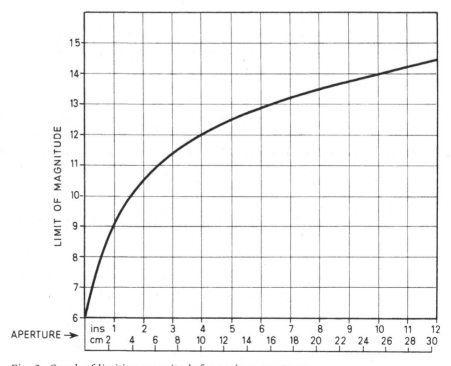

Fig. 3. Graph of limiting magnitude for various apertures

Table 5. *The Messier Objects in order of magnitude*

M. No.	Mag.	M. No.	Mag.	M. No.	Mag.	M. No.	Mag.
45	1.6	3	6.4	81	7.9	26	9.3
44	3.7	4	6.4	94	7.9	57	9.3
42	4.0	16	6.4	75	8.0	59	9.3
7	4.1	21	6.5	110	8.0	84	9.3
24	4.6	25	6.5	51	8.1	85	9.3
41	4.6	12	6.6	107	8.1	105	9.3
31	4.8	19	6.6	56	8.2	65	9.5
39	5.2	62	6.6	58	8.2	89	9.5
47	5.2	64	6.6	68	8.2	61	9.6
6	5.3	10	6.7	78	8.3	70	9.6
35	5.3	33	6.7	106	8.3	101	9.6
34	5.5	23	6.9	1	8.4	86	9.7
48	5.5	17	7.0	30	8.4	72	9.8
13	5.7	29	7.1	49	8.6	109	9.8
22	5.9	9	7.3	32	8.7	90	10.0
8	6.0	28	7.3	104	8.7	108	10.0
15	6.0	52	7.3	66	8.8	63	10.1
46	6.0	54	7.3	82	8.8	83	10.1
93	6.0	38	7.4	69	8.9	99	10.1
67	6.1	103	7.4	77	8.9	74	10.2
92	6.1	18	7.5	20	9.0	88	10.2
5	6.2	27	7.6	43	9.0	91	10.2
37	6.2	53	7.6	71	9.0	95	10.4
2	6.3	55	7.6	73	9.0	100	10.6
11	6.3	14	7.7	96	9.1	98	10.7
36	6.3	80	7.7	60	9.2	97	12.0
50	6.3	79	7.9	87	9.2	76	12.2

culated by dividing the focal length of the objective by the effective focal length of the eyepiece. Thus a reflector of 48 inches focal length used with an eyepiece of $\frac{1}{2}$-inch E.F.L. would give a magnification of $48 \div \frac{1}{2} = 96$.

Some of the Messier Objects, such as M 8, M 31, M 42, M 44 and M 45 cover more than one degree in at least one of their dimensions and so a fairly large field-of-view capability is an advantage. This can be obtained by having either a short focal length objective or a long focal length eyepiece. Astronomical eyepieces are not usually available with focal lengths much greater than about $1\frac{1}{2}$ inches but objectives can be obtained of very short focal length. The so-called 'rich-field' telescopes often used by comet observers usually have a focal ratio (ratio of focal length to aperture) of f.4 or even less, but this kind of telescope requires much greater accuracy in the figure of the objective to give satisfactory results, is more difficult to make and is consequently more expensive.

The telescope: eyepieces

The function of the eyepiece is to collect and magnify, with as little distortion as possible, the image already formed by the principal optics of the telescope.

These — the mirror and diagonal flat or prism in the case of a Newtonian reflector, the achromatic objective alone in a refractor — are the most important components of the instrument. Although eyepieces could be designed to correct some errors of the objective, this is hardly likely to happen by chance and if the objective is a poor one, not even the best of eyepieces will improve its performance. On the other hand, an inferior eyepiece may well fail to exploit the qualities of a good objective, so it is worth while paying some attention to the choice of this item of equipment.

The observer is not likely to come upon the 'perfect' eyepiece: many different requirements have to be met and any design must represent something of a compromise — hence the large variety of 'practical' eyepieces available.

To find and recognise objects such as nebulae and star clusters, which are often faint, large and diffuse, the observer will first require one good low-power wide-angle eyepiece which, with his particular telescope, will provide a field of view of at least one degree. The few Messier Objects which are larger than this, M 7, M 8, M 44, M 45, for example, may be viewed in the finder. To provide a field of one degree in telescopes of moderate focal ratio — say f.8 or f.10 — an eyepiece of about $1\frac{1}{2}$ inches effective focal length would probably be satisfactory. The design of the eyepiece, however, has also to be taken into account.

The Huyghenian eyepiece is a very old pattern — it was invented by Christiaan Huyghens in the 17th century — and its long existence is proof of its good qualities. However, it tends to produce serious distortion towards the edges of the field in telescopes of focal ratio less than about f.12 and so it is not really adequate when used in most modern amateur reflectors.

The Ramsden eyepiece is another well-tried design and is found frequently in modern prismatic binoculars. One disadvantage to this design is its small eye-relief so that the eye must be placed very close to the lens in order to view the whole available field. For spectacle wearers this is a decided objection.

The Kellner eyepiece is an improvement on both these types: it has improved eye-relief and provides a wide flat field combined with excellent colour correction. It has long been popular with comet observers whose needs are similar to ours and this type of ocular is usually easy to obtain and, an important consideration, inexpensive.

War-surplus, wide-angle, Erfle eyepieces can be excellent for low-power work but these may need an adapter to fit them to the standard $1\frac{1}{4}$-inch telescope drawtube and their generous field may be somewhat restricted by this modification.

For moderate magnification — and this can be applied with advantage to most of the Messier Objects once they have been picked up — an eyepiece of about $\frac{5}{8}$–$\frac{3}{4}$-inches E.F.L. would provide the next convenient step. As this eyepiece is the one which is most likely to be used for general work, it might be as well to obtain the best one can afford. The 'orthoscopic' design, of which there are several variants, provides a wide, flat field and excellent contrast and although more expensive than the equivalent Huyghenian or Ramsden, it could be considered a wise investment. The 'monocentric' design is also admirable, especially when used with telescopes of small focal ratio where demands are most exacting, but this may be more expensive still.

A higher power will be necessary when observing detail in small objects like the 'Ring' Nebula, M 57, and for the resolution of the globular clusters. Several of the Messier galaxies, too, such as M 51, M 64 and M 104 may well reveal a fair amount of detail under high magnification in good conditions.

The least expensive, and in many ways the most satisfactory method of obtaining high magnification is to use a Barlow lens in conjunction with either of the two oculars mentioned previously. The Barlow is a plano-concave lens which, when placed inside the prime focus of the objective, decreases the convergence of the rays, increases its effective focal length and thus provides higher magnification with the same eyepiece. The Barlow should be achromatic, i.e. a cemented doublet, of focal length about minus 2 to 3 inches and mounted in a short tube so that the distance between it and the eyepiece can be varied. This combination provides a kind of variable-magnification eyepiece of great flexibility and good performance. The Barlow can be purchased already mounted but for the observer who wishes to do this for himself, details of the method of mounting, including some ingenious alternative arrangements, can be found in works mentioned in the Bibliography.

'Zoom' eyepieces of variable magnification can also be obtained and although these may be frowned upon by some observers, the convenience of being able to vary the magnification by a simple twist of the thumb and forefinger, and without losing sight of the object in view, is not to be despised.

Prismatic, turret-type, multi-eyepiece arrangements may be used with refractors with some gain in convenience but, as with 'zoom' eyepieces and other such sophistications, penalties may well accompany their complexity. Indeed, it is possible to interpose so much glass between the eye and the object, that, in marginal cases, the light may well get lost on the way. It may be noted with profit that Wm. Herschel adopted the 'front view' arrangement for his great reflectors with the one object of avoiding the light loss due to the secondary mirror of the Newtonian design. He also employed a single-lens eyepiece for all except wide-field work in order to eliminate the losses in multiple lens oculars. Modern eyepieces with coated lenses are, however, much superior in this respect to those available in Herschel's time and few people use single-lens oculars today.

The telescope: filters
The use of filters in visual observing is a complex – and often contentious – subject. With a telescope of moderate aperture and clear, dark sky conditions, all the Messier Objects can be picked up with little difficulty. However, the increasing spread of light pollution from city and suburban areas makes the detection of low surface brightness nebulae more and more difficult. The employment of eyepiece filters which can block a good deal of mercury and sodium street lighting may make all the difference in many cases. Those with a wide band-pass, usually marketed under the name of 'Deep-Sky' or 'Nebular' filters may cost about £50 or so, and are often effective in enhancing the visibility of most nebulous objects. More selective, narrow band-pass filters can be obtained for use in detecting particular classes of objects such as planetary or some gaseous nebulae and are sometimes called 'Ultra High Contrast' (UHC) filters. The price range for these may well be around £100. For

the observer attempting to view the objects of the Messier catalogue for the first time it would be preferable to try to find a reasonably dark location for his telescope and to dispense with any kind of filter – at least to begin with. Certainly, before expending such sums as those mentioned above, it would be wiser to consult a friendly owner of such equipment for advice.

The telescope: the finder

A finder telescope is not a luxury; it is as important as the principal telescope itself: it may be simple, even crude, but it is vitally necessary. It should, of course be of smaller focal ratio than the main instrument so that a larger area of sky can be examined. It should be fitted with cross-hairs and be accurately aligned with the optical axis of the main telescope and, most important of all, should have an adequate aperture. I think it may be safely said that the finders supplied with most 'off-the-shelf' telescopes are too small and so are inadequate for the very important purpose of locating objects which are in any way 'difficult' to pick up, as many of the Messier Objects are. The objective of the finder should be not less than a quarter of the diameter of the objective of the telescope: one-third would be better still. A simple ring-and-bead or rifle type sight, fitted to the finder, may also help with the preliminary sighting.

The telescope: the mounting

Whatever type of mounting is used, it must be steady and movements about both axes should be smooth: if these two requirements are fulfilled there should be little difficulty either in finding or following any of the Messier Objects.

An equatorial mounting is naturally more convenient than an alt-azimuth, and a clock, or electric drive for the former is a refinement which is certainly desirable especially when using higher magnifications. Anyone with a little mechanical ability can build an excellent mounting out of old motor car components and there are plenty of these available at the breaker's yards. An old record-player unit can be adapted to give a very suitable source of power for a slow-motion drive if electricity is available at the observing site.

An observatory with a rotating dome is, no doubt, a fine luxury but the narrowness of the slit can be a drawback for wide range viewing and a simple run-off shed or a structure with folding walls and roof-sections may prove to be more convenient and cheaper. Even the most primitive screen or shelter will be a great boon on cold and especially on windy nights: this could be of canvas and be made easily portable but it should be capable of being firmly anchored otherwise it may prove more of a nuisance than a help.

THE OBSERVATORY

The choosing of a site for the observatory is discussed briefly under *Altitude and Culmination*. The fittings and equipment may be considered here. A firm base for the telescope is the only essential requirement; this may be of concrete and should be as substantial as possible, well sunk into the ground to a depth of at least 18 inches and quite independent of the floor structure. If much heavy traffic runs nearby, 'land wave drag' may cause unwelcome vibration even in a solid concrete base and in this case a cure may be found by mounting the telescope on heavy rubber 'shock absorbers' fastened firmly to both telescope and base.

Even on an open site for a portable telescope it is very necessary to have small concrete blocks set up in a permanent position so that the instrument can be aligned quickly as well as being provided with firm support. Here, too, a light, portable plywood floor can be an inestimable boon when dew and frost are present.

There is practically no limit to the amount of gadgets one may accumulate in the observatory but perhaps the most useful item is a comfortable, adjustable observing chair. This not only reduces fatigue during long observing sessions but also, by the same token, improves the quality of observations. Just as the telescope is provided with a firm base, the eye should be given one also and the head is steadier when one is comfortably seated.

THE DARK SKY

It is not always realised that the 'darkness' of the sky can be a very variable quantity. The period after sunset until the Sun's centre has descended to 6° below the horizon is called *evening civil twilight* and even at the end of this period the sky is quite light although a few of the brightest stars may be visible. A similar interval, of course, occurs before sunrise but this may be a less familiar sight to all but the most dedicated observers.

The period during which the Sun travels through another 6° until it is 12° below the horizon is called *nautical twilight**: during this time all the brightest stars will become visible while there is still enough light in the sky for the horizon to be distinctly seen, enabling the nautical navigator to take his sextant observations.

The subsequent interval while the Sun travels a further 6°, becoming 18° below the horizon, is called *astronomical twilight* and it is not until the end of this period that all trace of light in the sky directly derived from the Sun can be said to have gone.

The duration of astronomical twilight is not a constant value but varies both with the observer's latitude and with the sun's changing declination throughout the year. (The remaining factor, the observer's elevation above sea-level, is very small even for heights of several thousand feet.)

The largest factor is, by far, the observer's latitude: at the equator, the total of civil, nautical and astronomical twilight lasts for about 70 minutes and varies only a little throughout the year. At the poles, on the other hand, astronomical twilight alone is continuous throughout the 24 hours when the

---

*   Strictly speaking, the whole period counting from sunset is assessed as nautical twilight but it is more convenient here to separate the intervals. Similar remarks apply to astronomical twilight.

Sun's declination is between 12° and 18° below the Equator – a period of nearly 3 weeks in both spring and autumn.

Graphs of the duration of astronomical twilight, together with the local mean times of sunrise and sunset for 32°, 42° and 52° N latitude are included below (Figures 4, 5 and 6).

At 32° N – the latitude of Savannah, Georgia – it will be seen that the total period of twilight varies only a little from about 80 minutes in spring and autumn, to about 1 hour 40 minutes in midsummer, when the length of the 'dark' night is about $6\frac{1}{2}$ hours.

At 42° N (Chicago) the range is from 1 hour 30 minutes for the shortest period to 2 hours 15 minutes for the longest, which leaves an interval of complete darkness, even in June, of more than $4\frac{1}{4}$ hours.

At 52° N which passes through the southern parts of Canada and the British Isles, the position is much more difficult. Here, the period of twilight amounts to a minimum of about 2 hours in January, increasing gradually until about the middle of May, when the end of evening astronomical twilight merges with the beginning of morning astronomical twilight, so that no part of the night is completely dark. Near the end of July a short interval of 'darkness' begins again at local midnight and this lengthens gradually until January, when it is again at its longest.

Observations *can* be made during astronomical twilight, especially in that part of the sky away from the bright horizon, but conditions will not be at their best and for some of the fainter and more diffuse objects, the presence of only a trace of twilight may render them quite invisible in the telescope. The months of June and July are especially difficult for observers in the higher latitudes. The diffuse nebulosity in M 8 and M 20, both of which culminate

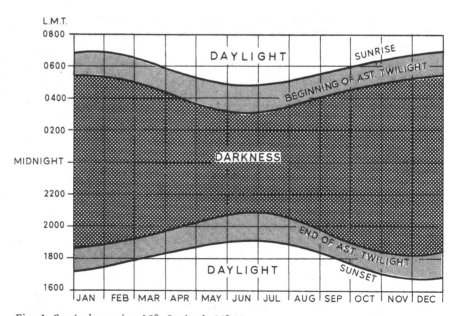

Fig. 4. Sun's depression 18°. Latitude 32° N

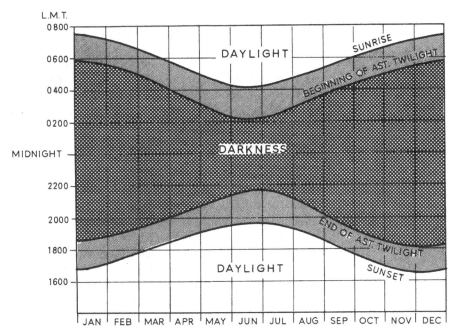

Fig. 5. Sun's depression 18°. Latitude 42° N

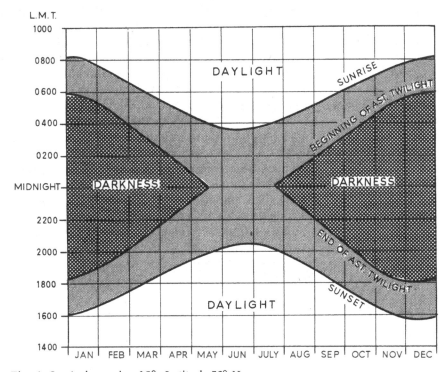

Fig. 6. Sun's depression 18°. Latitude 52° N

41

near midnight in these months, can be very elusive, both by the presence of twilight and by reason of their low altitude at culmination.

Another intruder into the dark sky is, *pace* lunar observers, the Moon. For at least five or six days either side of full Moon the viewing of Messier Objects may well be abandoned: the fainter galaxies and diffuse bright nebulae will probably be quite invisible and even the bright galactic clusters, flooded by moonlight, will look pale and insignificant. Again it is unfortunate that the period of July and August when the important group of Messier Objects in the Sagittarius region becomes accessible, is also doubly difficult as regards the full Moon.

At this time of the year the Moon when full has a south declination and has a low altitude for northern hemisphere observers: so also has the Sagittarius region. Furthermore the full Moon is then changing its declination northwards at a fairly rapid rate so that the amount by which it should rise later each successive night is reduced. The observer is thus confronted by a large Moon which is reluctant to leave the most wanted part of the sky.

This phenomenon, which in September is called the *Harvest Moon*, may well be a boon to the farmer but it represents a combination of celestial circumstances very unfavourable to the hunter of Messier Objects. Naturally, this situation cannot be altered: it merely means that every opportunity has to be taken during the short Moon-free periods to keep up-to-date with one's observations; otherwise the occasion may be lost and will not recur for another year.

The other forms of natural light in the sky such as *air-glow*, caused by light-emitting molecules and scattered starlight, although detracting from the darkness of the night, represent merely the penalties for living at the bottom of a deep atmosphere. The *Zodiacal Light* and *Gegenschein* have their causes outside our atmosphere, being derived from sunlight reflected and diffused by small meteoric particles in space. As phenomena, they are perhaps as interesting to observe as the stars themselves and it is only those devoted to astronomical photography who complain of this kind of interference.

The presence of artificial light, however, is for many observers the worst nuisance of all. The glare of nearby street lamps and the wide-angle, diffused light from towns even 10 or 15 miles away can seriously impair good seeing and it is an interference which does not wane with the seasons or lunar period. From the point of view of the astronomer the continued spread of street lighting to the countryside is to be deplored.

There is little one can do to mitigate this except to move house – which may prove an expensive remedy. If only one or two distant lights offend, a length of fencing or evergreen hedge, strategically placed, may screen some of the direct glare but the diffused light will still prove an astronomical hazard.

METEOROLOGICAL CONDITIONS

It may seem absurd to warn against trying to make astronomical observations in cloudy conditions but not all cloud is eight-eighths strato-cumulus: a thin layer of alto-stratus or cirro-stratus which may be quite evident during the daytime may not, at first glance, be at all apparent during the night. If, when the sky looks quite clear and the stars are all in view, the observer finds that some of the more diffuse objects like M 1 or M 33 are tantalisingly elusive, he should not immediately suspect his failing eyesight. A thin, almost invisible layer of very high cloud may well be to blame and this can often be detected by looking at the brightest object visible to the naked eye. Venus or Jupiter or Sirius under these conditions will exhibit a slight but unmistakable faintly coloured halo or corona and the whole sky, on closer examination, may have a slightly milky appearance.

Mist, forming by radiative cooling as it often does on autumn nights, has a similar dimming effect but this is much more noticeable and is confined to objects of fairly low altitude.

As an example of the effect of atmospheric conditions on the viewing of Messier Objects, the two galactic clusters M 46 and M 47 in Puppis may often provide a striking demonstration. These two clusters lie within $1\frac{1}{2}°$ of each other: M 47 is of mag. 5.5 and is an open cluster containing only a few bright stars while M 46 (quoted as mag. 6 but probably a magnitude fainter) consists of a large number of faint stars in a close cluster. If there is any sort of atmospheric obscurity, M 46 may be quite invisible while M 47 shines bright and clear with no appreciable dimming. On a clearer night M 47 may appear almost unchanged except that a few more faint stars may appear among the bright ones, but the difference in M 46 may be spectacular. Now hundreds of minute stars will be seen scattered over an area of nearly 30' diameter and seeming incredible that none of this had been visible before.

The meteorological situations which give the best viewing conditions for faint objects are often those in which the atmosphere is unstable, giving rise to large cumuliform clouds and showers during the day but which die away during the night. The instability allows dust and smoke in the atmosphere to escape upwards and the rain also has a marked cleansing effect on the air.

Anticyclonic situations in the summer, although providing cloudless skies and welcome sunbathing during the day, are often poor at night owing to the accumulation of heavy haze especially towards the horizon. Winter anticyclones tend to produce sparkling, crystal-clear nights (and very cold observatories) but the rapid cooling of the ground often produces varying density gradients in the atmosphere which result in a wavering or 'boiling' of the image especially under high magnification. This may be more of a nuisance to planet observers, straining to make out fine detail, than it is to 'deep sky' watchers who require transparency of atmosphere above all else, but it may still prove troublesome when detail is required.

A north-westerly air stream, often found after the passage of a cold front, is always unstable to some degree and so invariably produces good visibility.

The moderate wind and turbulence also tend to smooth out some of the discontinuities of density at least in the lower layers of the atmosphere and so give excellent seeing conditions – providing, of course that the air is not so unstable as to cause too frequent showers or thunderstorms during the night.

On occasions when the atmosphere has a high relative humidity and surface temperatures fall rapidly on clear, calm nights, dew forming on the optical surfaces of the telescope and eyepiece can be a major nuisance. When this occurs the moisture should never be removed by wiping or rubbing as damage to the mirror or lens will inevitably follow. A most effective tool for dispersing dew is an old hair-dryer or vacuum cleaner adapted to 'blow' rather than 'suck'. Even without a heating element in the circuit, the copious current of air clears the condensation in a matter of seconds. One word of warning: no oil should be given to the motor bearings for some of this may evaporate and be carried by the air-stream to make matters worse instead of better. If electrical power is not available in the observatory, some kind of reasonably large hand bellows, however crude, will do almost as well.

ALTITUDE AND CULMINATION

Whatever the state of the atmosphere, there is bound to be some loss in a ray of light passing through it and the amount of dimming will be directly proportional to the length of the path taken. A ray from a star at the observer's zenith will traverse the shortest path through the atmosphere but at lower altitudes the route will be more oblique and a loss in brightness will be apparent.

This loss is roughly proportional to the cosine of the altitude and so the brightness of an object will be reduced to about 75 per cent of maximum at 50° altitude and to 50 per cent at altitude 30°. Below 30° the loss becomes severe, the brightness being about 25 per cent at 15° altitude while at 6° above the horizon only 10 per cent of the light is transmitted.

As the maximum altitude of a body occurs at *culmination* (i.e. when it is on the observer's meridian and bears either due north or south) this is naturally the best time for observations.

To observe the maximum number of the Messier Objects the observer will need to have his view to the south as unobstructed as possible. All observatory sites are imperfect in some respect or other and the best position is generally arrived at through compromise but, if at all possible, it may well be an advantage if the SE aspect is kept fairly clear. This enables the observer to obtain a preliminary view of the object before it has reached culmination knowing that later on – in time or date – the altitude will be higher and so more favourable.

Finally, if one's own trees or fences obstruct the view they may well be demolished at will but if it is a neighbour's property which offends it is necessary to exercise the utmost restraint. Action may constitute trespass, requests looked upon as infernal impudence and even a mere suggestion may be a cause for lasting hostility. Instead, the man-next-door might be invited in to look at Saturn or M 104 at a carefully arranged time when the obstruction is just in the way. If the hint isn't made too obvious it may work wonders.

OBSERVING METHOD

Finding The easiest way to find an object in the heavens is to have a telescope accurately aligned with the meridian, fitted with properly calibrated hour angle and declination circles and linked to a precision sidereal time clock. The co-ordinates of the body are then set up and a glance through the eyepiece should then reveal the object in the centre of the field.

This is perhaps ideal and probably very few amateurs are so luxuriously equipped so here we shall deal with less sophisticated methods, assuming that the observer has only a basic telescope plus a simple, correctly aligned finder.

The first requirement is to estimate as accurately as possible the size of the field of view on one's lowest-power eyepiece. This is very simply done: merely direct the telescope on to any easily visible star on or near the celestial equator. Suitable stars are: α *Ceti*, any one of the stars of Orion's belt, *Procyon*, γ and ζ *Virginis*, λ and γ *Ophiuchi*, θ *Aquilae* and α *Aquarii*; but in fact any star, faint or bright between about 5° N and 5° S declination will do provided it can be seen clearly in the eyepiece.

Keeping the telescope stationary, the passage of the star should now be timed across the full diameter of the field of view, using a stop-watch if possible. As the Earth rotates through 15° in 1 hour or 1° in 4 minutes of time, the angular field of view in minutes of arc will equal the number of seconds of time taken in the transit, divided by 4. (Note: the 'time' measured should strictly be that of sidereal time but the difference from mean solar time for these small intervals may be ignored.) It is advisable to repeat the operation several times to get an average value and the calibration may well be applied to any other eyepieces in frequent use and the figures noted inside the cover of the observatory log for reference.

Stars of higher declination may be used with equal facility but in this case the value obtained must be multiplied by the cosine of the declination to obtain the correct figure.

An even quicker method can be used when the Pleiades are visible by identifying the individual stars visible at the extremes of the field. From the large-scale map of M 45 included as Map 9, use the scale of declination to give the diameter of the field in minutes. This method will also provide a little practice in recognising the brighter stars in this important galactic cluster.

The L.P. eyepiece with its field of view determined can now be used to step off the required intervals along N–S and/or E–W co-ordinates from the 'reference' star as given on the individual pages for each Messier Object. It is important to get into the habit of working systematically from the very beginning: taking a quick 'stab' at it may work once or twice, if you are lucky; generally it will not, and weaving round and round the vague position of the

45

object will turn out to be a waste of time and a considerable and unnecessary strain on one's temper.

The visibility of objects

The different classes of object in Messier's list requires slightly different techniques for observing them to the best advantage and these details are given below.

Gaseous nebulae

These are mostly large and some, like M 16 and M 20, are fairly diffuse: low power should be employed to give a wide field of view, and opportunity should be taken during the best seeing conditions. These remarks also apply to M 1. M 17 is considerably bright in parts and will stand higher magnification but the fainter nebulosity is best seen in a wide field. M 42 is easily visible in any conditions and there is something different to be seen in all ranges of magnification of this famous and spectacular nebula.

Galactic clusters

As these objects stand out most clearly by contrast with a generally sparse background, a wide field of view is the most effective. The larger clusters, like M 7, M 44 and M 45, extend beyond the field of medium focal ratio telescopes even on lowest power and a 'rich-field' instrument is needed to give the finest view. M 24 is a 'difficult' cluster but this is discussed in detail under its heading. Most galactic clusters are bright and may be seen even in moonlight but the more numerous, fainter stars may be lost in anything but a dark sky.

Planetary nebulae

M 57, like many planetaries, is small but bright and requires medium to high power to reveal its disc and higher power still to show the annulus. M 27 is an easy object and will show plenty of detail on high power: on the other hand, M 76 and M 97 are rather faint and diffuse and require low power and the very best conditions to be seen at all.

Globular clusters

As a class, these objects are the easiest to pick out as they are mostly bright and of reasonably large angular diameter. When found, they should be examined with the highest power they will bear when the outer regions, at least, should be capable of resolution into individual stars. A comparison of their different degrees of concentration as classified by Shapley, can prove a very interesting study.

Elliptical galaxies

These are among the smaller objects but they usually have a fairly high surface brightness and often present a characteristic 'pearly' appearance. Higher powers can generally be employed but apart from revealing an elliptical figure in some cases, no real detail can be made out even in quite large telescopes, and, for the amateur observer they are less interesting on this account. Similar remarks apply to the two SO galaxies in Messier's list.

Spiral galaxies

These objects exhibit a wide range of size and brightness: M 31 and M 33 can be seen with the naked eye while others like M 74, M 83 and M 98 are among the faintest of the Messier Objects. In nearly all cases it is only the central nucleus which is first seen in the telescope; the fainter, diffuse outer areas require low power and the use of averted vision to render them visible.

The edge-on galaxies such as M 82 or M 104 appear more distinct as they have a clearer outline and the 'spindle' shape is conspicuous. On the other hand, those galaxies which are presented face-on or nearly so are much more diffuse and the fainter objects of this type can appear very 'pale'. The Sc galaxies which lack a concentrated, bright nucleus can be the most difficult of all to pick up and it is for this reason that M 33, although having a total integrated magnitude of about 6, can disappear completely in anything but a dark sky. M 74, a face-on Sc galaxy in Pisces of mag. 10.2, is one of the most elusive of all the Messier Objects and it requires optimum conditions to be seen clearly in a small telescope.

The spiral galaxies in Messier's list can provide a great deal of interest for the amateur observer for many different types are represented: considerable detail can be made out in many of them and each one has its own recognisable characteristics which are easily seen and remembered with a little practice.

Dark adaptation It is a common experience that it takes a little time for the eyes to become adjusted to seeing 'in the dark'. The adjustment at first is quite rapid: after 10 minutes or so the improvement in night vision is most marked but it is not generally realised that this improvement continues, at a slowly decreasing rate, for a long time. Experiments have shown that even after 10 hours or more the eye has not reached the limit of its adaptation. This acquired adjustment to low-level illumination is, of course, completely lost if the eyes are exposed, even momentarily, to a bright light and, for this reason, all lamps used in the observatory for consulting star maps, etc., should be fitted with some kind of dark red filter.

If a bright light *has* to be used for some purpose or even if one has to go indoors for a short period, the dark adaptation of *one* eye at least can be preserved by closing it and doing what is necessary in a one-eyed manner. One's appearance may be thought a trifle odd but the accumulated adaptation of an hour or more is a valuable physical asset worth preserving.

This low-level sensitivity of the eye is not due only to the opening of the iris (this is an almost instantaneous reflex), but more to the gradual increase in the sensitivity of a small area around the centre of the retina. This property of dark-adjusted sight can be exploited when looking in the telescope for faint objects. If the gaze is focused, not directly at the object, but slightly to one side of it, the image will fall on the more responsive, outer portion of the retina, and a distinct improvement in perception will result. This use of 'averted vision' takes a little practice to acquire but it is a very valuable and necessary technique and soon becomes automatic.

Another physical quality of the eye may be used in the case of the very faintest objects. Many animals are able to detect, with great acuity, objects which are moving but seem unaware of them if they 'freeze'. Some of the mechanism for detecting objects in motion is in the retina itself and does not need the brain to interpret the information. In man, this faculty is of a very low order but it can be effective in marginal conditions. It may be exploited when looking for an object just on the threshold of detection by gently tapping the telescope so that all objects in the field of view are given a slight oscillation. This may well disclose whether the object is there or not and will also eliminate from consideration any spots or marks on the lenses of the eyepiece which will, of course, show a relative movement. This faculty is very slight and it may vary considerably with different observers but there is little doubt that the method works.

There are many factors which affect night vision. Visual fatigue due to long exposure to over-bright and over-active television images may persist for periods of an hour or more. Tobacco and alcohol are both deleterious drugs in this context, but on the other hand, aviation medicine has established that regulated doses of carotin (derived from carrots) greatly enhance the dark adaptation of night fighter pilots.

OBSERVING TECHNIQUE

The amount an observer may 'see' is governed, not only by physical qualities – the telescope, the atmosphere or eyesight – but also by mental or psychological qualities which may be summed up as the ability to 'observe'. This very important attribute, like 'traffic sense' or a 'good eye' in ball games, is difficult to define; to some extent it may be inborn but the ability can be acquired or greatly improved through training and experience.

What, to the casual eye, appears to be a mere 'blob' in the field of view of the telescope, may, with critical and organised examination, reveal a large amount of detail which, although faint, may be established with certainty. Much of the success in dealing with the apparently featureless 'blob' may be achieved through the medium of a number of self-examining questions thus:

   (i) Is it round, or oval, or spindle-shaped or what?
   (ii) If not round, in which direction is it elongated?
  (iii) Is it brighter or less bright in the centre?
  (iv) Is the brightness gradient gradual or abrupt?
   (v) Are there any dark markings?
  (vi) Are there any star-like points of light involved?
 (vii) Is there any trace of an outer ring?
(viii) What are its dimensions? (Compare known size.)
  (ix) What is its magnitude? (Compare known mag.)
   (x) Does it resemble any other known object?
      Etc, etc.

These questions are only examples and one may devise many more. The important thing is to ask them – all of them, all the time – until the process becomes automatic. When this stage has been reached one has become a 'good observer' . . . and still capable of further improvement.

There remains the matter of objectivity. It may well be that one is looking for an object already well described and photographed and vision is notoriously coloured by preconceived ideas. In this case it is even more vital to repeat the same catechism and be even more critical with the answers. If some of the observed features are distinctly at variance with a previous description then either one is looking at the wrong object, and this might prove embarrassing, or the original description was faulty and one can have the smug satisfaction of correcting it. There are probably more than a few errors of long standing in the description of quite common objects which have been perpetuated by a series of careless observations.

Naturally, it is always easier to make out a feature which has been seen and described before: like the 4-minute mile or climbing Everest – once it has been done, it is easier to repeat. Sir William Herschel once remarked, somewhat sourly: 'When an object is once discovered by a superior power, an inferior one will suffice to see it afterwards.'

The observer, however, should attempt to train his observational powers by viewing suitable 'difficult' objects *before* reading the previous description and, using the method outlined above, should write down a detailed description of his own. On comparing this with the observations of the authority with the 'superior power' he may well be surprised, and gratified, with the quality of his results.

In pursuing this course, the successes will inevitably be matched by a corresponding number of 'failures' and one's own weaknesses soon become apparent and will require correction. However, with perseverance and a continual, conscious effort to try to improve, the rewards of a highly developed observational sense may prove more valuable than buying bigger and better equipment, and, of course, considerably cheaper.

It might be fitting to conclude with another quotation from that great observer W. Herschel; a modest, and much more characteristic claim. 'Seeing is in some respect, an art which must be learnt. Many a night I have been practising to see, and it would be strange if one did not acquire a certain dexterity by such constant practice.'

RECORDING OBSERVATIONS

The importance of maintaining a meticulous record of all observing sessions cannot be stressed too firmly. A first requirement is the purchase of a simple, hard-cover, ruled exercise book to serve as an 'Observatory Log'. The pages can be ruled into columns to include entries of date, beginning and ending times of observation, weather conditions (temperature, wind velocity, cloud cover, sky transparency, etc.) objects observed (or looked for even if not seen) and a fairly generous space for 'Remarks'.

If you are going to keep a written record of your observations while at the telescope itself, you will need a separate clip-board to hold your current working sheets and a dim red light source. Some observers find it more convenient to dictate their observing remarks into a miniature tape-recorder, transcribing the aural record later. If, however, you intend to make drawings at the telescope (and all observers, however doubtful of their 'artistic' ability, should attempt this from the beginning) you will need the clip-board and red light anyway.

Drawing at the telescope

For sketching at the telescope first prepare a suitable stock of drawing blanks on which the field-of-view is outlined by a circle (say 10 cm in diameter) within which is another concentric circle of half that size. Upon this both the vertical and horizontal diameters should be drawn to complete the graticule. Before drawing is begun it is vital to mark the field orientation with at least an 'N' (for North) and either an 'E' (for East) or an 'f' (for 'following'). A clear indication of the field diameter for the eyepiece in use should also be made.

Once the desired object has been found and verified the observer should commence with putting in a few of the brightest field stars as markers, preferably one or two in each quadrant. If a mistake is made in plotting a star it is safer to mark the error firmly with an 'x' so that it can be eliminated at the redrawing stage.

Star clusters containing no nebulosity can be drawn with an ordinary B or HB pencil using graduated sizes of 'dots'. The brighter and richer clusters such as M 44, M 45 and M 67 cover a wide field and may have to be dealt with by first adding a larger (say 15 cm diameter) circle to the drawing template and manipulating the telescope to cover the greater area.

Drawing globular clusters is generally a much less onerous task. There is little point in attempting to plot the individual stars in the immediate boundary of the globular, but it is possible to build up, by a controlled pattern of small dots, a recognizeable portrait of each one which would serve to distinguish it from other members of its class.

All the planetary and gaseous nebulae are interesting subjects for portraiture, varying from the tiny yet precise oval of the 'Ring' Nebula (M 57) to the immense and diffuse Orion Nebula (M 42, M 43). By any criterion this latter object is unique and the region is best exploited by covering the whole complex in two or even three overlapping fields.

The extra-galactic nebulae (galaxies) also differ widely in the demands they impose upon the draughtsman. Most of the elliptical galaxies appear as fairly featureless ovals or circles, and the main differences lie in the light gradient between centre and periphery and in the orientation of the major axis.

Among the spiral galaxies there are many gradations of form which need to be delineated and these are best analysed by reference to the 'catechism' set out on page 48. When one turns to the 'showpiece' galaxies such as M 31, M 33 or M 101, the best approach is to use a fairly high magnification and let the object drift through the field of view in successive strips in declination, building up the picture by 'scanning'.

Finally, each drawing made at the telescope, however carefully done, is bound to have numerous blemishes solely due to the difficult conditions endured by the observer. As long as adequate written annotations are made as marginal notes, the sketches can always be redrawn at leisure, but the revision should follow the original with as little delay as possible.

4 *The Messier Objects in detail*

INTRODUCTION

On the following pages each of the Messier Objects is described in detail: the notes below may serve to explain the arrangement and derivation of the information given.

1. The objects are arranged in their order as published in Messier's catalogue. Their corresponding numbers in the *New General Catalogue* of Dreyer are also given except for M 45 which was not listed in the NGC. Those objects which were seen by Lacaille in 1751–52 are given his classification also; those observed by de Chéseaux, the numbers in the list of 1746, and those by Bode, their numbers in his 1777 catalogue.

2. 'Popular' names are given for a few well-known objects, e.g. the 'Crab' Nebula for M 1, etc.

Each object is given its classification according to type, as follows:

(i) Gaseous nebulae: 'emission' or 'reflection'.
(ii) Planetary nebulae: their classification according to Prof. Vorontzov-Veljaminov as detailed in Chapter 2, page 26.
(iii) Galactic clusters: Shapley's classification as explained in Chapter 2, page 20. R. J. Trumpler's classification is also given in the text.
(iv) Globular clusters: Shapley's classification of the degree of concentration, class I representing highest concentration and class XII, the lowest.
(v) Galaxies: Hubble's earlier classification as explained in Chapter 2, page 28.

3. Positions of the objects are for the epoch 1950.0. Positions for epoch 2000.0 are given in Appendix 1.

4. Angular dimensions, visual magnitudes and distances are quoted for each object. For most objects, the angular apparent size will vary with circumstances as better atmospheric transparency and larger aperture will enable fainter and fainter portions to be seen. The figures given here are intended as a guide to the size of the object when seen in a 6–8-inch reflector in good average conditions.

The apparent magnitudes have been obtained from a variety of sources. I have selected those values which seemed to me to agree best with my own observations. Some comments, where applicable, are included in the text.

Distances are also variously derived. In general, the most recent estimates have been used, but especially for extra-galactic systems, the values quoted should be considered only the best approximations available.

5. Next follows a description of each object by its discoverer (if known) and by Messier's own original notes and date of observation. The descriptions by later observers such as Smyth and Webb are also included and have been selected to show the development of observation since Messier's time.

Messier's descriptions have been taken directly from an original copy of the *Connaissance des Temps* for 1784, in a somewhat free translation made by myself which, I hope, has not strayed too far from literal accuracy. A more detailed account of the *Connaissance des Temps*, including Messier's introduction to his catalogue, is given in *Notes on Sources* in Chapter 5. Admiral Smyth's descriptions and comments have been selected from his *Cycle of Celestial Objects* of the later, augmented edition by G. F. Chambers, 2nd edition, Oxford, 1881. Those of Webb are from the 6th edition of *Celestial Objects for Common Telescopes*, revised by T. E. Espin, Vol. II, London, 1898.

The descriptions under the heading, *Lick XIII*, are from the Lick publication, *Studies of the Nebulae* by Heber Doust Curtis. Further information about this authority is also given in *Notes on Sources* in Chapter 5.

Brief biographies of many of the early observers of nebulae and of other astronomers connected with Messier are included in Chapter 5.

6. More recent astrophysical data and information of interest are given for most of the objects. The amount of modern information available varies greatly from object to object. Some, such as M 1, M 31 and M 42, have received almost continuous attention from astronomers while others, such as the galactic clusters M 18 and M 24, the globulars M 10 and M 68 and the galaxies M 59 and M 89 have been virtually 'neglected'. I have tried to keep this information as up-to-date as possible, but new developments, especially in radio astronomy, are so rapid that some, possibly important, new knowledge will, of necessity, be omitted at the date of publication.

7. Finally, I give my own descriptions and drawings of each of the objects. The descriptions I have tried to make as objective as possible: the purpose has been to assist the observer to identify the object and to point out features which I thought would be distinctive and interesting.

The drawings were all made at the eyepiece of my own 8-inch Newtonian reflector, supplemented by a portable 4-inch reflector of F.L. 3 feet. This latter instrument, although not of good quality, was useful in providing a view which approximated to that which Messier himself might have had.

For those objects which were not visible from my observatory in latitude 51° 26½' N I took the advantage of a visit to Johannesburg, South Africa, where I was able to use a fine 12½-inch Cassegrain reflector, built by D. Overbeek of Germiston, Transvaal.

The reader will note that the telescope drawings are not all to the same scale: the 65' field represents a magnification of 40×, the 40' field 100× and the 20' field 200×.

M 1 NGC 1952

The 'Crab' Nebula. A supernova remnant in TAURUS

R.A. 5 h 31.5 m.   Dec. N 21° 59' (1950).

Size 6'×4'   Mag. 8.4   Distance 2.0 kpc, 6500 lt-yrs.

Discovered by John Bevis in 1731 but found independently by Messier in 1758.

MESSIER   (Sept. 12th, 1758) 'Nebulosity above the southern horn of Taurus. It contains no star; it is a whitish light, elongated like the flame of a taper, discovered while observing the comet of 1758. Observed by Dr Bevis in about 1731. It is reported on the English *Celestial Atlas*.'
(A note in Messier's handwriting added in the margin of his copy of the *Connaissance des Temps* for 1784 reads: 'Seen by Dr Bevis in about 1731 according to his letter to me of 10th June 1771.')

BODE   (No. 11), Nov. 8th, 1774. 'A small nebulous patch without stars.'

J. H.   'A cluster of stars; barely resolvable.'

SMYTH   'A large nebula, pearly-white, of oval form with a major axis Np & Sf and the brightest portion toward the South.'

WEBB   'Oblong; pale; 1° N.p. ζ Tauri.'

ROSSE   (1844) 'A cluster . . . it is no longer an oval resolvable Nebula; we see resolvable filaments singularly disposed, springing principally from its southern extremity, and not, as is usual in clusters, irregularly in all directions. Probably greater power would bring out other filaments, and it would then assume the ordinary form of a cluster . . .'

LASSELL   (Dec. 15th, 1852. At Malta) 'With 160× it is a very bright nebula, with two or three stars in it, but with 565× it becomes a much more remarkable object . . . Long filaments run out from all sides and there seems to be a number of very minute and faint stars scattered over it.'

(Jan. 6th, 1853) 'The brightest parts are about 2' in length, while the outlying claws are only just circumscribed by the edge of the field of 6' diameter.'

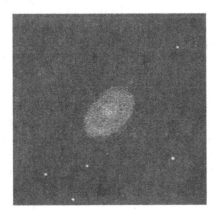

M 1 (Smyth) From A Cycle of Celestial Objects *(B.A.A.)*

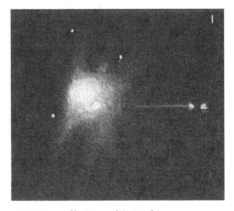

M 1 (Lassell) From his Malta observations (B.A.A.)

53

D'ARREST   (Sept. 30th, 1861) 'A wonderful phenomenon . . . a pale and unresolvable nebula.'

(Mar. 25th, 1865) 'Huge, rhomboidal. $5\frac{1}{2}' \times 3\frac{1}{2}'$. Brightens from the centre, especially towards the preceding part.'

LICK XIII   'Two stars of mag. 16 are close together near the centre but it is not certain that either of them is a central star. This very complex and interesting object is nearly $6' \times 4'$ in P.A. about 125°. It is not a typical planetary in form and it is doubtful whether it is properly to be included as a member of the class. The nebular matter is intrinsically quite faint.'

Although Messier's list was compiled with the object of preventing confusion between nebulae and comets, M 1 has since been frequently so mistaken. A notable occasion was in August 1835 when Halley's comet, on its second predicted return, passed close to the nebula.

M 1 is almost certainly a remnant of the supernova observed by Chinese and Japanese astronomers in AD 1054. It was visible to the naked eye for 2 years and at its maximum was as bright as Venus and visible, even in daylight, for 23 days.

If the original supernova was an arresting phenomenon 900 years ago, its remains, now known as the 'Crab' Nebula, are no less so today, for it is undoubtedly one of the most remarkable and interesting objects in the heavens and is quite unique in many ways.

It seems to consist of two kinds of matter, each giving off a different type of radiation. The bright filaments detected by Lord Rosse as 'fringes' give an emission-line spectrum like that found in gaseous nebulae while the remaining amorphous nebulosity emits a highly polarised light. This latter also produces *'synchrotron radiation'* – an optical and radio emission caused by high-energy electrons moving in a weak magnetic field. This process is also known to occur in other sources such as Cygnus 'A' – a distant pair of interpenetrating galaxies – and also in M 82 (q.v.).

Synchrotron radiation of radio frequency only has been detected in the intense radio source Cassiopeia 'A', about 3400 parsecs away, which Shklovsky, Minkowski and Baade all consider to be the remnant of another supernova, probably one which occurred in AD 1700.

Each year the Sun passes a little to the north of M 1 on or about June 14th, its nearest approach being to within about $2\frac{1}{2}°$. In 1952, Machin and Smith of the Cavendish Laboratory, Cambridge, measured the refractive effect of the Sun's corona on radio emission from the nebula. Even at a distance of 5° a reduction in the intensity of the M 1 emission was observed, and at the closest approach the emission declined to about 25 per cent of its normal value.

In 1964 observations with high-altitude rockets revealed that M 1 is a strong source of X-rays, the first identification of an X-ray source with an optically visible object outside the solar system. This led to the adoption of the 'Crab' nebula as a 'standard candle' against which to measure the intensity of other celestial X-ray emissions.

In 1968 a pulsating radio source (pulsar) was detected in the central region. It was suspected that this source derived from one of the two faint stars which appear near the centre of the nebula. On Jan. 15th 1969, by a brilliant series of observations with the 36-inch reflector at Tucson, Arizona, Cocke, Disney and Taylor showed that the south-preceding component – a blue 16 mag. star – was pulsating in optical wavelengths also, and at the same frequency (0.033 s) as the radio pulsar, now designated as NP 0532.

The regularity of the pulses is constant to within a few parts in a hundred million, but even so a slow-down rate in frequency has been detected which amounts to about ten microseconds a year.

These and other observations soon led to the theoretical explanation which assigned a rapidly rotating, superdense neutron star as the driving mechanism for the prolific emission. Since then supernova remnants elsewhere in the sky have been studied intensively, and at least three classes are now recognised, the 'Crab' nebula being the exemplar of the *plerionic* type with a 'filled centre' form of emission in radio, optical and X-ray radiation, but without a marked shock-wave 'shell'.

Observation with the IRAS satellite in 1984 showed that the infra-red radiation comes from an area almost equal to its visible size, and is interpreted as due to thermal emission from dust particles at a temperature of 70 to 100 Kelvin.

The 'Crab' nebula continues to be a fruitful source for the study of complex astrophysical problems: in 1970 a faint 'jet' was observed some 3' north of the centre but not radially aligned with the pulsar, and its origin remains a mystery. Another fascinating phenomenon is the occurrence of momentary 'glitches' or sudden variations of the rotation rate of the pulsar, which Sir Francis Graham Smith, in 1986, interpreted as due to changes in magnetic coupling between the neutron superfluid and the solid crystalline crust of the star.

Establishing the distance of M 1 is a matter of considerable importance: the following summary was made by Dr Virginia Trimble in 1970.

From proper motions and radial velocities:

1. Line of sight parallel to long axis	1380 parseconds
2. Line of sight parallel to short axis	1990 parseconds
3. Matching distribution functions of above	1450 parseconds
4. Matching dispersions due to turbulence of above	1670 parseconds
5. If all features have equal total speeds	2170 parseconds

From physical properties of supernovae:

6. Brightness from Chinese records (Type I S/N)	1800 parseconds
7. From 21 cm emission and absorption	1580 + 700 parseconds
8. From soft X-ray absorption	2100 parseconds
9. From dispersion measures of pulsar	2200 + 1500 parseconds
10. From absorption of continuum (about 1 mag.)	2000 parseconds

The best 'acceptable' value seems to lie conveniently close to 2 kpc or about 6500 light-years. (For observing details, see next page.)

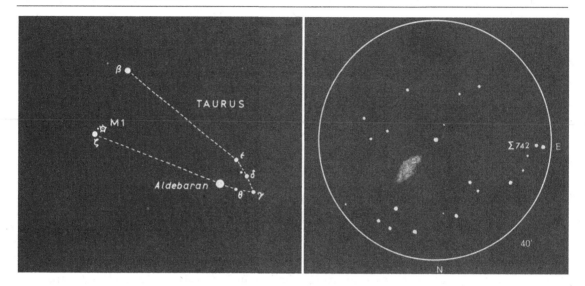

To observe   M 1 is easy to see when the sky is clear and dark, with no Moon present: in any but the very best conditions it may well be quite invisible.

From *Aldebaran* look ENE and find 3rd mag. ζ *Tauri* (the southern horn of the Bull). M 1 is 1° N and 1° W of ζ and just under ½° W of a 6th mag. star Σ 742 which will be in the same field of view in a L.P. eyepiece.

M 1 appears quite bright in good conditions but little more can be seen than a hazy oval patch, at first glance not unlike a tailless comet. No nucleus is apparent and the P.A. of the major axis is not easy to estimate except that it is approximately NW–SE. In very good conditions with 8-inch aperture it has a slightly more definite though irregular outline and a shape more lozenge than oval. The serrated outline and the bright filaments can be seen only in large telescopes.

The background is part of the Milky Way and very many faint stars can be seen. Σ 742 is a double star given by Webb as A, mag. 7.2 yellow, B, mag. 7.8 white. P.A. 258°, d. 3".3 (1890). 'A charming object.'

Objects near by   Σ 742 is easily split into its components; P.A. is now 272° (1964) and D. is 3".6.

Fig. 1. (Rosse 36-inch)

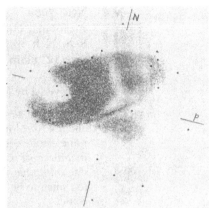

Fig. 2. (Rosse 72-inch)

The drawing of M 1 which gave rise to the 'Crab' description was made using the 36-inch reflector at Birr Castle in about 1844 under the direction of the 3rd Earl of Rosse. (Fig. M 1 (1)). As Dr D. W. Dewhirst of Cambridge Observatory has pointed out, when the 72-inch telescope was brought to bear on the nebula in 1848, a very different picture emerged, and Lord Rosse virtually repudiated the earlier sketch. In 1855, a new drawing, made by R. J. Mitchell, who was staff astronomer at the time, was approved by the 4th Earl as a 'best representation' of the object. (Fig. M 1 (2)). This accords well with modern optical descriptions.

It may seem odd that the epithet applied to the first drawing (which in any case resembles more a pineapple than a crab) should have stuck so firmly; but despite a suggestion by W. B. Somerville of University College London in 1988, that because of its even better resemblance to the map of Ireland, it should be renamed the 'Irish' nebula, it is now much too late in the day for such rationalist tinkerings with tradition.

M 2 NGC 7089
A globular cluster, class II in AQUARIUS
R.A. 21 h 30.9 m.   Dec. S 1° 03′ (1950).
Size 12′ diam.   Mag. 6.3   Distance 17.0 kpc, 55 000 lt-yrs.

Discovered by Maraldi in 1746 while looking for de Chéseaux' comet.

MARALDI (Sept. 11th, 1746) 'This one is round, well terminated and brighter in the centre, about 4′ or 5′ in extent and not a single star around it to a pretty large distance; none can be seen in the whole field of the telescope. This appears very singular to me, for most of the stars one calls nebulous are surrounded by many stars, making one think that the whiteness found there is the effect of the light of a mass of stars too small to be seen in the largest telescopes. I took, at first, this nebula for the comet.'

MESSIER (Sept. 11th, 1760) 'Nebula without star, centre brilliant; surrounded by a circular light resembling the nucleus of a comet. (Diam. 4′)'

BODE (No. 70, Oct, 22nd, 1775) 'West from star 24. Round; with a bright nucleus embedded in a nebula.'

W. H. 'A cluster of very compressed, exceedingly small stars.'

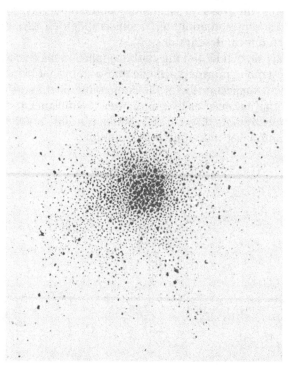

M 2 (Rosse)
From Collected Scientific Papers *(B.A.A.)*

PROFESSOR
VINCE,
who saw M 2 in William Herschel's 40-foot telescope in Sept. 1779, wrote afterwards, 'The scattered stars were brought to a good, well-determined focus, from which it appears that the central condensed light is owing to a multitude of stars that appeared at various distances behind and near one another. I could actually see and distinguish the stars even in the central mass.'

J.H.
'It is like a heap of fine sand.'

CACCIATORE
(In Palermo, 1814) 'The components were about as difficult to enumerate as the sand on the seashore.'

SMYTH
'A fine globular cluster. This magnificent ball of stars condenses to the centre and presents so fine a spherical form that imagination cannot but picture the inconceivable brilliance of their visible heavens to its animated myriads.'

WEBB
'Beautiful round nebula, diam. 5' or 6', having a granulated aspect in 3.7-inch — the precursor of resolution. In 9-inch, resolution evident; the margin seems to diffuse itself away, probably in rays.'

ROSSE
'Streams of stars branch out, taking the direction of tangents.'

LICK XIII
'Fine globular cluster, 7' in diam.'

M 2 has an ellipticity of 9 in P.A. 135° but this is not noticeable visually as it appears quite round. The cluster is composed mainly of red and yellow giant stars of abs. mag. −3.

It contains few variable stars compared with most globular clusters; about 17 are known to date, three of them 13th mag. classical Cepheids of periods 15.57, 17.55 and 19.3 days respectively. They were discovered by H. Arp and G. Wallenstein in 1961. It also contains one *R V Tauri* variable of period 67 days with a range of two magnitudes.

One of the brightest stars in the cluster was noted as variable by Chèvremont as early as 1897. It is somewhat irregular in period and was first thought to have a period of about 30 days but this later was found to be about 11 days, with a range of brightness a little more than one magnitude.

Chèvremont's variable is within the recognised bounds of the cluster but Shapley suspected that it may not, in fact, be a true member of it.

The cluster has a spectral type of F0 and its colour index is −0.06. Hogg gives the average magnitude of the 25 brightest stars as 14.61 and apparent diameter of the cluster, 11.7'. Becvar, in the *Atlas Coeli Katalog*, gives Shapley's apparent diameter of 8.2' and quotes the true diameter as 32 parsecs.

The age of M 2, according to Arp (1962) is 13 000 million years; similar to the globular clusters M 3 and M 5. (For observing details see next page.)

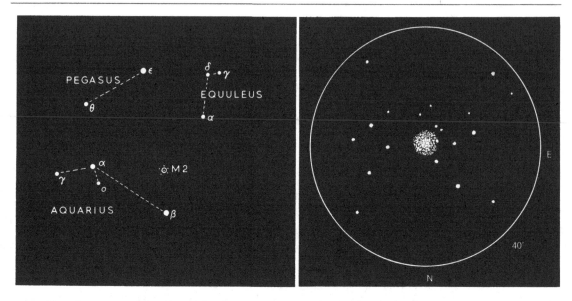

To observe   If you have just found M 15, move 1° E and then S for 13°, *or,* from ε *Pegasi* move S to β *Aquarii* (3rd mag). From β *Aquarii* move 5° N or until level with α *Aquarii.* M 2 forms almost a right-angled triangle to the north with α and β *Aquarii.*

A good bright globular cluster but very different in many ways from nearby M 15. There is not as much contrast between the centre and outer edges and M 2 can be seen to shade gradually fainter from the centre.

The outer edges are resolvable in an 8-inch but the stars are all small and the general outline is much rounder than M 15. Being a bright object, it will stand M.P. well and better resolvability may follow. Under some conditions it may seem to be surrounded by a faint, greenish-blue glow, but this effect, which can be found with some other globulars such as M 3 and M 15, is probably a spurious optical illusion due to the differential colour sensitivity of the eye. (See M 15.)

M 3 NGC 5272
A globular cluster, class VI in CANES VENATICI
R.A. 13 h 39.9 m.   Dec. N 28° 38′ (1950).
Size 6′ diam.   Mag. 6.4   Distance 15.0 kpc, 48 500 lt-yrs.

Discovered by Messier in 1764.

MESSIER   (May 3rd, 1764) 'Nebula without star; centre brilliant, gradually fading away; round. In a dark sky, visible in a telescope of 1-foot. Plotted on chart of comet of 1779. (Diam. 3′)'

W. H.   'A beautiful cluster of stars about 5′ or 6′ diam.'

J. H.   '11–15 mag. stars making lines and irregular rays.'

SMYTH   'A brilliant and beautiful congregation of not less than 1000 small stars. It blazes splendidly towards the centre and has outliers in all directions except S.f. where it is so compressed that, with stragglers, it looks like a "Medusa pelluceus".'

WEBB   'Blazing splendidly, that is, running up to a confused brilliancy towards the centre, with many outliers. 3.7-in hardly resolved. Buffham resolved centre in 9-inch "With" mirror.'

ROSSE   'Rays running out on every side from a central mass in which there are several small, dark holes.'

LICK XIII   'The main portion of this very beautiful globular cluster is about 8′ diam.'

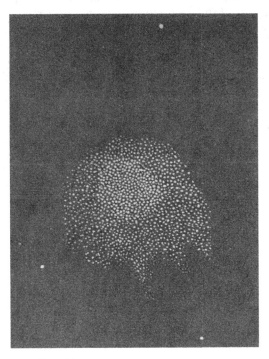

M 3 (Smyth)
The drawing, if not the actual cluster, does resemble a jellyfish (B.A.A.)

M 3 contains more variable stars than any other globular cluster: 189 of them have been observed to date.

It has also been studied more intensively than any other globular in order to estimate the age of clusters and the evolution of stars. Its age has been variously estimated as 5000 million years (Baade), 11 400 million years (Woolf), 20 000 million years (Arp) and 26 000 million years (Sandage). However, in 1958, Hoyle at Cambridge, using an electronic computer to evaluate different theoretical models, and comparing the results with the observed H–R diagram for this cluster, obtained an age of 6500 million years. Its diameter, estimated from the distance of the most distant cluster-type variables from the centre, is about 100 parsecs.

The brighter stars in M 3 are of remarkably uniform magnitude and in this respect it is very similar to M 22.

However, it was shown by Sandage in 1953 that one unusual feature of the cluster is that it contains a very young, blue star of spectral type O8 (see also M 13).

Shapley considered it to be the most asymmetrical of the globular clusters and gave its ellipticity as 8 which makes it one of the more oval globulars also.

Walter Scott Houston, writing in his *Deep Sky Wonders* column in *Sky & Telescope*, noted that the core of M 3 does not lie in the exact centre of the blaze.

H. S. Hogg, in her catalogue of clusters in the *Handbuch der Physik*, Vol. 53, gives a spectrum of F2 for the cluster and a colour index of −0.05. The average brightness of the 25 brightest stars is 14.23 mag.

In 1954, Sandage made a count of 44 500 stars of mag. 22.5 or brighter within 8′ radius of the centre of M 3. The total diameter of the cluster on photographs taken with the 200-inch Mt Palomar telescope is 20′ and the total mass has been calculated by Johnson and Sandage to be equivalent to 245 000 Suns. (For observing details see next page.)

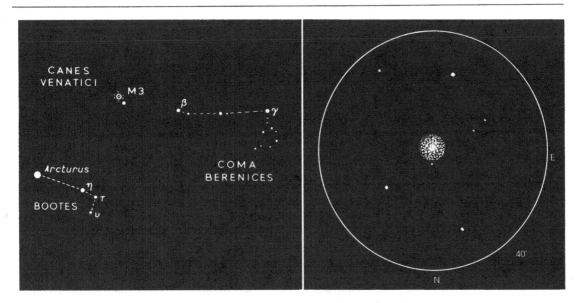

To observe Although an isolated object, it is not difficult to find, being very bright and distinct even in moderate conditions.

From β *Comae Berenicis* (4th mag.) move $\frac{1}{2}$° N and then 6° E. There is a $5\frac{1}{2}$ mag. star about 30′ to the SW.

A very fine globular cluster, very bright and appearing perfectly round to the eye. Even at 40× the edges have a mottled appearance and in a 6-inch at 100× about two thirds of the cluster can be resolved into small, faint stars. Higher magnification with an 8-inch resolves all but the bright, central nucleus.

The whole cluster has a bluish glow: it lies inside a triangle of 8–9 mag. stars but the background is otherwise sparse.

The distribution of stars appears to be very even and I could not make out any resemblance to Admiral Smyth's description of it being like a jellyfish.

M 4 NGC 6121
A globular cluster, class IX in SCORPIUS
R.A. 16 h 20.6 m.   Dec. S 26° 24' (1950).
Size 20' diam.   Mag. 6.4   Distance 3.0 kpc, 10 000 lt-yrs.

Discovered by de Chéseaux in 1746.

DE CHÉSEAUX   (No. 19) 'Close to Antares . . . it is white, round and smaller than the preceding ones. I do not think it has been found before.'

LACAILLE   (Cape of Good Hope, 1751–52) 'It resembles the small nucleus of a faint comet.' (Lac. I. 9.)

MESSIER   (May 8th, 1764) 'Cluster of very small stars: with an inferior telescope it appears more like a nebula; this cluster is situated near Antares and on the same parallel. Observed by M. de la Caille and reported in his catalogue. Reviewed 30th Jan. & Mar. 22nd 1781. Diam. $2\frac{1}{2}$'.'

BODE   (No. 31 in 1777 Catalogue) 'A nebulous patch like a comet nucleus.'

W. H.   In 1783 resolved it into stars 'having a ridge of 8 or 10 pretty bright stars running from the middle to N.f.'

SMYTH   'A compressed mass of small stars with outliers and a few small stellar companions in the field. It is elongated N–S and has the aspect of a large, pale, granulated nebula, running up to a blaze in the centre.'

WEBB   'Large, rather dim, resolvable, followed by a vacant, star-less space.'

M 4 was considered by Shapley to be probably the nearest of all the globular clusters: it is also one of the largest and most open globulars as its classification (class IX) indicates. Shapley also considered it to be very slightly elliptical (0.9 in P.A. 115°) but more recent measurements of 1205 stars down to mag. 17.5 show no ellipticity.

The total of stars down to mag. 19.3 had been counted as 10 300. M 4 was known to contain 33 variable stars in 1907 and in recent years, Baade, who studied this cluster closely, found 37 variables with a period of 0.24 days and 41 with a period of 0.25 days.

The average magnitude of the 25 brightest stars, according to Mrs Helen Sawyer Hogg, is 13.11.

In 1984 Richer and Fahlmann obtained a distance modulus of 12.5 for this globular cluster. This is equivalent to about 3 kpc or 10 000 lt-yrs.

A pulsar of millisecond frequency was discovered in M 4 in 1987, the first to be observed in a globular cluster. (Another was found in M 28 soon after.) The rapid rotation rate in such an old star is thought to be due to acceleration by interaction with a binary companion. (For observing details, see next page.)

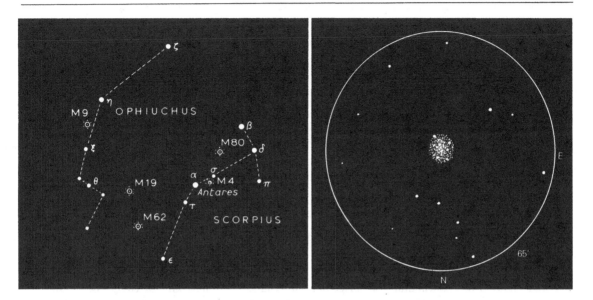

To observe M 4 is very easy to find, being only 1½° W of the bright star *Antares*.

Although Shapley said of M 4 that it was 'inconspicuous because of heavy intervening space absorption', in the amateur's telescope it appears large and bright and much more conspicuous than near-by M 80. It is easily resolved and stands higher magnification well. There are many curving strings of stars, especially to the south and a bright concentration of stars can be seen a little to the north of centre.

In small and moderate telescopes the cluster is distinctly elongated in an approximately N–S direction as Smyth observed, but as seen on a good photographic plate, with all the faint stars included, it appears quite circular.

To observers in the southern United States, M 4 is almost the equal of M 5 or M 13 but for British observers its brightness is a little dimmer by its low altitude. The brighter portion of the cluster is about 10' diameter but fainter areas can be traced up to 20' diameter in a 6-inch.

Several keen-eyed observers have detected the central 'bar' of 11th mag. stars running for about 2½' in P.A. 12° which Wm Herschel noted in 1783. The cluster itself can be seen with the naked eye in the best conditions.

Objects near by A globular cluster, NGC 6144, can be seen 50' to the ENE (30' NW of *Antares*). It is rather small and faint and *Antares* itself must be excluded from the field to avoid 'flooding' the cluster. Its magnitude is about 10½ and it is 3.3' in diameter.

M5 NGC 5904
A globular cluster, class V in SERPENS CAPUT
R.A. 15 h 16.0 m.   Dec. N 2° 16′ (1950).
Size 20′ diam.   Mag. 6.2   Distance 7.2 kpc, 23 500 lt-yrs.

Discovered by Gottfried Kirch. May 5th, 1702.

MESSIER (May 23rd, 1764) 'A fine nebula which I am sure contains no star. Round; seen well in a good sky in a telescope of 1-foot. Plotted on chart of comet of 1763. Diam. 3′.'

W. H. 'In May 1791 counted about 200 stars with the 40-foot reflector, 'although the middle was so compressed that it is impossible to distinguish the components'.

J. H. Compared it to a snowball and as if 'projected on a loose, irregular ground of stars'.

SMYTH A close cluster near 5 Librae (now 5 Serpentis). This superb object is a noble mass, refreshing to the senses after searching for faint objects, with outliers in all directions and a bright, central blaze which even exceeds M 3 in concentration.'

WEBB 'A beautiful assemblage of minute stars, 11–15 mag. Greatly compressed in the centre.'

ROSSE (May 26th, 1875) 'More than 7′ or 8′ diam. Very condensed part in middle about 1′ diam. Stars 12–15 mag., many going out from centre in curved lines.'

LICK XIII 'A beautiful bright globular cluster: the main portion is about 12′ diam.'

M 5 can be seen in photographs to be distinctly elliptical, being about 10 per cent elongated in P.A. 50°. It is also unusually asymmetrical for a globular cluster although M 62 shows an even greater irregularity of outline. The suspicion that some of the stars which make up the cluster were variable,

M 5 (John Herschel) From G. F. Chambers, Handbook of Descriptive and Practical Astronomy *(B.A.A.)*

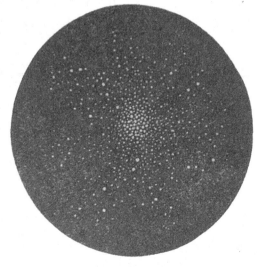

M 5 (Smyth) From A Cycle of Celestial Objects *(B.A.A.)*

was first noted by Dr A. A. Common in 1890 and by 1904 no less than 85 variables had been confirmed among the outlying stars. By 1955 the number of short-period variables observed had risen to 97.

The variable stars have been well studied and besides being used to obtain the distance of the cluster, they formed the basis of an important experiment by Harlow Shapley in 1917. Shapley used a series of photographs taken with the 60-inch Mount Wilson reflector in order to determine if there were any difference in the speed of light at different wavelengths.

Many of the variables in this cluster have a rapid rise in brightness of almost one magnitude in a period of 30 minutes: this enables the times of median magnitude for individual stars to be predicted very accurately. Shapley made exposures of the whole cluster in yellow light, lasting for 30 minutes and interrupted by an exposure of 1 or 2 minutes on ordinary blue-sensitive plates. By this method, the times of median magnitude for 14 stars were obtained in both yellow and blue light. In six cases the yellow light appeared to arrive first; in seven others, the blue, while for the remaining star, the difference in the arrival times was zero.

As the average difference found was only about 1 minute – after a total light journey of about 27 000 years – it was reasonably concluded that the possible difference of speed of the two wavelengths of 4500 Å for the blue and 5500 Å for the yellow, of only about one part in 20 000 million, really indicated that there was no difference at all.

Walter Baade studied this cluster intensively in his investigations into stellar populations. Using a different system from that of Shapley, he divided globular clusters into two classes. Those which contained more long-period variables and fewer short-period variables were put into class 1 and those in which long-period and short-period were about equally numerous, made up class 2. M 5, together with M 3, is in class 1.

The age of M 5 has been estimated by Arp to be about 20 000 million years, and by Sandage as 24 000 million years. Arp, in 1962, amended his estimate to 13 000 million years.

The cluster lies on the edge of a very dense cloud of galaxies which Shane at Lick Observatory has estimated to contain 200 galaxies per square degree, the cloud itself containing separate groups of galaxies.

Hogg gives the spectrum of M 5 as F7, the colour index as −0.05 and the average magnitude of the 25 brightest stars, 13.97. The true diameter of the cluster, according to Becvar, is 40 parsecs or about 130 light-years: this makes it the largest of the Messier globulars although several others, not seen by Messier, are larger still. The largest known is NGC 6522 with a diameter of 73 parsecs. (For observing details see next page.)

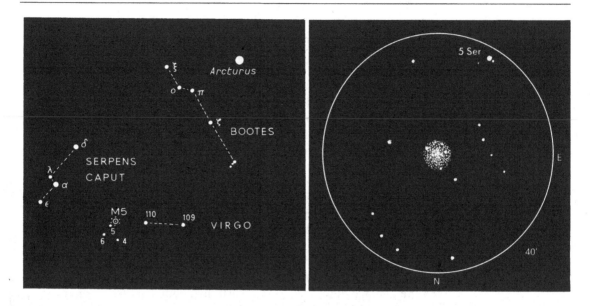

To observe   M 5 is best found by moving SE from *Arcturus* to the 4th mag. *109* and *110 Virginis*. 4° E of *110 Virginis* is a small triangle formed by *4*, *5* and *6 Serpentis*. M 5 is just NW of *5 Serpentis*, and being only 20' away, will appear in the same L.P. field.

M 5 is a fine sight in the smallest telescope being large and bright and easily resolved. It stands high magnification well, when curves of separate stars can be seen to the north and little knots of stars to the south. The whole arrangement of outlying stars appears to form a spiral pattern at times. Only M 13 is superior in the northern hemisphere.

5 Serpentis is a double star (Σ 1930); A, 5 mag. pale yellow, B, 10 mag. light grey. P.A. 37°, dist. 11".0. (1923).

Although M 5 appears to lie in a very barren region of the sky, the group of galaxies to the west contains at least half-a-dozen members which could be picked out in an 8-inch or 10-inch telescope.

M 6   NGC 6405
A galactic cluster, type 'e' in SCORPIUS
R.A. 17 h 36.8 m.   Dec. S 32° 11' (1950).
Size 25' diam.   Mag. 5.3   Distance 405 pc, 1300 lt-yrs.

Discovered by Hodierna before 1654. Counted 18 stars.

DE CHÉSEAUX	(No. 1) 'Here is a very fine cluster.'
LACAILLE	'An unusual cluster of small stars disposed in three parallel bands forming a lozenge 20'–25' in diameter and filled with nebulosity. (Lac. III. 12.)
MESSIER	(May 23rd, 1764) 'A cluster of small stars between the bow of Sagittarius and the tail of Scorpius. To the naked eye it resembles a nebula without a star but even a small telescope reveals it as a cluster of small stars. Diam. 15'.'
J. H.	'A fine, large, discrete cluster of stars. 10–11 mag. One star is 7 mag. one 7.8. Fills field.'
	(In catalogue of 1864) 'A cluster, large, irregularly round, not much condensed. A 7 mag. star in it but component stars generally range from 10 mag. down.'
J. E. GORE	'Stars in zig-zag lines.'
SPLENDOUR OF THE HEAVENS	'A most beautiful open cluster of galactic type but too South for satisfactory observation in the U.K. Somewhat irregular in shape, with a central rib of stars and resembles a butterfly with open wings.'
FLAMMARION	'Stars of 7–10 mag. very dispersed and arranged in a remarkable pattern. Three starry avenues leading to a large square.'

The yellow magnitudes for 132 stars in this cluster have been determined using an electric photometer attached to the 60-inch reflector at the Harvard Boyden observatory in South Africa. In this 1959 survey the German astronomers, K. Rohlfs, K. Schrick and J. Stock, give the magnitudes for the five brightest stars as: 6.2, 6.8, 7.2, 7.3, 7.9.

There are 10 stars with magnitudes 8 to 9, 22 with magnitudes 9 to 10, 27 with magnitudes 10 to 11 and a further 67 between 11 and 14 magnitude.

According to an earlier survey by Dr A. Wallenquist, this rich cluster contains 80 stars brighter than mag. 11 in an area 54' in diameter.

Trumpler's class for this cluster is II,3,m and he obtained a spectrum of B5 for the brightest star.

Wallenquist, in his 1959 survey of the properties of galactic clusters, calculated a density for the central part of 24.5 stars per cubic parsec and an average density for the cluster as a whole of 0.63 stars per cubic parsec. He gives the true diameter of M 6 as 6.4 parsecs or 21 light-years but Becvar, in the *Atlas Coeli Katalog*, gives only 4.1 parsecs or 13½ light-years. (For observing details see next page.)

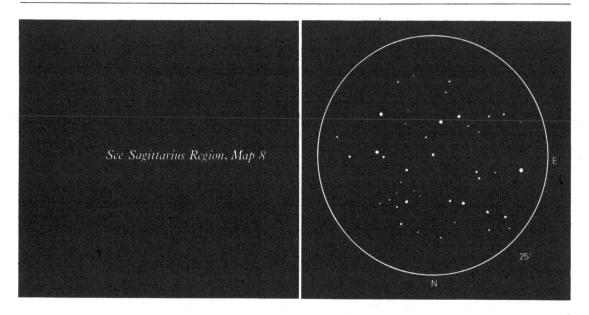

See Sagittarius Region, Map 8

25'

E

N

To observe   There are quite a number of clusters in this region and care has to be taken to pick out the right one. However, M 6 is large and bright and, when found, is quite unmistakable.

From *Antares* move 4° S and a little less than one hour of R.A. to the E and pick up 4th mag. *45 Ophiuchi*. M 6 is 2½° S and 2° E of this star.

M 6 is a brilliant and very rich cluster containing, among the very many bright stars, some pretty groups of small and faint stars. One of these, a little chain of seven stars of identical brightness, in the form of an inverted V, in the NW quadrant, is especially noticeable.

The cluster as a whole has some resemblance to a butterfly with open wings and curved antennae to the east, but this is better seen with a field of about 1°. A higher power, however, is better to bring out the fainter stars, many of which are in close groupings.

Lacaille's 'three parallel bands forming a lozenge' were not apparent and his 'nebulosity' was almost certainly due to the effect of the many faint, close stars not being resolved in his small telescope.

M 6 is not easily seen from latitudes higher than 50° N and the best views are obtainable from situations which are at least no higher than 40° N.

Objects near by   A somewhat scattered cluster, NGC 6383, mag. 5.5 and containing about 40 stars in a 5' field can be seen close to the SW.

M 7 NGC 6475
A galactic cluster, type 'e' in SCORPIUS
R.A. 17 h 50.6 m.   Dec. S 34° 48' (1950).
Size 50' diam.   Mag. 4.1   Distance 250 pc, 800 lt-yrs.

PTOLEMY (No. 567) 'A nebulous cluster following the sting of Scorpius.'

HODIERNA (Before 1654) counted 30 stars.

LACAILLE (Observing at Cape of Good Hope 1751–52) (Lac. II. 14.) 'Group of 15 or 20 stars very close together in a square'.

MESSIER (May 23rd, 1764). 'A cluster considerably larger than the preceding (M 6). It appears to the naked eye as a nebulosity; it is situated a short distance from the preceding, between the bow of Sagittarius and the tail of Scorpius. Diam. 30'.'

J. H. (Catalogue of 1864). 'A cluster, very bight, pretty rich, little compressed. The component stars from 7 to 11 magnitude.'

M 7 was the last item in the list of 16 'nebulous stars' published by Wm. Derham in 1733 and is there described as 'Lying between Scorpio's Tail and the Bow of Sagittarius.' Derham did not actually observe it but explains that he transcribed the list from Hevelius' *Prodromus Astronomiae* of 1690.

The last two items in Derham's list, however, were derived not from Hevelius, who would have been quite unable to observe them from Dantzig (lat. 54° N), but from Halley's *Catalogus Stellarum Australium* of 1678. M 7 is No. 29 in Halley's catalogue. (It was not included in his list of six nebulae published in 1715.)

M 7 contains 80 stars brighter than mag. 10 in a field 1°.2 in diameter according to Dr Wallenquist.

Its age has been estimated by Koelbloed at about 70 million years which compares with M 44 of about 400 million years and M 45 of only 20 million years.

In 1959, Dr Wallenquist, using a system of star-counts in successive concentric shells, classified galactic clusters according to their degree of central concentration. He assigned numerical values to this system, in which 0 represents an even density throughout, and higher numbers a more rapid gradient of concentration to the centre. On this scale, M 67 had a small gradient of 1.08 while M 7, with a value of 11.81, had the highest degree of concentration for any of the clusters investigated by him.

Trumpler classified M 7 as I,3,m and gave the spectrum of the brightest star as B5.

The true diameter of the cluster, according to Wallenquist, is 5.3 parsecs or $17\frac{1}{2}$ light-years but the *Atlas Coeli Katalog* gives a greater value of 6.6 parsecs or $21\frac{1}{2}$ light-years. (For observing details see next page.)

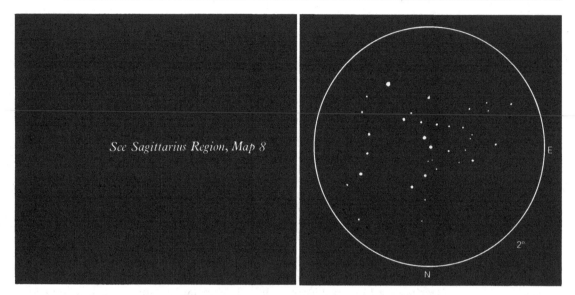

See Sagittarius Region, Map 8

To observe From M 6 move $2\frac{1}{2}°$ S and $2\frac{1}{2}°$ E, *or*, from *Shaula (λ Scorpii)* move $2\frac{1}{2}°$ N and 4° E.

This cluster is so large it is best observed through the finder telescope: it is at least 1° in total diameter and there are many more stars nearby which add to its beauty in a wide field.

There is a bright, fairly close pair near the centre and many of the stars are arranged in curves. To the south of the centre is a distinct, smooth arc of stars ending in a yellow star to the west.

A higher magnification brings out much more detail but the cluster appearance is lost in a field less than 1° diameter.

Some of the brighter stars form almost a hollow square which tallies with Lacaille's description.

Objects near by Nearly 1° NW of the centre of M 7 is a small globular cluster, NGC 6453. It can be seen almost equidistant between two small stars lying E–W. It is fairly faint however.

45' to the SE is a galactic cluster, H 18, containing one 10 mag. star and 20 to 30 fainter members, some of which are arranged in a W pattern. It is not very bright but worth looking for.

M 7 is best seen from latitudes lower than about 40° N. It is not really visible from the British Isles and in the northern U.S.A. is too low to be seen at its best.

M 8   NGC 6523 (with NGC 6530)
The 'Lagoon' Nebula. A galactic cluster and gaseous emission nebula in
SAGITTARIUS
R.A. 18 h 00.1 m.   Dec. S 24° 23' (1950).
Size 90'×40'   Mag. 6.0   Distance 1.5 kpc, 4850 lt-yrs.

FLAMSTEED   (No. 2446) 'Nebulosum antecedentam Arcum' about 1680.

DE CHÉSEAUX   (No. 8) 'Cluster in Sagittarius' bow.' 1745–46.

LE GENTIL   (1747) 'A small nebulosity like the tail of a comet with numerous stars . . . like the more transparent and whitish localities of the Milky Way.'

LACAILLE   'Three stars enclosed in a trailing nebulosity parallel to the equator.' (Lac. III. 13.)

MESSIER   (May 23rd, 1764) 'A cluster which looks like a nebula in an ordinary telescope of 3 feet but in a good instrument one observes only a large number of small stars. A fairly bright star nearby is surrounded with a very faint glow: this is *9 Sagittarii*, 7 mag. The cluster appears elongated NE–SW. Diam. 30'.'

W. H.   'A set of milky streaks and loops.'
(*Phil. Trans.* 1785) 'An extensive milky Nebulosity divided into two parts; the north being the strongest. Its extent exceeds 15'; the southern part is followed by a parcel of stars which I suppose to be the 8th of the *Connaissance des Temps.*'

J. H.   'A collection of nebulous folds and matter surrounding and including a number of dark, oval vacancies and, in one place, coming to so great a degree of brightness as to offer the appearance of an elongated nucleus. Superimposed upon this nebula and extending in one direction beyond its area, is a fine and rich cluster of scattered stars which seems to have no connection with it as the nebula does not, as in the region of Orion, show any tendency to congregate about the stars.'

SMYTH   'A very singular nebula.'

WEBB   'A splendid galaxy object, visible to the naked eye. In a large field we find a bright, coarse triple star followed by a resolvable luminous mass, including two starry centres and then by a loose, bright cluster; a very fine combination.'

Although Sir John Herschel thought the cluster separate from the nebulosity, they are, in fact, intimately connected and many of the stars can be seen in photographs to be linked by curving wisps of nebulous matter, as in the Pleiades (M 45).

M 8 contains a large number of irregular variable stars and also some 'flare' stars similar to those found in M 42 and M 45. These strange stars exhibit very sudden bursts of brilliance, lasting only a very short period. The best known of these is *UV Ceti*, discovered by Luyten of the University of Michigan in 1948, which has been seen to increase its brightness from mag. 10.3 to 6.8 in only 20 seconds.

The star which excites the nebulosity of M 8 to glow was thought by Baade to be a very hot O-type object lying deep in the darker regions and not optically visible. In 1969, however, the very visible star *9 Sgr.* (mag. 5.9, type O 5) was itself identified as the source, with perhaps the fainter star Herschel 36, located on the edge of the 'hour-glass' portion of the nebula, also contributing to the excitation. Indeed, according to A. D. Thackeray of Radcliffe Observatory, there may be more heavily reddened O-type stars in the region which may add to the total output.

Among the glowing stars and gas, many small, dark globules of obscuring matter have been detected by Bart J. Bok of Harvard in 1946. These have a diameter of about one-hundreth of a light-year or so, and are thought to be sufficiently dense to be gravitationally stable against the disruptive forces of the Galaxy and nearby stars. These globules may continue to contract and eventually become visible as 'protostars', shining by their own radiation.

M 8 is an emission-type bright nebula with a gaseous spectrum and contains a ratio of helium to hydrogen of 0.110.

It is also known to be a radio source in the 9.4 cm band.

The cluster, NGC 6530 is classed by Shapley as type 'e' (intermediate rich), mag. 6.3, diam. 10' containing 25 stars. This is a young cluster in which star-formation is thought to be taking place at the present time: its physical association with the gaseous nebula is hardly to be doubted. (For observing details see next page.)

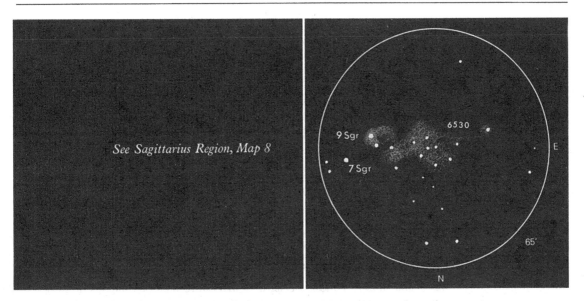

See Sagittarius Region, Map 8

To observe   M 8 is most easily found from the 'little dipper' in Sagittarius. Continue the curve of three bright stars, ζ, φ and λ for another 5° in the WNW, *or*, if you have just found M 20, move 1½° S and 15' E.

Even from the higher latitudes of the northern U.S.A. and Great Britain, M 8 can easily be seen with the naked eye, despite its low altitude at culmination.

In binoculars or small telescopes it appears much as Messier saw it: a cluster of many stars looking slightly nebulous as a whole. These are the stars involved in the gaseous nebula and the associated cluster, NGC 6530.

A larger aperture or a lower latitude are necessary to bring out the beauty of the bright diffuse nebulosity and a rich-field telescope is best of all.

With good transparency of atmosphere and a moderate aperture, a small, distinct patch of bright light can be seen to the west around the star *9 Sagittarii* and, using averted vision, further areas of fainter diffuse nebulosity can be seen to the east, surrounding many of the stars in the cluster. The star to the extreme east appears slightly nebulous and on the extreme west, the 5½ mag. *7 Sagittarii* has a bluish tinge of nebulosity around it.

M 8 is a most interesting visual object in almost any instrument and, in the best conditions, careful examination can reveal a wealth of fascinating detail.

M 9 NGC 6333
A globular cluster, class VIII in OPHIUCHUS
R.A. 17 h 16.2 m.   Dec. S 18° 28' (1950).
Size 6' diam.   Mag. 7.3   Distance 7.9 kpc, 25 800 lt-yrs.

MESSIER (May 28th, 1764) 'Nebula without star on the right leg of Ophiuchus. Round and faint. Reviewed Mar. 22nd, 1781. (Diam. 3')'.

W. H. Resolved it in his 20-foot telescope in 1784. He thought it (like M 10) 'a miniature of M 53'.

ROSSE (June 3rd, 1851) 'The outline not round; on south side is an outlying portion separated from the chief portion by a dark passage.'

SMYTH 'A globular galaxy cluster with a coarse telescopic double in N.p. quadrant. This fine object is composed of a myriad of minute stars, clustering to a blaze in the centre and wonderfully aggregated with numerous outliers seen by glimpses.'

WEBB 'Small, apparently resolvable.'

D'ARREST 'Large and shining: almost circular although not regularly terminated. Resolved into stars mingled with nebula on magnification 147×. Appears almost bi-central, N–S.'

LICK XIII 'Bright globular cluster, 3' in diameter.'

M 9 was investigated by Baade for variable stars. Among the shorter variables, he found 10 with periods of 2.4 days.

It is one of the nearer of the globular clusters to the galactic centre being a little less than 2 kpc from the nucleus. Shapley gives its ellipticity as 9 and estimates its distance as 20.8 kpc but later measurements have greatly reduced this figure. The currently adopted distance is about 8.0 kpc.

In Helen Sawyer Hogg's catalogue of clusters in the *Handbuch der Physik*, the spectral class of M 9 is F2, its colour index +0.06 and the average magnitude of the 25 brightest stars, 15.5 mag. The true diameter of the cluster, according to Becvar, is 15.0 parsecs or about 49 light-years. (For observing details see next page.)

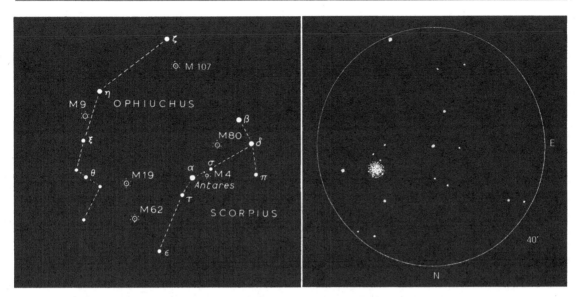

To observe   From η *Ophiuchi* (2½ mag.) move 3° S and 2° E.

A small globular cluster of typical appearance, having a bright, flat-looking centre and fainter outer areas which are resolvable on M.P. in an 8-inch. It appears perfectly circular with a diameter of about 5' although, with averted vision, it can be seen to be slightly larger, perhaps extending to 7' or 8'.

Most of the faint stars nearby are to the east. A large, dark cloud obscures the stars to the west and south.

Objects near by   A very small but bright globular cluster, NGC 6356, lies 80' to the NE.

77

M 10 NGC 6254
A globular cluster, class VII in OPHIUCHUS
R.A. 16 h 54.5 m. Dec. S 4° 02' (1950).
Size 8' diam. Mag. 6.7 Distance 5.0 kpc, 16 300 lt-yrs.

MESSIER (May 29th, 1764) 'Nebula without star in the belt of Ophiuchus. Fine and round. Seen in 3-foot telescope. Reported on 2nd chart of track of comet of 1769.* (Diam. 4')'.

BODE (No. 33, Aug. 14th, 1774) 'Nebulous patch without stars, very pale.'

W. H. Resolved it into 'a beautiful cluster of extremely compressed stars, resembling M 53.'

J. H. 'A fine, large cluster . . . diameter 5', with stragglers, several of which are of larger size to about 12'; all resolve into stars 11–15 magnitude; very compressed.'

SMYTH 'A rich globular cluster of compressed stars of a lucid white tint, somewhat attenuated at the margin and clustering to a blaze in the centre. Easily resolvable by moderate means.'

WEBB 'Bright cluster. Beautiful group lies f. lucida bright or.'†

ROSSE (May 30th, 1851) 'A dark lane above the centre quite across, or rather the upper one-sixth of cluster is much fainter than the rest.'

BRODIE 'A fine cluster but inferior to M 92, Hercules.'

LICK XIII 'Fine bright globular cluster, diameter about 8'. Central brighter part about 2'.'

Baade found very few variables in this cluster. Its ellipticity, according to Shapley, is 9. He also calculated its distance to be 11.2 kpc but Kinman, in his survey of the distances of globular clusters, makes the distance much less at about 5 kpc and this is supported by the figure given by Becvar in the *Atlas Coeli Katalog*.

Helen Sawyer Hogg gives the spectral class of M 10 as G0, its colour index +0.03 and the average magnitude of the 25 brightest stars as 14.06 mag.

The true diameter of the cluster, according to Becvar, is 26 parsecs or about 85 light-years. (For observing details see next page.)

---

* A manuscript note in the margin of Messier's copy reads: 'Reviewed 6th March, morning, 1781, "Always very fine"'; 'Reviewed 10th March, morning, 1790, Night-glass of Rebour'.

† 'Or' is the heraldic term for gold.

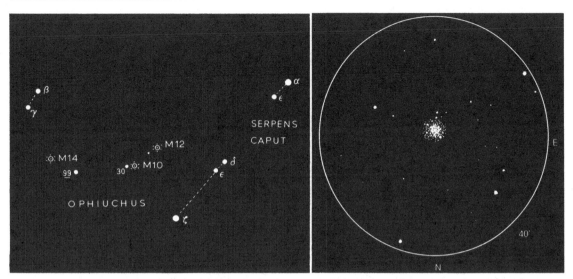

To observe   If you have just found M 12, move $1\frac{1}{2}°$ SE to a 6 mag. star and continue in the same direction for another $2\frac{1}{4}°$. A 5th mag. red star, *30 Ophiuchi* lies 1° to the E, *or*, first find α and ε *Serpentis* and then move about 10° SE and pick up δ and ε *Ophiuchi* which are almost in the same straight line. From ε, then move 10° E to M 10.

This globular can be compared with M 12 which is $4\frac{1}{2}°$ to the NW. M 10 is of similar size but can be seen to be more concentrated especially towards the centre (M 10 is class VII while M 12 is class IX). M 10 seems slightly brighter than M 12 although it is generally given as 1/10 mag. fainter.

The outer portions of M 10 are easily resolved on M.P. in a 6-inch or 8-inch but the central portion is very dense.

Except for *30 Ophiuchi*, 1° to the ESE, there are few bright stars near but the background is studded with a large number of very faint stars.

M 11 NGC 6705
'The Wild Duck'. A galactic cluster, type 'g' in SCUTUM
R.A. 18 h 48.4 m.   Dec. S 6° 20' (1950).
Size 12' to 30' diam.   Mag. 6.3   Distance 1.7 kpc, 5500 lt-yrs.

Discovered by Kirch in 1681.

KIRCH   'A small, obscure spot with a star shining through and rendering it more luminous.'

DERHAM   First resolved it into stars in a telescope of 8 feet F.L. 'It is not a nebula but a cluster of stars somewhat like that which is in the Milky Way.'

DE CHÉSEAUX   (No. 14) 'A prodigious cluster of small stars about 4' diam.'

LE GENTIL   (July 19th, 1749) 'A prodigious cluster of very small stars, forming a large, white cloud; six of the principal stars form a large letter V somewhat similar to the Hyades but with the opening towards the south.' (Le Gentil also thought that it contained 'a true nebula in its northern extremity'.)

MESSIER   (May 30th, 1764) 'Cluster of a great number of small stars which can be seen in a good telescope. In a 3-foot (F.L.) instrument it looks like a comet. The cluster is mingled with a faint light. 8 mag. star in cluster. Seen by Kirch in 1681. Reported on the English *Great Atlas*.'

W. H.   '11 mag. stars divided into 5 or 6 groups noted independently in a 5½-inch. An 8 mag. star is a little within the apex. Just visible to the naked eye.'

SMYTH   'A splendid cluster of stars on the dexter chief of Sobieski's* shield. This object, which somewhat resembles a flight of wild ducks in shape, is a gathering of minute

*   The full title of the constellation 'Scutum' was 'Scutum Sobieskii' or 'Clypeus Sobieskii' and was formed by Hevelius in honour of John Sobieski, the third King of Poland, famed for his heroic part in the relief of the siege of Vienna by the Turks in 1683.

M 11 (Kirch) From
Ephemeridum Motuum
Celestium ad annum 1682
(British Museum)

M 11 (Smyth) The cluster itself is much
more angular than shown in the sketch
(B.A.A.)

stars with a prominent 8 mag. star in the middle and two following; these are decidedly between us and the cluster.'

BRODIE   'Smyth's sketch is utterly unlike the original in my 8½-inch refractor in 1864, for the general outline is rather square. Smyth's rounded apex is to me nearly a right-angle.'

WEBB   'Noble fan-shaped cluster.'

D'ARREST   'A magnificent pile of innumerable stars. Irregular and as if divided into several agglomerations.'

HELMERT   At Hamburg observatory in 1870 made a catalogue of 200 of the component stars.

LICK XIII   'Bright, rather open cluster, 6' in diameter. Not globular.'

M 11 is one of the most compact and dense galactic clusters known and contains about 500 stars brighter than mag. 14 and 600 brighter than mag. 14.8. Shapley considered it 'apparently a condensation in the rich background of the Milky Way' and put it into his class 'g' of dense clusters.

The density of stars in the cluster has been estimated at about 83 stars per cubic parsec near the centre and about 10 stars per cubic parsec within a half-radius. This gives an average separation of less than 1 light-year between stars in the central region. (Compare M 36, M 37 and M 45.)

Trumpler, who was one of the foremost investigators of galactic clusters, said: 'at the centre of M 11 the observer would find about 40 stars with parallaxes of 2" or more and which would appear from 3 to 50 times as brilliant as Sirius shines in our sky.'

Besides being very dense, M 11 contains many more yellow giant stars of abs. mag. − 1.0 than most clusters, being similar to M 67 in this respect. It is probably not a condensation in the Scutum star-cloud against which it appears, but is closer to the Sun. Its age, according to Johnson, Sandage and Wallenquist, is about mid-way between M 44 and M 45, making it a comparatively young cluster.

The magnitudes and colours of 40 stars in the cluster were used to obtain its distance and a further 1000 stars have also been classified for magnitude and colour.

Trumpler's classification for M 11 is II,2,r and he gave the spectral class of the brightest star as B8. Its true diameter if 5.5 parsecs. (For observing details see next page.)

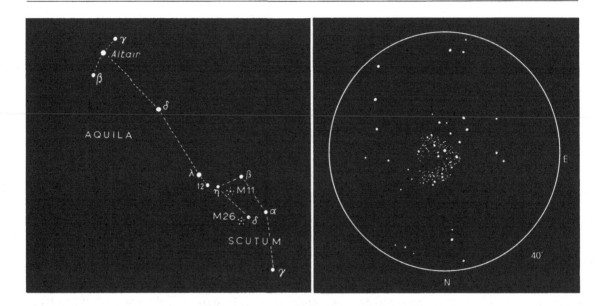

To observe   M 11 is a great favourite of amateur observers for it shows up well even in a small telescope. To find: from *Altair* move SW to δ, λ and *12 Aquilae* and then to η *Scuti* (5th mag.). M 11 is 2° W and slightly S of η *Scuti*. It is visible to the naked eye in good conditions.

A very beautiful and impressive, compact cluster. The densest part of the cluster – and it is very dense – is in the shape of a fan or arrow-head with a bright 8 mag. star just inside the apex. To the east (ahead) of the arrow-head there is another, almost symmetrical, shallow 'V' of stars which probably prompted Smyth's 'flight of wild ducks' description.

The angle at the apex appears quite sharp, as Brodie remarks, and Smyth's sketch, which shows a distinctly rounded 'nose', is somewhat unreal.

The dense portion of the cluster is quite small, about 10' or 12' in diameter, but the whole object is much larger, being nearly 30' diameter overall.

Objects near by   A double star (ADS 11719), mags. 6.2 and 8.7, P.A. 171°, d. 113", lies 40' to the NW.

One degree to the NW is the variable star R.Scu., an RV Tauri type, semi-pulsating variable of period 140 to 146 days with a range of 4.8 to 6.0 mag. It also exhibits an exceptionally low minimum of less than 8.0 mag. every 4th or 5th cycle.

M 12   NGC 6218
A globular cluster, class IX in OPHIUCHUS
R.A. 16 h 44.6 m.   Dec. S 1° 52′ (1950).
Size 9′ diam.   Mag. 6.6   Distance 5.8 kpc, 19 000 lt-yrs.

MESSIER   (May 30th, 1764) 'Contains no star; round, faint; near by is a 9 mag. star. Reported on 2nd chart of comet of 1769.' (Diam. 3′.)

BODE   (No. 32, Aug. 14th, 1774) 'A new nebula without stars, very pale.'

W. H.   Resolved it in 1783. 'A brilliant cluster 7′ or 8′ in diam. – the most compressed parts about 2′.'

SMYTH   'A fine, rich globular cluster with a cortège of bright stars and many minute, straggling outliers. The resolvable mass is greatly condensed toward the centre with several very bright spots.'

WEBB   'Resolvable. (J. H. 10–16 mag.).'

ROSSE   'Long straggling tentacles of a slightly spiral arrangement. Finely grouped.'

LICK XIII   'Fine globular cluster; central part about 2′, outer part about 8′ in diam. Apparently less compact than most globular clusters.'

For some years M 12 was considered to be an intermediate form between the compact galactic clusters such as M 11 and the more easily resolved of the globulars. We now know that the two kinds of clusters are completely different in age and type of star population. M 12 is not very concentrated and Shapley included it in his class IX while M 56 and M 68 (class X) with M 55 (class XI) are less concentrated still and M 71 is so open that it was given a classification with the galactic clusters.

H. S. Hogg's catalogue of clusters gives the spectrum of M 12 as F7, its colour index as 0.0 and the mean magnitude of the 25 brightest stars as 13.97.

Shapley used a distance of 11.0 kpc for his cluster but Kinman and Becvar quote distances close to 6 kpc.

The true diameter of the cluster, according to Becvar, is about 30 parsecs or about 100 light-years.

Sandage has found 13 variable stars in M 12. (For observing details see next page.)

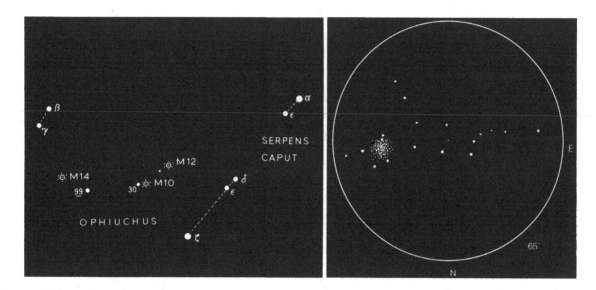

To observe From M 10 move 2° N and 2° W, *or*, from δ *Ophiuchi* move 2° N and 8½° E.

Although it is not as concentrated as most globular clusters, M 12 is more interesting on that account as it can be resolved with ease even in a small telescope.

The cluster itself seems entangled in a long skein of stars lying roughly E–W (perhaps this is Smyth's 'cortège') and the whole ensemble has a distinctly pleasing appearance.

The total outline of the cluster extends to at least 10′ and is far from being perfectly circular. The slightly denser centre also seems to be flattened on the SW. Compared with nearby M10, it is much less concentrated towards the centre.

M 13   NGC 6205
A globular cluster, class V in HERCULES
R.A. 16 h 39.9 m.   Dec. N 36° 33' (1950).
Size 23' diam.   Mag. 5.7   Distance 8 kpc, 26 000 lt-yrs.

The best-known of all globular clusters, which Halley discovered by a chance observation in 1714.

HALLEY   'This is but a little patch but it shows itself to the naked eye when the sky is serene and the Moon absent.'

MESSIER   (June 1st, 1764) 'A nebula which I am sure contains no star. Round and brilliant; centre brighter than the edges. Near two 8 mag. stars. Reported on chart of comet of 1779. Diam. 6', reported on the English *Celestial Atlas*.'

BODE   (No. 30, Sept. 9th, 1774) 'A very distinguishable nebula which appears as a pretty lively and round nebulous patch. In the centre is a bright nucleus.'

W. H.   'A most beautiful cluster of stars exceedingly compressed in the middle and very rich. Contains about 14 000 stars.'

J. H.   'Exhibits hairy-looking, curvilinear branches.'

SMYTH   'An extensive and magnificent mass of stars with the most compressed part densely compacted and wedged together under unknown laws of aggregation.'

WEBB   'Spangled with glittering points in 5½-inch achromat and superb in larger telescopes.'

ROSSE   'More distinctly separated and brighter than anticipated; singularly fringed appendages to the globular figure branching out into the surrounding space.'

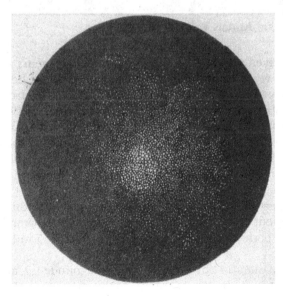

M 13 (John Herschel) From G. F. Chambers' Handbook *(B.A.A.)*

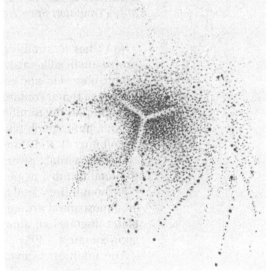

M 13 (Rosse) Drawn by Bindon Stoney (B.A.A.)

D'ARREST   'On magnification 95×, beautifully resolved with arm-like extensions.'

DR J. P. NICHOL   'No plate can give a fitting representation of this magnificent cluster . . . Perhaps no one ever saw it for the first time without uttering a shout of wonder!' (Derham said that he had observed M 13 but merely remarked that it looked 'much alike' the other nebulae. De Chéseaux included it in his list as No. 21 but did not observe it as he did not have its position.)

MITCHEL   (*Stellar Worlds* (1869), pp. 256–261, Wm Collins, Glasgow) 'No one can behold this magnificent object for the first time without exclamations of wonder and astonishment.'

M 13 (Trouvelot) From Harvard Annals *Vol. VIII (B.A.A.)*

M 13 has naturally received a great deal of attention from amateurs and professionals alike as it is very bright, readily accessible to northern hemisphere observers and easily resolved, even in small telescopes. Wm. Herschel estimated that it contained about 14 000 stars but J. E. Gore, writing in 1904, thought that the number should be much less 'or it would be brighter than it is'. In a photograph taken at the beginning of this century by the American astronomer H. K. Palmer, 5482 stars were counted. Shapley estimated that it contained a total upwards of 100 000 stars and more recently, Otto Struve put the total number as possibly nearer half a million.

Although the visual object shows many different forms and often exhibits an almost spiral arrangement of branches, the full outline on photographic plates discloses an almost perfectly circular outline of ellipticity 9.5 in P.A. approximately 130°.

The brightest individual stars in the cluster are of magnitude 12 and, excluding the five brightest stars, the average photographic magnitude of the 25 next brightest stars is 13.75.

M 13 contains only about 15 variable stars compared with the 187 in M 3 and nearly 100 in M 15. The distance of M 13, obtained from these 15 variables, works out about 7 kpc.

Some 50 years ago it was thought that the whole of space was filled with light-obstructing matter and that this would have the effect of making the more distant stars redder than they should be for their spectral type. However, in about 1915, Harlow Shapley observed that the stars in M 13 were, if anything, bluer than average, showing that the reddening was ineffective. The galactic latitude of M 13 is fairly high, being about 40° N and it was suspected that the obscuring matter might be more thickly distributed toward the galactic plane. To determine whether this was the case, Hubble and Seares at Mount Wilson in 1920, made similar measurements of other clusters in various galactic latitudes and at various distances. These included M 3, M 5, M 11, M 15, M 22, M 35, M 36, M 38 and M 50 and the results showed no appreciable reddening up to a distance of 13 kpc.

M 13 contains a young, blue star of spectral type B2: this is Barnard No. 29 and its membership of the cluster has been confirmed by radial velocity measurements. Why so old a cluster should include such a young star is still a mystery (see also M 3).

M 13 is a fairly open globular – it is in Shapley's class V – and its high galactic latitude means that the major axis of its orbit is large. This shows that for most of its long history it has kept well away from the disrupting effect of the galactic nucleus. In H. S. Hogg's catalogue it is given a spectral class of F2 and a colour index of – 0.03. The average magnitude of the 25 brightest stars is 13.75.

Becvar, in the *Atlas Coeli Katalog*, gives the apparent magnitude as 4.0 photographic and 5.7 visual. The linear diameter of the cluster is 30 parsecs or nearly 100 light-years and its distance, 6.9 kpc, according to the same authority. Shapley's earlier distance estimate was 10.3 kpc and T. D. Kinman, in his revision of globular cluster distances, gives a figure of 7.4 kpc.

The age of M 13 has been calculated as 24 000 million years (Sandage) and 17 000 million years (Arp), the latter revising his estimate to 14 000 million years in 1962.

From radial velocity measurements (150 miles/s) the distance of M 13 has been estimated as about 8 kpc or 25 000 light-years.

Using a radial-velocity spectrometer on the 200-inch Hale telescope at Mt Palomar in 1971, Dr R. F. Griffin was able to measure the motions of 62 stars in the cluster. He found that the axis of rotation was close to P.A. 20°, which is also the direction of the minor axis of the elliptical outline. Dr Griffin also remarked that seeing M 13 in the 200-inch was 'rather like looking into a bowl of sugar!' (For observing details see next page.)

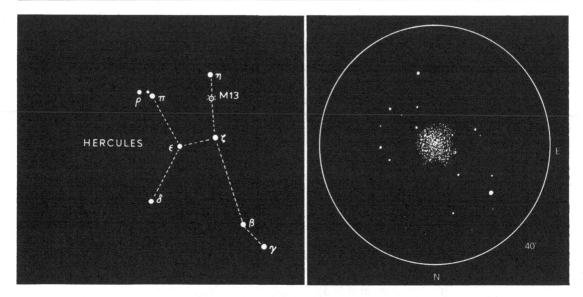

To observe   M 13 is easily seen with the naked eye on a good night. It is $2\frac{1}{2}°$ S of η *Herculis* or about one-third the distance from η to ζ *Herculis*.

This globular is a fine, show-piece object for small telescopes, being large, bright and open enough to be well resolved into stars even in a 4-inch. It appears to be slightly flattened on the east side and is sometimes seen to consist of long, curving strings of stars extending from a bright nucleus.

It stands magnification well and on 250× in an 8-inch telescope it can be resolved practically to the centre.

Objects near by   About 40′ to the NE is an 11 mag. spiral galaxy, NGC 6207. In a 6-inch or 8-inch this can be seen just to the west of a pair of 9–10 mag. stars as a faint, slightly elongated nebula about 2′×1′ with a brighter nucleus.

M 14   NGC 6402
A globular cluster, class VIII in OPHIUCHUS
R.A. 17 h 35.0 m.   Dec. S 3° 13' (1950).
Size 6' diam.   Mag. 7.7   Distance 7.4 kpc, 24 000 lt-yrs.

MESSIER   (June 1st, 1764) 'Nebula without star, not large, faint, visible in 3½-foot telescope. Round; near 9 mag. star. Reported on chart of comet of 1769. (Diam. 7').'

W. H.   Resolved it in 1783. 'Extremely bright, round, easily resolvable: with 300× I can see the stars. This cluster is considerably behind the scattered stars as some of them are projected upon it.' Later, in 1791: 'It resembles the 10th *Connaissance des Temps* (M 10) which probably would put on the same appearance as this if it were removed half its distance farther from us.'

J. H.   'A most beautiful and delicate globular cluster; not very bright, but of the finest star dust; all well resolved ... excessively rich. All the stars equal, 15 or 16 mag.'

ROSSE   'A cluster: stars small and very close together.'

SMYTH   'A large globular cluster of compressed, minute stars. A fine object of a lucid white colour and very nebulous in aspect. In a splendid field of stars, the lustre of which interferes with it. By diminishing the field under high powers, the cluster loses its definition.'

WEBB   'Large, glimpses of resolution.'

D'ARREST   'Elegant, comet-like, almost round, irregularly terminated, resolved on 226×.'

ROBERTS   'Curves and lines of stars radiating in all directions outwards from the dense cluster; vacancies in the centre.'

Shapley gives M 14 an ellipticity of 9 in P.A. 110°. His early distance estimate was about 19.7 kpc but in T. D. Kinman's *Revision Catalogue of Globular Cluster Distances*, and also in Becvar's *Katalog*, the distance is reduced to about 7.4 kpc.

Helen Sawyer Hogg quotes the spectrum of the cluster as G0 and its colour index is fairly large at +0.11, showing that it is considerably reddened. The average magnitude of the 25 brightest stars is 15.44.

In 1964, the first nova ever to be photographed in a globular cluster was discovered in M 14 by Amelia Wehlau of the University of Western Ontario. The nova actually brightened in 1938 and was found at the later date from plates taken by Helen Sawyer Hogg between 1932 and 1963, using the 74-inch reflector of the David Dunlap Observatory and the 72-inch of the Dominican Astrophysical Observatory. 255 photos were obtained on 124 nights during this period and the nova appears on only eight of them, these being taken on June 21–28th, 1938.

The brightest that the nova appeared on the plates was only 16.0 mag. which explains why it had not been noticed before. Mrs Hogg estimates that this is equivalent to an absolute magnitude of about −1.5 but she considers that it may have been as bright as mag. 10 or −7.5 abs. mag.

The only other nova found in a globular cluster was *T Scorpii*, in M 80 which was discovered in 1860 when it reached a visual magnitude of 7.

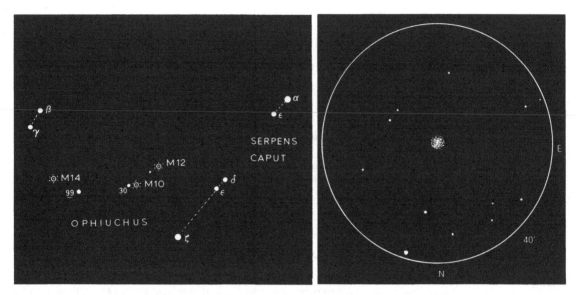

To observe   M 14 is a little isolated and perhaps it is best approached from M 10. From M 10 move 1° S and then sweep 7° E to the star *99 Ophiuchi* (4½ mag.). M 14 is 2° N and 3° E of *99 Ophiuchi* (*99 – Piazzi – in Norton = 47 – Flamsteed – in Becvar*).

M 14 seems fainter than the given magnitude of 7.7 and at first sight looks rather more like an elliptical galaxy than a globular cluster. It appears as a perfectly round, hazy patch in which the brightness gradually tapers off towards the edges. A trace of resolution just appears at the periphery using 100× in an 8-inch and it was not bright enough to stand higher magnification. The whole area is more or less homogeneous and shows no definite central condensation.

Admiral Smyth seems to have had similar difficulty in trying to resolve it but Sir Wm Herschel, using 300× with his 20-foot F.L. 18.7-inch aperture reflector, found it easily resolvable. It evidently needs a large aperture to resolve well.

M 15 NGC 7078
A globular cluster, class IV in PEGASUS
R.A. 21 h 27.6 m.   Dec. N 11° 57' (1950).
Size 12' diam.   Mag. 6.0   Distance 15.1 kpc, 49 500 lt-yrs.

Discovered by Maraldi on Sept. 7th, 1746 while looking for de Chéseaux' comet.

MARALDI   'A nebulous star, fairly bright and composed of many stars.'

MESSIER   (June 3rd, 1764) 'Nebula without a star; it is round and brilliant in the centre. M. Maraldi reported this nebula in the Memoirs of the Academy in 1746. Diam. 3'.'

BODE   (No. 71, Sept. 23rd, 1774) 'Round; involved in a dense nebula; containing no stars.'

W. H.   In 1783 resolved it into stars. Considered it a good test object.

D'ARREST   'Most magnificent cluster, very bright, with beautiful arm-like extensions apparent. The nucleus is eccentric, being displaced to the east.'

SMYTH   'Although this noble cluster is rated globular, it is not exactly round and under the best circumstances is seen with stragglers branching from the central blaze. Under M.P. there are many telescopic and several brightish stars in the field but the accumulated mass is completely insulated.'

WEBB   'Bright and resolvable; blazing in centre. A glorious object in a 9½-inch "With" mirror. Very fine specimen of a completely insulated cluster. Buffham, with a 9-inch speculum finds a dark patch near the middle with two faint, dark lines or rifts like those in M 13, unnoticed by J. H. or D'Arrest.'

ROBERTS   Photograph shows the stars arranged 'in curves, lines and patterns and the centre nebulous.'

LICK XIII   'A bright, unusually beautiful globular cluster, 8' diam.'

M 15 has been closely studied by Shapley, Baade and others as it is rich in variable stars: with nearly 100 variables, it is the third richest after M 3 and the bright southern globular, ω *Centauri*.

A faint planetary nebula was discovered to be a member of the cluster in 1957. A spectrogram was obtained which showed three emission lines typical of the gaseous envelope of planetary nebulae.

Shapley gives the ellipticity of M 15 as 8 in P.A. 35°. In H. S. Hogg's catalogue of clusters, M 15 has a spectrum of FO, a colour index of −0.08 and the magnitude of the 25 brightest stars, on average, is 14.31. Its true diameter, according to Becvar's *Atlas Coeli Katalog*, is 27 parsecs or about 88 light-years.

The distance of the cluster, in Shapley's earlier estimate, was 13.1 kpc: Becvar gives 15.1 kpc and T. D. Kinman, 16.0 kpc.

The planetary nebula, K648 was found to be a weak radio source in 1980, and in 1974 the cluster itself was found to be an X-ray emitter. By 1973 a few more variables, mostly *RR Lyrae* type, were discovered, to bring the total of variables to 112. One classical Cepheid, Type II with a period of 17.11 days has also been found.

In 1972, D. J. Stickland made a detailed study of the colour profile of M 15 using the historic 36-inch Crossley reflector of Lick Observatory. Although the cluster appeared to have a distinct bluish tinge in its visual image, the spectrum scanner revealed that the maximum brightness is found well into the red region of the spectrum as should be expected from the great age of the stars of which the cluster is composed. The bluish cast seen visually is therefore illusory. (See M 2 and M 3.)

In 1988, French and British astronomers succeeded in pinpointing the X-ray source in M 15. It was observed as a bright object, variable in both ultraviolet and X-ray emission with a period of about nine hours and which is thought to be produced by a normal-plus-neutron star binary system. An additional feature was obtained from radial velocity measures which showed the binary to be moving out of the cluster itself with a speed of some 150 km/s.

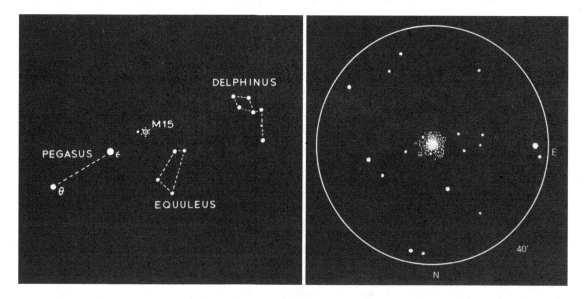

To observe   Although in a fairly barren region of the sky, it is not difficult to find as the 2nd mag. star, ε *Pegasi*, is quite close. From ε *Pegasi* move $3\frac{1}{2}°$ W and $2\frac{1}{4}°$ N. There is a 6 mag. star 20' away to the east.

M 15 is a fine, bright globular cluster with a smallish, exceptionally bright centre, surrounded by a large area of resolvable stars. The total size is not less than 8' diam. As Admiral Smyth noted, the outline is distinctly non-circular and it is not difficult to make out many small protuberances and inlets in the outer region.

It is a fine object for small telescopes, being hardly inferior to M 5 or M 13.

The planetary nebula K648 involved in the cluster is very small (3" diam.) and at 13.8 mag. is a difficult object to locate. Its coordinates, determined by Adams, Penn and Seaton of UCL in 1980, are R.A. 21 h 27 min 34 s and Dec. +11° 57' 14" (1950).

Objects near by   Just over 1° to the SSW is the double star, Σ 2799. Both components are 7.4 mag., P.A. 289°, d. 1".5 (1925).

M 16 NGC 6611
A gaseous emission nebula and associated galactic cluster in SERPENS
R.A. 18 h 16.0 m.   Dec. S 13° 48' (1950).
Size 8' (diam. of cluster)   Mag. 6.4   Distance 1.8 kpc, 5870 lt-yrs.

Discovered by de Chéseaux in 1746.

DE CHÉSEAUX (No. 4) 'A cluster of stars between the constellations of Serpens, Sagittarius and Antinous.'

MESSIER (June 3rd, 1764) 'A cluster of small stars enmeshed in a faint glow. In an inferior telescope it appears like a nebula. (Diam. 8').'

J. H. 'A cluster containing at least 100 stars.'

SMYTH 'A scattered but fine, large, stellar cluster. As the stars are disposed of numerous pairs among the more evanescent points of more minute components, it forms a pretty object in a telescope of moderate capacity.'

WEBB 'A grand cluster.'

ROBERTS Of a photograph taken by him in Aug. 1875. 'A large bright *nebula* with a cluster apparently involved in it.'

BARNARD Also described it as a nebula in 1915.

The gaseous nebula has been photographed with the 120-inch Lick and 200-inch Palomar telescopes and has been seen to contain many bright-edged, dark lanes which are strongly visible in red, blue and ultra-violet light. The nebulosity, which appears brighter in the blue and ultra-violet, shows a striated structure somewhat similar to that photographed in the Pleiades, M 45.

The cluster contains a large number of faint red stars which may be reddened by absorption in the gaseous matter surrounding them.

M 16 also contains many small, black globules, especially in the SE portion of the nebula. There are compact dust clouds similar to those discovered in M 8 and studied by Bok who thinks that, after a slow, gradual contraction, they become 'protostars', visible by their own radiation.

Strong turbulence has been detected around the dark 'elephant trunk' region in the north of the nebula. Fred Hoyle has described the phenomenon as 'a hot cloud of gas expanding into a cooler region like an exploding bomb'. M 16 is very useful as a starting point from which to explore the many Messier Objects in Sagittarius, Ophiuchus and Scorpius. (For observing details, see next page.)

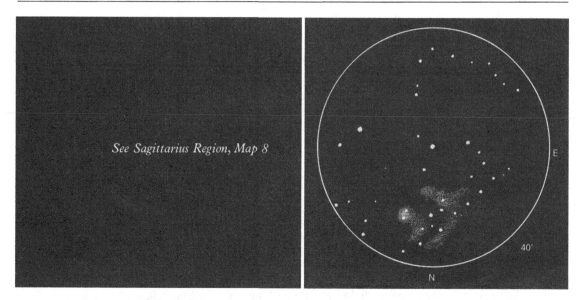

See Sagittarius Region, Map 8

To observe   M 16 is very useful as a starting point from which to explore the many Messier Objects in Sagittarius, Ophiuchus and Scorpius.

From *Altair* move SW through δ and λ *Aquilae* to 4th mag. γ *Scuti*. M 16 lies 1° N and 2½° W from γ *Scuti*. Use low power and confirm the position by moving 2½° S and very slightly E when M 17 will be seen – an unmistakable object (q.v.).

The cluster is fine and conspicuous and situated at the northern end of a large, distinct, curving 'S' of bright stars. The nebulous part is faint but distinctly glowing and of varying intensity, being brightest around a fairly close pair of bright stars at the tail of the 'S'. Averted vision increases the area of nebulosity that can be seen and at 8-inches aperture a dark intrusion can be made out at the extreme north of the cluster.

Objects near by   A faint galactic cluster H 19 lies 40' to the NW, diam. 5', mag. 12.2, class 'g', contains 20 stars.

94

M 17 NGC 6618

The 'Omega' or 'Horse-shoe' Nebula. A gaseous emission nebula in
SAGITTARIUS
R.A. 18 h 17.9 m.   Dec. S 16° 12′ (1950).
Size 46′×37′   Mag. 7.0   Distance 1.8 kpc, 5870 lt-yrs.

Discovered by de Chéseaux in 1746.

DE CHÉSEAUX (No. 20) 'A nebula which has never been discovered: it has a shape quite different from the others: it has the perfect form of a ray or the tail of a comet, 7′ long and 2′ wide. Its sides are exactly parallel and quite well terminated. The centre is whiter than the edges. It makes an angle of 30° with the meridian.'

MESSIER (June 3rd, 1764) 'A train of light without stars, 5′ or 6′ in extent, in the shape of a spindle, a little like that in Andromeda's belt (M 31) but the light very faint. In a good sky, seen very well in a 3½-foot telescope. Diam. 5′.'

W. H. (*Phil. Trans.* 1785) 'A wonderful extensive Nebulosity of the milky kind. There are several stars visible in it but they can have no connection with that nebulosity and are, doubtless, belonging to our own system scattered before it. It is the 17th of the *Connoissance des Temps.*'

J. H. 'The chief peculiarities are, first, the resolvable knot in the following portion of the right branch which is, in a considerable degree, insulated from the surrounding nebula and, secondly, the smaller and feebler knot as the N.p. end of the same branch where the nebula makes a sudden bend at an acute angle.'

MITCHEL (*Stellar Worlds* (1869), pp. 256–61, Wm Collins, Glasgow) 'A bright condensed mass of light, nearly round ... encircled by a perfectly dark vacuity on the outside of which nebulous light is again seen.'

SMYTH 'The horse-shoe or omega nebula: a magnificent and irresolvable luminosity occupying more than a third of the area in a splendid group of stars, principally from 9 to 11 mag. reaching more or less all over the field.'

WEBB 'Horse-shoe nebula, visible in finder, 1° N of M 18.'

HUGGINS 'Spectrum gaseous.'

FLAMMARION 'Like a smoke-drift, fantastically wreathed by the wind.'

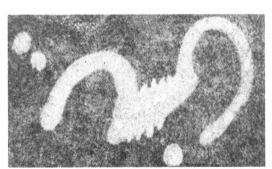

M 17 (Lewis Swift, Director of Warner Observatory, Rochester, N.Y.) As drawn from memory in 1885 (B.A.A.)

M 17 (Chambers) From his Handbook *1889 (B.A.A.)*

CHAMBERS   'Like a swan, floating on water.'

LICK XIII   'Very bright, very large, diffuse nebulosity, showing a wealth of detail, filling an area about 26'×20'.'

HOLDEN   who studied the many observations of M 17 made between 1833 and 1875, thought that it had changed position with regard to the neighbouring stars and, according to G. F. Chambers, Schröter also suspected some change in its structure.

Many early observers made drawings of this object, including Lord Rosse, Lassell, Holden, Trouvelot and Smyth.

M 17, like many of the bright, diffuse nebulae, appears to consist of both bright filaments and dark, obscuring matter. However, the absorption regions were considered by Duncan to be different from those in M 8 and M 16. The nebula also appears to contain dark filaments radiating from a point in the centre of the southern portion of the nebula.

The brightest point of the nebula, which is roughly rectangular in shape, has been estimated to have a mass about 800 times that of the Sun. This is a good deal greater than the figure for the Orion Nebula (M 42).

Although Wm Herschel thought the stars that it seemed to contain had no connection with it, there are, in fact about 35 stars which are actually embedded in the gaseous nebula.

M 17 is also a radio source on 9.4 cm and is within 2° of the X-ray source Sgr XR-2.

Infra-red observations in 1979 revealed a 29 per cent polarisation in M 17, which seems to indicate the existence of 'bipolar nebulae' or lobes of scattering material in the polar regions. A twin-lobed radio source was also detected extending from a central infra-red cluster. (For observing details see next page.)

See Sagittarius Region, Map 8

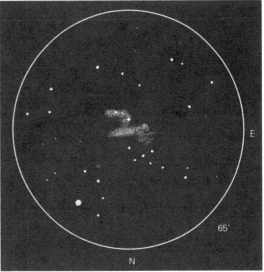

To observe   From γ *Scuti* move 1½° S and 2° W. There is a 6 mag. star 30' to the NW of the centre of M 17 and will be seen in the same field of view of a L.P. eyepiece.

At first sight, only the bright 'bar' may be seen in the centre of an open cluster of stars. This bar glows strongly and, in poor conditions or at low altitude could be mistaken for an 'edge-on' galaxy except that it has no central condensation. On further examination, the two stars close to the south can be seen to be embedded in nebulosity also. With averted vision, faint patches of light can be seen in a curve to the NE and the whole nebula then looks something like a figure 3. With patience, a great deal of fine detail can be made out over an area nearly 40' in diameter.

Chambers' description of 'a swan, floating on water' is a good impression of the brightest portion which may be all that is visible in a small telescope.

When found, this object, being easily recognised, makes a convenient 'sky-mark' from which one can find one's way to the numerous other objects in this populous region.

M 18 NGC 6613
A galactic cluster, type 'd' in SAGITTARIUS
R.A. 18 h 17.0 m.   Dec. S 17° 09' (1950).
Size 7' diam.   Mag. 7.5   Distance 1.5 kpc, 4900 lt-yrs.

Discovered by Messier 1764.

MESSIER (June 3rd, 1764) 'A cluster of small stars, a little below M 17; surrounded by slight nebulosity. Easier to see than M 16. Appears nebulous in 3½-foot telescope: with a better telescope, the stars can be seen. (Diam. 5')'.

SMYTH (Commenting on Messier's description) 'The means of that very zealous observer did not quadrate with his diligence.' (On M 18) 'A neat double star, A9, B11, both bluish, P.A. 133°, d. 248" (1879) in a long and straggling assemblage of stars. The whole vicinity is very rich and there are several splendid fields about a degree to the South.'

WEBB 'Glorious field in a very rich vicinity. South, lies a region of surpassing splendour.'

This galactic cluster does not seem to have received much attention from astronomers: it is not included in Wallenquist's 1959 *Catalogue of Galactic Clusters* but in Helen Sawyer Hogg's list in the *Handbuch der Physik* it is described as having an apparent diameter of 7', integrated magnitude of 8.0, and its brightest star having a magnitude of 10.9.

Trumpler classified it as II,3,p – i.e. having a fairly high concentration, with a large range of magnitude among the cluster stars and poor – with less than 50 stars. In the *Atlas Coeli Katalog* it is credited with having no more than 12 stars among its members.

The description in the *New General Catalogue* of J. L. E. Dreyer is also brief: 'Cl, P, vlC' – A cluster, poor, very little compressed.

Photographs taken with the 48-inch Schmidt telescope at Mt Palomar reveal a region of faint nebulosity surrounding the whole cluster. (For observing details see next page.)

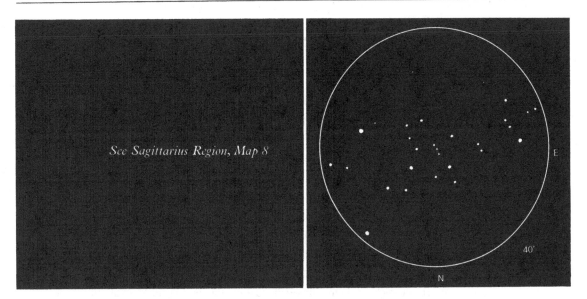

See Sagittarius Region, Map 8

To observe   From M 17 move 1° S and slightly W, *or*, from γ *Scuti* move 2½° S and 2° W.

M 18 is an open and rather inconspicuous cluster with one or two brightish stars and with the remaining, fainter members forming something of an 'S' shape.

Like several of the other 'poor' clusters in Messier's list, such as M 21, M 26, M 29 and M 103, this object is, paradoxically, better distinguished in a poor telescope – or at least, one with a wide field and using a low magnification. With a field of, say 2°, the smaller scale makes the cluster appear more compressed and almost 'nebulous' as Messier described it.

In the amateur's average 6-inch or 8-inch Newtonian of about f.8 or so, the cluster appears very poor and scattered and difficult to pick out from the many other groups round about.

Objects near by   There is a patch of bright, diffuse nebula about 45' to the NW which may be seen in a good rich-field telescope.

The 'splendid' and 'glorious' fields to the south, mentioned by Smyth and Webb are, of course, portions of the Milky Way and really include the object which Messier catalogued as No. 24. The whole area is best seen with a very wide field (see M 24).

M 19   NGC 6273
A globular cluster, class VIII in OPHIUCHUS
R.A. 16 h 59.5 m.   Dec. S 26° 11' (1950).
Size 5' diam.   Mag. 6.6   Distance 6.9 kpc, 22 500 lt-yrs.

Discovered by Messier in 1764.

MESSIER   (June 5th, 1764) 'Nebula without star. Round, easily seen in 3½-foot telescope. Nearest neighbouring star is 28 *Ophiuchi*. (Diam. 3').'

W. H.   Resolved it in 1784.

J. H.   'Superb! A globular cluster, very bright, round, diameter 10'. Resolved into stars of 14, 15 and 16 mag.'

SMYTH   'A fine, insulated globular cluster of small and very compressed stars of creamy white tinge and slightly lustrous in the centre. Near the large opening or hole, about 4° broad, in the Scorpion's body, which W. H. found to be almost devoid of stars.'

There are many dark nebulae in this part of the Milky Way, the one which W. H. mentions as being almost devoid of stars is close to M 80 about 10° away from M 19. See M 80.)

WEBB   'A mass of stars; large; fairly bright, but very low.'

M 19 is one of the most elongated of globular clusters. Shapley gives it an ellipticity of 6 in P.A. 15° and estimated that there could be counted twice as many stars along the major axis as along the minor.

It is also one of the nearer of the Messier globulars to the galactic centre, being a little over 1 kpc away.

In H. S. Hogg's list in the *Handbuch der Physik*, the spectrum of M 19 is F5, its colour index +0.04 and the average magnitude of the 25 brightest stars 14.8 mag.

Its true diameter, according to Becvar, is 20 parsecs or about 65 light-years. Shapley's earlier estimate of its distance was 16.3 kpc but later figures put it at about 7.0 kpc. (For observing details see next page.)

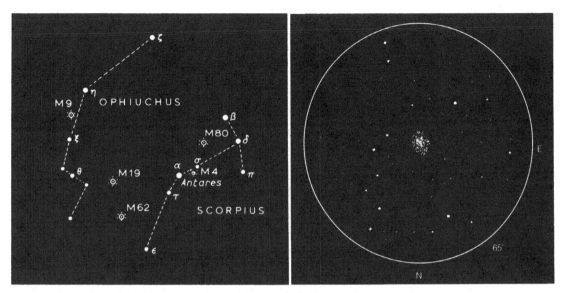

To observe   M 19 is easily located, being 8° due E of *Antares* and embedded in the Milky Way.

It appears as a small globular mass, rather like M 80 but slightly brighter and having a less concentrated centre. Its ellipticity is easily apparent, though not as much as the true (photographic) ratio of 6. It appears to be about 10 per cent to 15 per cent longer N–S than E–W.

The edges are fairly easily resolved on M.P. but for observers in higher latitudes it is not really bright enough to stand higher magnification and resolution to the centre.

It is accompanied by a long, N–S zig-zag of stars to the west and the background is filled with the faint dusting of the Milky Way.

Objects near by   There is a globular cluster, NGC 6293, 1½° to the ESE, 8.4 mag. 1.9′ diam. and another, NGC 6284, 95′ to the NNE, mag. 9.7, 1.5′ in diameter.

M 20 NGC 6514

The 'Trifid' Nebula. A gaseous emission nebula in SAGITTARIUS
R.A. 17 h 58.9 m.   Dec. S 23° 02' (1950).
Size 29'×27'   Mag. 9.0   Distance 700 pc, 2300 lt-yrs.

Discovered by Le Gentil before 1750. (Bigourdan.)

MESSIER   (June 5th, 1764) 'Cluster of stars a little above the Ecliptic between the bow of
Sagittarius and the right foot of Ophiuchus. Reviewed 22 Mar. 1781.'

W. H.   Catalogued it in 4 portions (41 H. IV and 10, 11 and 12 H. V).

J. H.   First to call it 'trifid'. 'Singularly trifid, consisting of 3 bright and irregularly formed
nebulous masses, graduating away insensibly externally but coming up to a great
intensity of light at their interior edges where they enclose and surround a sort of
3-forked rift or vacant area, abruptly and uncouthly crooked and quite void of
nebulous light. A beautiful triple star is situated precisely on the edge of one of
those nebulous masses just where the interior vacancy forks into two channels.'

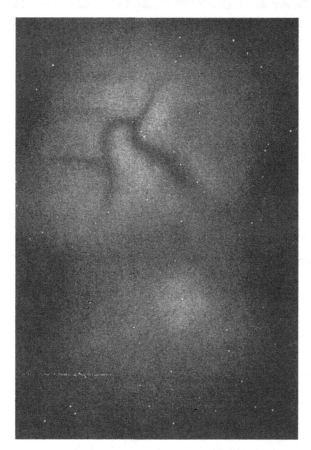

*M 20 (Trouvelot) Made using the 15-inch refractor of Harvard College Observatory
in 1874 (B.A.A.)*

SMYTH   'After examining M 23 I lowered the telescope a couple of degrees and gazed for
the curious Trifid nebula, 41 H. IV, but though I could make out the delicate triple
star in the centre of its opening, the nebulous matter resisted light of my telescope
so that its presence was only indicated by a peculiar glow.'

WEBB   'The Trifid nebula, closely f. a cruciform group. Very curious object: pair with
minute *comes* where 3 ways meet; dark rifts through nebulosity. Nebula imperfectly
seen; rather low. Grand region.'

LICK XIII   'Very bright diffuse nebula covering an area 24'×20' with remarkable dark lanes.'

M 20 is a typical emission nebula, being a large mass of glowing gas excited
to emit radiation by a hot star or stars within it. It is very symmetrical in shape
and decreases in density rapidly from the centre outwards.

The 'trifid' division is due to dark, obscuring matter closely connected with
the nebula itself: a dark cloud in the SE part of the nebula has a typical bright
edge like those seen in M 8 and M 16.

M 20 is also a radio source on 9.4 cm. This high-frequency radiation and
absence of evident radiation on longer wavelengths leads to the conclusion
that the radio emission is 'thermal' in origin and is the result of collisions
between electrons and protons. For this source of energy, temperatures of
about 10 000 K are necessary. Similar radiation can be found in M 8 and
M 42.

The real diameter of the nebula has been estimated at about 30 light-years.
The gas contains a ratio of helium to hydrogen of about 0.094.

As noted by J. H. the centre of the nebula contains a conspicuous triple star
(ADS 10991). These are all O-type giants with abs. mag. of about −5.2. The
components are A, 76 mag, B, 10.7 mag, C, 8.7 mag, but four fainter members of the system have also been found. (For observing details see next page.)

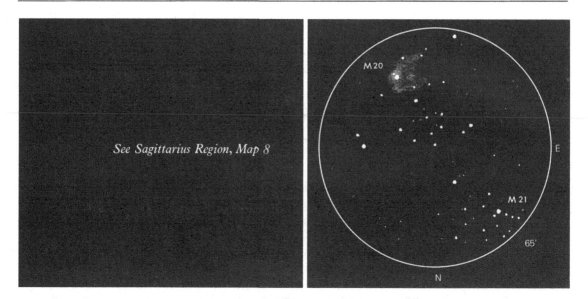

See Sagittarius Region, Map 8

To observe   From μ *Sagittarii* (mag. 4) move 2° S and 2½° W (M 21 will be seen 50' to the NE of M 20).

As both Smyth and Webb noted, the nebula is not a very conspicuous object in the latitudes of the British Isles and the same remarks apply to the northern parts of the U.S.A. However, even here, a good aperture and a clear sky will reveal a fair amount of the nebulosity.

In lower altitudes, as in the southern U.S.A., where the nebula rises higher in the sky, the chances of seeing it as 'trifid' are much more favourable. A fine, transparent atmosphere and an instrument of wide aperture and short focal length are the ideal.

In less favourable circumstances the brightest portion of the nebula is to the SW surrounding two brightish stars close together. From this point, averted vision can be used to build up the fainter portions of the nebula.

Its total area of about 29'×27' is almost as large as the full Moon but the whole of this amount would be found only by photography.

M 21 NGC 6531
A galactic cluster, type 'd' in SAGITTARIUS
R.A. 18 h 01.7 m.   Dec. S 22° 30' (1950).
Size 12' diam.   Mag. 6.5   Distance 1.3 kpc, 4250 lt-yrs.

Discovered by Messier in 1764.

MESSIER   (June 5th, 1764) 'Star cluster near M 20: the brightest star is 11 *Sagittarii*, 7 mag. The stars of both clusters are 8–9 mag. and enveloped in nebulosity.'

SMYTH   'A course cluster of telescopic stars in a rich gathering galaxy region. About the middle of a conspicuous pair, A, 9 mag. yellowish; B, 10 mag. ash-coloured; P.A. 317°, d. 30".9 (1875). Messier included some outliers in his description and what he mentions as nebulosity must have been the grouping of minute stars in view.'

WEBB   'In a lucid region.'

M 21 is a cluster which shows, on photographic plates, a strong concentration toward the centre. In 1953, S. N. Svolopoulos of Norman Lockyer Observatory, examined 109 stars within a radius of 6'.8 of the centre of the cluster. Of these, 57 were found to be members and the distance modulus obtained was 9.34 mag. which is equivalent to a distance of 680 parsecs.

The stars are distributed according to brightness as follows:

Visual Mag.	8	9	10	11	12	>12.5
No. of stars	1	4	4	6	18	28

Trumpler's classification for M 21 is I,3,p and he made the spectrum of the brightest star, BO.

In Wallenquist's catalogue the mean magnitude of the five brightest stars is 8.2 and 36 stars were counted in a diameter of 12'. The central density of the cluster Wallenquist gives as 9.36 stars per cubic parsec with an overall density of 0.52 stars per cubic parsec.

Becvar gives the true diameter as 2.6 parsecs but Wallenquist's figure is 5.1 parsecs or about 17 light-years.

M 21 is considered to be a comparatively young cluster. (For observing details see next page.)

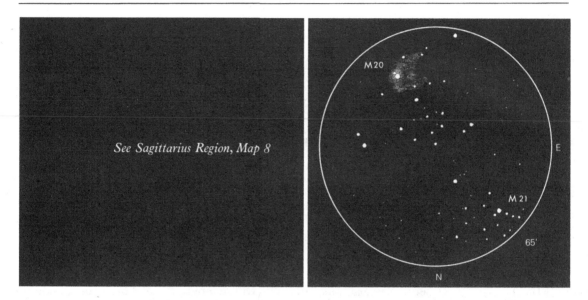

See Sagittarius Region, Map 8

To observe   M 21 is a little less than 1° NE of M 20 or 2½° SW of μ *Sagittarii*.

It is difficult to decide which star Messier referred to as the brightest. The one he names, *11 Sagittarii*, is several degrees away to the SE and is 5 mag. The brightest of the stars in both M 20 and M 21 is GC 24526, mag. 5.7, on the northern edge of M 20.

M 21, which can just be seen in the same field of view with M 20 on L.P. appears as a smallish, fairly compact cluster at the north-east end of a double string of fairly bright stars which connect it with M 20. It is somewhat sparse and formless and contains no nebulosity at all. Lower power and wide field will give the best views of this cluster and M 20.

M 22 NGC 6656

A globular cluster, class VII in SAGITTARIUS
R.A. 18 h 33.3 m.   Dec. S 23° 58' (1950).
Size 17' diam.   Mag. 5.9   Distance 2.4 kpc, 7800 lt-yrs.

The first globular cluster to be discovered: Hevelius is supposed to have seen it sometime before 1665 but Halley, de Chéseaux and Messier all credit its discovery to Abraham Ihle in 1665.

HALLEY   1716. 'This nebula is near the ecliptic ... not far from the point of the Winter Solstice. This, it seems, was found in the year 1665 by a German gentleman named Abraham Ihle whilst he attended the motion of Saturn, then near his aphelion. This is small and luminous and emits a ray like the former.'

DE CHÉSEAUX   (No. 17) '5' diam., round and of a reddish colour.'

LE GENTIL   drew it in Aug. 1747 as seen in a telescope of 18-feet F.L. and saw it as 'very irregular, long-haired and spreading some kind of rays of light all around its diameter.'

LACAILLE   (1751–52) 'It resembles the small nucleus of a comet.' (Lac. I. 12.)

MESSIER   (June 5th, 1764) 'A round nebula without star near 25 Sag. (7 mag.). Seen in 3½-foot telescope. Position found from lambda Sagittarii. Abraham Ihle, a German, found it in 1665 while observing Saturn. M. Le Gentil observed it in 1747 (Diam. 6') reported on the *English Atlas*.'

W. H.   Was the first to resolve it into stars and described it as a circular cluster.

J. H.   Recommended it as a test object for telescopic resolving power. 'A magnificent globular cluster: gradually brighter in the middle but not to a nucleus. All the stars of two sizes; 10 and 11 mag. as if one shell over another; larger ones ruddy.'

SMYTH   'A fine globular cluster: consists of very minute and thickly condensed particles of light with a group of small stars preceding by 3 min. somewhat in a crucial form.'

WEBB   'Beautiful bright cluster, very interesting from visibility of components; largest 10 and 11 mag., makes it valuable as a clue to the structure of many more distant or difficult nebulae.'

LICK XIII   'A beautiful globular cluster 12' or more in diameter.'

Harlow Shapley paid a lot of attention to this globular cluster: he and Pease counted 70 000 stars in it. He gives its ellipticity as 8 in P.A. 25° and estimates that it contains about 30 per cent more stars along that axis than in the minor axis. Shapley also confirms J. H.'s estimate that it has considerable uniformity in the magnitudes of its brighter stars.

It is one of the nearest of the globular clusters to the Sun and, with a galactic latitude of 9° S is very close to the galactic plane and in the midst of the great star-clouds of the Milky Way.

The long-period variable of 199½ days discovered by Bailey in 1920 is now considered not to be a member of the system. However, more than 20 short-period variables are established as belonging to the cluster.

In Helen Sawyer Hogg's catalogue, M 22 is given a spectrum of F6, a colour index of +0.03 with the mean magnitude of the 25 brightest stars, 12.93.

Its true diameter, according to the *Atlas Coeli Katalog*, is 34 parsecs or about 110 light-years.

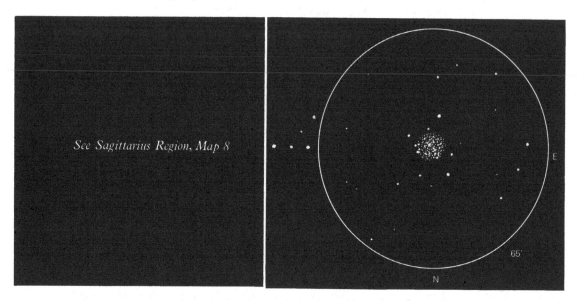

See Sagittarius Region, Map 8

To observe   From λ *Sagittarii* (mag. 3) move 1½° N and 2° E. It is large, bright and unmistakable.

A fine, large globular cluster of noticeably even brightness over an area a little greater than 15′. Quite easily resolvable even in the centre with apertures of 6 inches or more. It is distinctly non-circular but the P.A. of the major axis is not the same as that of the photographic alignment (25°) but seems to lie more E–W.

Although of low altitude at culmination in northern regions it is still a fine object and in lower latitudes it is magnificent. Being a comparatively loose cluster, class VII, it is easier to resolve than M 13 and nearly as bright.

There is a conspicuous group of bright stars to the west which Smyth noted as cruciform.

Objects near by   There is another cluster, NGC 6642, about 70′ to WNW mag. 7.9 and less than 1′ diameter which is sometimes classed as a galactic cluster, type 'd' but which is probably a globular.

Note   Those observers who are fortunate enough to live in low latitudes and can observe the magnificent ω *Centauri* cluster may notice the resemblance between it and M 22. They are of similar classes of concentration (ω *Centauri* is class VIII) and both can be easily resolved. M 22 is a little smaller and about two magnitudes fainter.

M 23 NGC 6494
A galactic cluster, type 'e' in SAGITTARIUS
R.A. 17 h 54.0 m.   Dec. S 19° 01' (1950).
Size 35' diam.   Mag. 6.9   Distance 660 pc, 2150 lt-yrs.

Discovered by Messier in 1764.

MESSIER   (June 20th, 1764) 'A star cluster very near 65 *Ophiuchi*. The stars of this cluster are very close to one another. Position determined from µ *Sagittarii*. Diam. 15'.'

J. H.   'About 100 stars, 9–10 to 13 mag.'

SMYTH   'A loose cluster: an elegant sprinkling of telescopic stars over the whole field under moderate magnification. The most clustering portion is oblique in a direction S.p. and N.f. Precedes a rich outcropping of the Milky Way.'

WEBB   'Grand low power field.'

FLAMMARION   'On a photograph, six stars in a circle can be seen near the centre and above and to the left, nine stars in an arc.'

Trumpler classified M 23 as I,2,r and gives the spectral class of the brightest star as B9.

Wallenquist counted the probable number of cluster stars as 129 in a diameter of 34' and gives the mean magnitude of the five brightest stars as 9.4. The density of the cluster in the central part is 31 stars per cubic parsec and an average density for the whole cluster, 1.2 stars per cubic parsec. It is considered by Wallenquist to be one of the older galactic clusters.

M 23 was examined in 1953 by S. N. Svolopoulos at Norman Lockyer Observatory for colours and magnitudes. He observed a total of 333 stars and found 149 members with a diameter of 27'.2.

The distribution of stars in magnitude was as follows:

Visual mag.	10	11	12	13	>13.5
No. of stars	12	17	24	20	96

(For observing details see next page.)

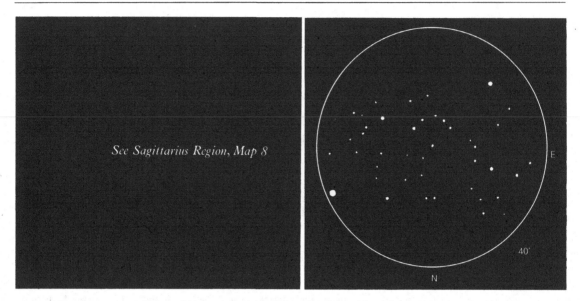

See Sagittarius Region, Map 8

To observe   From μ _Sagittarii_ move 2½° N and 3½° W _or_, from M 9 move ½° S and then 8° E.

M 23 is a fine, bright open cluster of mainly 10–13 mag. stars. There is a bright 6 mag. star to the NW and some observers have described this star as being like a diamond in a ring of stars formed by the cluster. To me, however, the main pattern of the cluster converges on an 8 mag. star to the NE forming a distinctive fan-like shape, having three or more concentric curves of stars, all centred on the 8 mag. star.

The diameter of the cluster has been given variously as 25′ to 50′ and it is a little difficult to decide exactly where the boundaries of the cluster end. The main portion, however, fits neatly inside a 40′ field.

M 24   (NGC 6603)

A Milky Way star-cloud. (A galactic cluster, type 'g') in SAGITTARIUS
R.A. 18 h 15.5 m.   Dec. S 18° 26′ (1950).
Size 1½° (4.5′) diam.   Mag. 4.6 (11.4)   Distance (6603) 5 kpc, 16 000 lt-yrs.

Star-cloud discovered by Messier in 1764.

MESSIER   (June 20th, 1764) 'Cluster on the parallel of the preceding and near the tip of
Sagittarius' bow, in the Milky Way: a large nebulosity in which there are many
stars of different magnitudes: the light which is spread throughout this cluster is
divided into several parts. The position is given for the centre of the cluster. Diam.
1° 30′.'

BODE   (No. 51) 'A nebulous star cluster above mu in the bow of Sagittarius.'

J. H.   'Pretty large, very rich, with stars of 11–20 mag.' (The entry in J. H.'s *General
Catalogue* for No. 4397 and later included in the NGC as No. 6603, reads: '!Cl, vRi,
vmC, R, st 15, M.W.' A remarkable object, a cluster, very rich, very much
compressed, round, stars of 15 mag., in the Milky Way.)

SMYTH   'A beautiful field of stars: the whole is faintly resolvable, though there is a
gathering spot with much star-dust. A double star, H & S 264, follows in S.f.
quadrant and a wider one, H & S 263, S.p.'

WEBB   'Magnificent region visible to the unaided eye as a kind of protuberance in the
Galaxy and so considered by J. H. It is accompanied by two little pairs; 2° N of μ
Sagittarii.'

ROSSE   'Unresolved nebulous light in it.'

J. E. GORE   (writing in 1904) 'A photograph taken in Feb. 1894 shows the stars free from
nebulosity. On this photograph the streams of stars surrounding the centre seem to
be arranged in spirals and strongly suggest that the cluster has been evolved from a
spiral nebula.'

A. M. CLERKE   (in 1905) 'Visible to the naked eye as a dim cloudlet near μ *Sagittarii* and named by
Fr. Secchi as "Delle Caustiche" from the peculiar arrangement of its stars in rays,
arches, caustic curves and intertwined spirals.'

FLAMMARION   (1918) 'Several arrangements of stars worthy of attention can be observed here. A
little to the right of the centre there is a little nebulous patch. It is difficult to see
this cluster separate from the rest of the deep sky in the middle of the Milky Way.'

J. L. E. DREYER   In his notes to the first part of the NGC Dreyer adds this supplementary comment
by J. H. 'h. 2004 = M 24. W. H.'s two observations hardly consist with this
description and their deviation of nearly +3 m from Messier's place makes it very
doubtful whether he really saw this object.'

The Messier Object, M 24, has a magnitude of 4.6 but is not a true galactic
cluster, being a detached portion of the Milky Way about 1½° in diameter as
Messier reported in his 1781 catalogue. The galactic cluster, NGC 6603, is
embedded in this rich star field.

Wallenquist does not include it in his 1959 catalogue of galactic clusters but
Trumpler classified NGC 6603 as I,2,r.

In the *Handbuch der Physik*, Helen Sawyer Hogg describes the actual cluster (under M 24) as being of magnitude 11.4 and with an apparent diameter of 4.5', the magnitude of the brightest star being 13.0.

The confusion about this object seems to have arisen by giving the magnitude of the large Milky Way portion seen by Messier (mag. 4.6) to the small, true cluster, NGC 6603, whose diameter is only 4'.5. It is thus described in the *Atlas Coeli Katalog* and similarly in several other lists of Messier Objects but I have not been able to trace where the error first began. Even J. H. was not quite sure which was which.

The true diameter of NGC 6603, according to Becvar's *Katalog*, is 5.8 parsecs or about 19 light-years.

See Sagittarius Region, Map 8

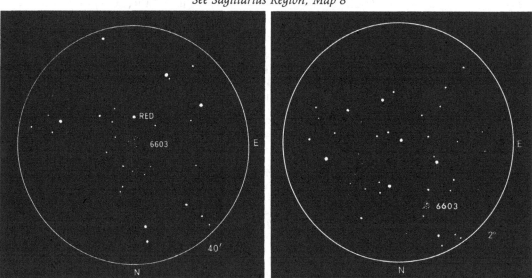

To observe (M 24)   For the young observer, making his first exploration of the Messier clusters and nebulae, M 24 probably proves more troublesome than any other object. This is, no doubt, due to the incorrect identification of Messier's No. 24 with the galactic cluster, NGC 6603 by many of the later 'editors' of his catalogue.

As mentioned above, M 24 is large and bright, while NGC 6603 is both small and faint.

M 24 proper can be seen quite easily with the naked eye on a clear night, as a bright, detached portion of the Milky Way about two-thirds of the way between γ *Scuti* (4½ mag.) and μ *Sagittarii* (4th mag.).

Binoculars will provide a fine view when it will appear as a large, bright star-group, not unlike M 44 ('Praesepe') in general appearance, although not quite as brilliant as that famous cluster.

In a telescope which can provide a field of view of about 2° the object may be picked up from the easily recognised M 17. It will appear in the SW of the field when M 17 is quitting the northern edge. If it is then centred in the field it will appear as a brilliant clustering of fairly bright stars, including a diamond-shaped group, with very many fainter stars in the background, the whole filling most of the 2° field of view.

In any field less than about $1\frac{1}{2}$° diameter, the object tends to lose its 'identity' altogether and merely looks like any other part of the rich background of the Milky Way.

To observe (NGC 6603) NGC 6603, which is probably the 'gathering spot' described by Admiral Smyth, is not likely to be seen on low power or with a small aperture. It may be distinguished in a 6-inch in the best conditions by observers in the lower latitudes but, for most observers, an 8-inch, used with a power of about 150× would be best.

NGC 6603 may be seen to the NE of the 'diamond-shaped' group in M 24, immediately to the north of a distinct red or orange-tinted star. The cluster is only about 4' or 5' in diameter and is pretty faint so that it seems quite 'nebulous' at first appearance. In the best conditions a higher magnification can be used when it may be seen to form a luminous, dense cluster of small stars containing several narrow, curved dark lanes.

Any observer who searches for, and finds, this small, faint cluster will realise that it must have been far out of reach of Messier's crude telescopes, while the large, bright, Milky Way cloudlet is quite as he describes it.

Of the descriptions given by the earlier observers, quoted above, that of Flammarion is the most helpful.

Objects near by Barnard's dark nebula, B92 lies close to the west. This is about 15'×9' in extent and has one 12 mag. star near its centre. It may be seen in a rich-field telescope but is better revealed by photography.

113

M 25　IC 4725
A galactic cluster, type 'd' in SAGITTARIUS
R.A. 18 h 28.8 m.　Dec. S 19° 17' (1950).
Size 35' diam.　Mag. 6.5　Distance 630 pc, 2060 lt-yrs.

Discovered by de Chéseaux in 1746.

DE CHÉSEAUX　　(No. 5) 'A star cluster between the bow and head of Sagittarius.'

MESSIER　　(June 20th, 1764) 'A cluster of small stars in the neighbourhood of the two previous clusters. The nearest star is 21 Sagittarii, 6 mag. (Flamsteed). The stars seen with difficulty in 3½-foot telescope. No nebulosity seen. Position determined from μ Sagittarii. Diam. 10'.'

BODE　　(No. 56) 'A nebulous star cluster, northerly in a triangle with μ and λ Sagittarii.'
M 25 was omitted from J. H.'s catalogue of 1864 and also from the first part of the NGC. It was rediscovered by Schmidt in 1866 and added to the *Index Catalogue* in 1908 using a position found by Bailey.

SMYTH　　'A loose cluster of large and small stars: the gathering portion of the group attains an arched form and is thickly strewn in the South where a pretty knot of minute glimmers occupies the centre, with much star-dust around.'

WEBB　　'Coarse and brilliant.'

J. E. GORE　　'Closely following this nebula is the short-period variable star *V. Sagittarii* which varies from 7.0 to 8.3 mag. with period of about 6¾ days.'

This cluster was examined by Irwin in 1955 and by Sandage in 1960 for the colours and magnitudes of its members. Two giant M-type and two giant G-type stars had their radial velocities measured in 1960. The two G-type giants were found to be members of the cluster but the two M-type are probably not members.

The distance modulus obtained was 8.9 plus or minus 0.2 magnitudes which gives a distance of about 630 parsecs.

Near the centre of M 25 is a classical Cepheid variable, *U Sagittarii*, first discovered by J. B. Irwin in 1956 and found to have a period of 6.74 days. M. W. Feast of Radcliffe Observatory obtained spectra of 42 stars in the cluster and measured their radial velocities. 35 of the member stars had a radial velocity of +4 km/s and so, also, had the Cepheid, showing that it was almost certainly a true member of the cluster. This, together with a similar discovery in NGC 6087 (*U Normae*) gave an independent measurement of the absolute magnitude of a Cepheid variable as the distance of the cluster was estimated by other means.

Trumpler gave M 25 a classification of IV,3,r and quoted the spectrum of the brightest star as B4.

Dr Åke Wallenquist, in his 1959 Catalogue, *Some Structural Properties of Galactic Clusters*, counted 86 stars in a diameter of 34' and estimated the mean brightness of the five brightest stars as 8.6 mag. He also estimated the density of stars in the central portion of the cluster to be 18.4 stars per cubic parsec and 0.70 stars per cubic parsec for the cluster as a whole.

114

The true diameter of the cluster is 6.2 parsecs in Wallenquist's list and 6.4 parsecs in Becvar's *Katalog*; this is equivalent to a diameter of about 20 light-years.

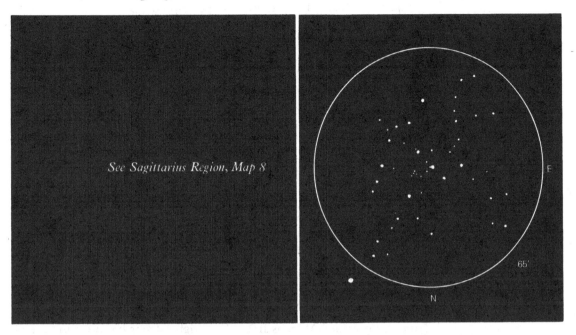

See Sagittarius Region, Map 8

To observe   From μ *Sagittarii* move 2° N and 4° E, *or*, from M 17 which is immediately identifiable, move 3° S and 2° E. The position can be checked by a star (5 mag.) 1° N.

Like several other galactic clusters in and near the Milky Way, M 25 may be a little difficult to separate from other groups round about. However, the position can easily be confirmed from the reference stars quoted above.

It is a fine, though rather open cluster with a dark background. There is a bright, slightly yellow star near the centre, and to the West of this, a delicate, hook-like pattern of stars catches the eye. Many of the other stars are arranged in long, curving tendrils, making a very pretty group altogether.

M 26 NGC 6694
A galactic cluster, type 'f' in SCUTUM
R.A. 18 h 42.6 m. Dec. S 9° 27' (1950).
Size 9' diam. Mag. 9.3 Distance 1.5 kpc, 4900 lt-yrs.

Possibly discovered by Le Gentil before 1750. (Bigourdan.)

MESSIER (June 20th, 1764) 'A cluster near η and o in Antinous (now α and δ Scuti) in between which there is one brighter. Not distinguished with a 3½-foot telescope and it needed a better instrument. Contains no nebulosity. Diam. 2'.'*

J. H. 'Large and pretty rich, stars 12–15 mag.'

SMYTH 'A small and coarse but bright cluster of stars in a fine condensed part of the Milky Way. The principal members of this group lie nearly in a vertical position with the equator and the place is that of a small pair in the south of the field.'

WEBB 'Coarse cluster.'

James Cuffey, of Kirkwood Observatory, Indiana University, reported in 1940 that a striking feature of M 26 is a well-defined zone of low star density in a region 3'.1 in diameter, immediately surrounding the nucleus.

In this area, the density of stars was about 13 per cent less than in adjacent areas. This may be due either to a spherical shell of less dense star concentration or, what is much more likely, a marked zone of dark, obscuring matter in this region of the cluster.

Trumpler classified M 26 as II,2,r. The brightest star has a magnitude of 11.9 and a spectral class of B8.

Wallenquist counted 94 stars in a diameter of 11' and estimated the density of the central portion of the cluster to be 15.8 stars per cubic parsec and 1.57 stars per cubic parsec for the whole cluster.

Its true diameter is 4.9 parsecs or 16 light-years, according to Wallenquist but Becvar, in the *Atlas Coeli Katalog*, gives 7.9 parsecs or nearly 26 light-years for this dimension. (For observing details see next page.)

---

* Messier added, in a manuscript note to his copy of the catalogue of 1784: 'I have seen it very well with a Gregorian telescope magnifying 104×. The cluster contains no nebulosity.'

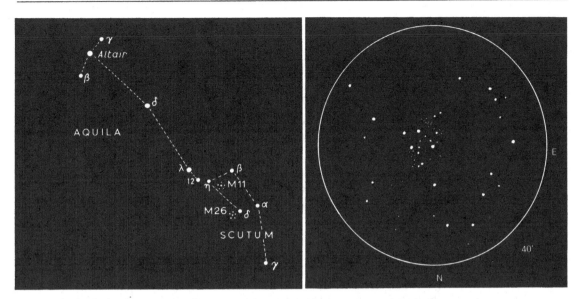

To observe   From δ *Scuti* (4½ mag.) move 1° ESE, *or*, from M 11 move 3° S (the bright globular cluster, 47 H. I, NGC 6712, may be encountered at 2½° S) then 1½° W to M 26.

M 26 is not a very impressive cluster but it contains some fairly dense patches of faint stars. At the centre, the four brightest stars form a small kite shape with the apex to the SW. Two curving arms of faint stars can be traced to the N and S and this gives, as Smyth noted, a distinct elongation of the cluster in the meridian.

The diameter of the denser portion is about 10′ or a little less. It is a pretty enough cluster but unfortunately it appears rather formless and faint by comparison with nearby M 11.

Objects near by   δ *Scuti* (1° WNW) is double. A, 5.5 mag.; B, 10 mag. P.A. 130°, d. 52″.5.

47 H. I, NGC 6712 (2° E and 1° N) although listed by William Herschel as a bright nebula, is a 10th mag. globular cluster. It is about 3′ in diameter and has a soft outline with a brighter centre.

117

M 27   NGC 6853

The 'Dumb-bell' Nebula. A planetary nebula, class IIIa in VULPECULA
R.A. 19 h 57.4 m.   Dec. N 22° 35' (1950).
Size 8'×4'   Mag. 7.6   Distance 300 pc, 975 lt-yrs.

Discovered by Messier in 1764.

MESSIER	(July 12th, 1764) 'Nebula without star. Seen well in 3½-foot telescope. Appears oval and contains no star. Recorded on chart of comet of 1779. Reviewed Jan. 31st, 1781.' (Diam. 4'.)
W. H.	(*Phil. Trans.* 1785) 'Though its appearance is not large, it is probably a double stratum of stars of a very great extent, one end of which is turned towards us. That it is thus situated may be surmised from its containing, in different parts, nearly all the three nebulosities; *viz.* the resolvable, the coloured but irresolvable and a tincture of the milky kind.'
J. H.	Saw the faint luminosity which fills the lateral concavities of the body and converts them into protuberances so as to render the general outline of the whole nebula a regular ellipse, having its shorter axis the common axis of the two bright masses of which the body consists; i.e. the longer axis of the oval form in which it was imperfectly seen by Messier. (Smyth.)
SMYTH	'This magnificent and singular object is situated in a crowded vicinity where field after field is very rich. M 27 is truly one of those splendid enigmas which, according to Ricciolus, are proposed by God but never to be subject to human solution. The axis of symmetry is not less than 5'. P.A. of axis 32°.'
ROSSE	'Not actually resolvable but symptoms of it. The 6-foot (diam.) reflector surrounds it with an external ring like a chemical retort.'
D'ARREST	'Very large and shining; two objects blending into one another. In the more luminous S.p. portion is a brighter region placed eccentrically: the N.f. part is of fairly even brightness. Beautiful appearance on magnification 95×.'
WEBB	'Just visible in a 1¼-inch finder. In a rich field: two hazy masses in contact, with p. the brighter. Minute stars in it, 2 or 3 (5½-inch achromat.) 7 (Denning, 10-inch speculum), 18 (Ingall, 5½-inch Dialyte.).'
J. E. GORE	'Photos by Schaeberle found it a great counter-clockwise spiral at least ½° in extent.'
LICK XIII	'Central star of mag. 12. 8'×4' in P.A. 125°. One of the giants of the planetary class and of great importance in theories of planetary (nebula) structure because of the easy visibility of its intricate details. Radial velocity − 37.2 km/s.'

M 27 is undoubtedly the finest of the planetary nebulae in the whole heavens.

However, it is neither the largest nor the brightest of the planetary nebulae but as the larger objects of this class are all faint, and the brightest ones very small, M 27 is the most easily seen and reveals the most detail. (For observing details see page 120.)

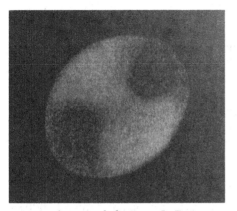

M 27 (John Herschel) From G. F. Chambers' Handbook *(B.A.A.)*

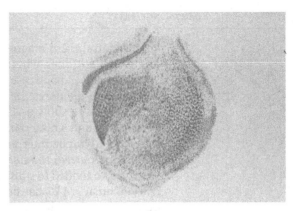

M 27 (Rosse) Drawn by Johnstone Stoney (B.A.A.)

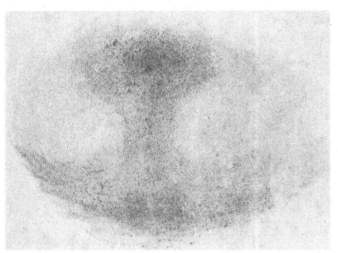

M 27 (Rosse) From Collected Scientific Papers *(B.A.A.)*

M 27 (Rosse) From Collected Scientific Papers *(B.A.A.)*

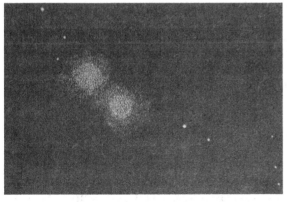

M 27 (Smyth) From A Cycle of Celestial Objects *(B.A.A.)*

M 27 (Rosse) Drawn by Bindon Stoney (B.A.A.)

The central, 12th magnitude star has a continuous spectrum without any bright or dark lines. Photographs of the nebula in different ranges of the spectrum show in one case a granular structure while, in another, it is seen to contain thin filaments.

This seems to show a very complex form of stratification and Vorontzov–Veljaminov and Kramer, in 1937, thought that its structure is so complicated that it could not be interpreted by any unique theoretical model. L. H. Aller, in his book *Gaseous Nebulae*, writes, 'subsequent observations with large telescopes have tended to substantiate these conclusions'. Like many other planetary nebulae, M 27 has been found to be expanding. The rate is about 6".8 per century and indicates that the original explosion took place about 3000 to 4000 years ago.

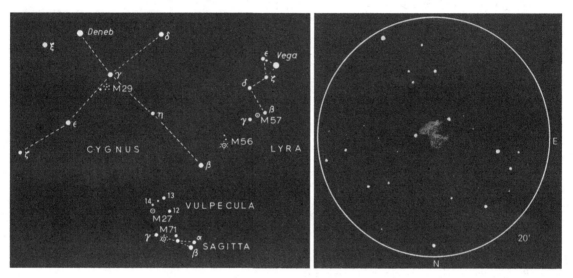

To observe   M 27 is very easy to find, being 3° N of γ *Sagittae* and quite unmistakable even in a small telescope.

It is a fine object, very bright and distinct against a dark background very rich in stars. The waisted 'dumb-bell' is quite clearly seen in telescopes of 6-inch aperture and looks like two truncated cones, apex to apex in P.A. about 30°–210°, the 'cut-off' being quite sharp in this direction and more vague at right angles.

With a larger aperture, a soft, bluish-green glow can be seen to surround the 'dumb-bell' and in large apertures and in photographs it appears to have its major axis in P.A. 125°.

A 12 mag. star can be distinctly seen very close to the western edge but this, and several other faint stars noted by Webb and others, are unconnected with the nebula. The 12 mag. central star is quite difficult to pick up.

Objects near by   Only a 5th magnitude star, *14 Vulpeculae* 25' to N.

M 28   NGC 6626
A globular cluster, class IV in SAGITTARIUS
R.A. 18 h 21.5 m.   Dec. S 24° 54′ (1950).
Size 15′ diam.   Mag. 7.3   Distance 5.8 kpc, 19 000 lt-yrs.

MESSIER   (July 27th, 1764) 'Nebula containing no star. Round, seen with difficulty in 3½-foot telescope. Position determined from λ Sagittarii. Reviewed 20th March 1781. Diam. 2′.'

W. H.   First resolved it and classed it as a cluster.

J. H.   'Very bright, round, very much compressed, resolved into stars 14–15 mag. A fine object.' He also recommended it as a test object for the resolving power of telescopes.

SMYTH   'A compact globular cluster of very minute stars. Not very bright and preceded by two telescopic stars in a vertical line.'

WEBB   'Not bright, 1° N.p. λ.'

LICK XIII   'Bright globular cluster, 4′ diam.'

Shapley gives the ellipticity of M 28 as 9 in P.A. 50°. His early distance estimate of 16.6 kpc has been much reduced by later observations, Kinman and Becvar both giving about 4.8 kpc. Whelan and Hogg in 1984 obtained 5.8 kpc (19 000 lt-yrs).

In H. S. Hogg's list of globular clusters in the *Handbuch der Physik*, its spectral class is given as F9, the colour index as +0.09 and the average magnitude of the 25 brightest stars as 14.73.

The true diameter of the cluster given in the *Atlas Coeli Katalog*, is 23 parsecs or about 75 light-years. Becvar quotes Shapley's figure of 4′.7 for the apparent diameter of the cluster while Hogg gives 15′.

M 28 was found by Alfred H. Joy in 1949 to contain a long-period variable star. Its period was 17 days and it had a light curve similar to the *W Virginis* variables.

Long-period variables – as opposed to the *RR Lyrae* type with periods of less than a day – are not unknown in globular clusters. About 70 have been found altogether and many of these are similar to *W Virginis*. Classical Cepheid, long-period variables do not appear to exist in globular clusters.

Another long-period variable of 90 days with a range of two magnitudes has been found recently, and up to 1973 the tally of *RR Lyrae* variables amounted to 18.

In 1987 a millisecond pulsar was detected in this cluster and also in M 4 (See M 4). (For observing details see next page.)

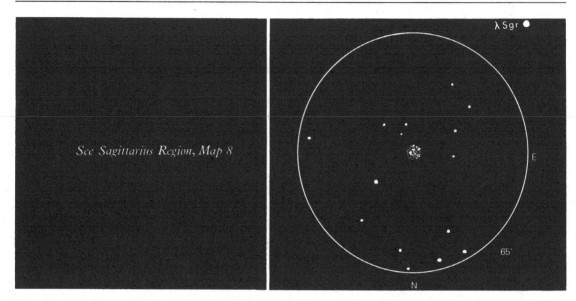

See *Sagittarius Region, Map 8*

To observe   From λ *Sagittarii* move 1° NW, *or*, from M 22 move 1° S and 2½° W.

M 28 looks a very compact cluster, having a bright, glowing centre, fading off rapidly toward the edges. From latitudes of 50° N and higher it appears smaller and fainter than its vital statistics suggest but in lower latitudes, where it gets to a more respectable altitude above the horizon, it glows more strongly, although still small.

It is quite difficult to resolve except at the extreme edges and then only in good conditions and with a wide aperture. In any case it is a less rewarding object than the magnificent M 22 which is close to the SW.

M 29   NGC 6913
A galactic cluster, type 'd' in CYGNUS
R.A. 20 h 22.1 m.   Dec. N 38° 22' (1950).
Size 7' diam.   Mag. 7.1   Distance 1.2 kpc, 4000 lt-yrs.

Discovered by Messier in 1764.

MESSIER   (July 29th, 1764) 'A cluster of 7 or 8 very small stars. Looks like a nebula in 3½-foot telescope. Found from γ Cygni. Reported on chart of comet of 1779.'

BODE   (No. 69) 'A star cluster under γ in the breast of Cygnus.'

SMYTH   'A neat but small cluster of stars in the preceding branch of the Milky Way. In S.p. portion are two stars, treated here as a pair. A, 8 mag. yellow; B, 8½ mag. dusky; P.A. 225°, d. 66" (1879). Messier's description, though fair, his declination is much out.'

FLAMMARION   'Resembles a miniature archipelago in this opulent stellar region.'

M 29 was not included in Wallenquist's 1959 catalogue of galactic clusters but Trumpler classified it as III,3,p. H. S. Hogg, in the *Handbuch der Physik*, gives its diameter as 7', its integrated magnitude as 9.0 and the brightest star as 9.4 mag.

Becvar's *Atlas Coeli Katalog* describes the cluster as containing only 20 stars and gives the true diameter as 3.3 parsecs or about 11 light-years.

In 1954, W. A. Hiltner of Yerkes Observatory, made a study of this cluster and obtained values for the intersellar polarisation of the light of its brightest members. A fair correlation was obtained between the amount of polarisation and the estimated values of interstellar absorption.

Harris, in the same year, also found that the cluster stars exhibited an irregular obscuration. (For observing details see next page.)

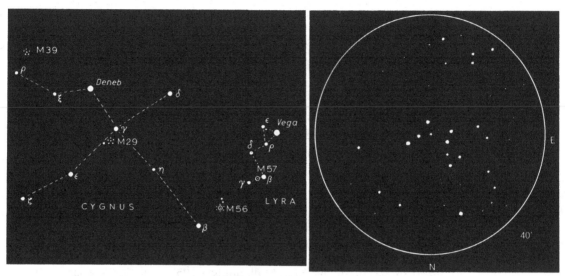

To observe   M 29 is a little less than 2° S and a little E from γ *Cygni*.

A very sparse cluster, merely consisting of seven or eight fairly bright stars, the four brightest forming a quadrilateral and another three, a small triangle to the north. There are only a few other faint stars in the vicinity and the cluster appears fairly isolated.

The cluster appeared nebulous in Messier's 3½-foot telescope owing to its poor light-grasp and lack of definition, but in fact M 29 is surrounded by a portion of the bright diffuse nebula which is widely strewn in this part of the Milky Way. This, however, is only detectable by photography.

Observers with small telescopes may find this cluster more in correspondence with Messier's description. A simple 4-inch reflector with a wide field will show it more compressed and perhaps almost 'nebulous'.

M 30 NGC 7099
A globular cluster, class V in CAPRICORNUS
R.A. 21 h 37.5 m.   Dec. S 23° 25' (1950).
Size 9' diam.   Mag. 8.4   Distance 12.6 kpc, 41 000 lt-yrs.

Discovered by Messier in 1764.

MESSIER   (Aug. 3rd, 1764) 'Nebula discovered near 41 Capricorni (Flamsteed). Seen with difficulty in 3½-foot telescope. Round, contains no star. Found from ζ Capricorni. Reported on chart of comet of 1759. Diam. 2'.*

W. H.   Resolved it in 1783 into 'a brilliant cluster with two rows of stars, 4 or 5 in a line which probably belong to it.'

J. H.   'A globular cluster; bright, 4' long by 3' broad; all resolved into stars, 16 mag. besides a few 12 mag.'

SMYTH   'A fine, pale white cluster. This object is brighter and from the straggling streams of stars on its N edge, has an elliptical aspect with a central blaze: few other stars in the field.'

WEBB   'Moderately bright, beautifully contrasted with 8 mag. star beside it. Comet-like on 64×. With higher powers, resolvable.'

ROSSE   'A spiral arrangement of its branches.'

LICK XIII   'Bright globular cluster, 5' diam.'

In Shapley's list of globular clusters, M 30 is given an ellipticity of 9: he found it contained only three variable stars among 480 stars examined.

Helen Sawyer Hogg quotes a spectral class of A7 for the cluster in the *Handbuch der Physik* and gives the colour index as −0.11 which indicates a fairly large 'blue' tendency. The average magnitude of the 25 brightest stars is 14.63.

The true diameter of the cluster in the *Atlas Coeli Katalog* is 23 parsecs or about 75 light-years. Shapley's figure for the distance of M 30 was 14.6 kpc but Becvar gives 12.6 kpc which lies within the limits of Kinman's modulus of 15.5 ±0.5 mag.

Eight more variables were added up to 1973. A point-like ultra-violet source discovered in 1983 was found to be a post-red giant star beginning to evolve towards the white dwarf cooling sequence. (For observing details see next page.)

*   Messier's manuscript note reads: 'Seen only with difficulty in an ordinary telescope of 3½-feet. It is round and I saw no star there having observed it with a good Gregorian telescope of 104×.'

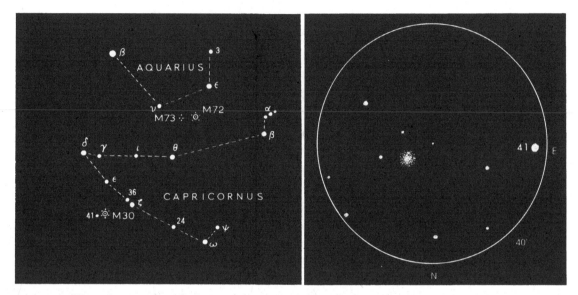

To observe   M 30 is quite easy to find. From ζ *Capricorni* (4th mag.) move ¾° S and 3° E. A 5½ mag. star, *41 Capricorni*, lies about 25' ESE.

A small globular cluster with a good bright nucleus, 1' to 2' diam. surrounded by a glowing area out to about 5' diam. Outside this are some faint individual stars. The outer areas are easily resolved on M.P. but the centre is quite dense. For northern observers, its low altitude detracts from its brightness for, in lower latitudes it is seen to glitter brilliantly.

There is an 8 mag. star about 8' away to the west and faint stars in strings appear to the north and west.

Objects near by   *41 Capricorni*, 25' to the ESE is double, A, 5.3 mag., B, 13.0 mag.; P.A. 203°, d. 5.3" (the *comes* at 13 mag. makes it difficult for small apertures).

126

M31   NGC 224
The 'Great Nebula'. A spiral galaxy, type Sb in ANDROMEDA
R.A. 0 h 40.0 m.   Dec. N 41° 00' (1950).
Size 160' x 40'   Mag. 4.8   Distance 680 kpc, 2.22 million lt-yrs.

With the other, very different, Great Nebula (M42 in Orion) M31 shares the honour of being the most famous and most closely studied of the Messier Objects or, for that matter, of any celestial object outside the solar system.

Being quite easily visible to the naked eye, it was well known to ancient astronomers and the first written record of it seems to have been that of Al-Sufi, who included it in his *Book of the Fixed Stars* for AD 964 as a 'little cloud' and a familiar object which had been observed as early as AD 905.

SIMON MARIUS,   a contemporary of Galileo, was the first to examine it through the telescope on Dec. 15th, 1612 and described it in the preface to his *Mundus Jovialis* as, 'like the

M31 The constellation of Andromeda as depicted on an early Persian MS copy of Al-Sufi's Book of the Fixed Stars. *The nebula is shown by a faint, dotted circle at the snout of the Fish (British Museum)*

M31 Another representation of Andromeda with the nebula more clearly marked. This figure, taken from a 15th-century Persian MS, is shown reversed. The lettering is a 17th-century addition. Taken from Ishmael Boulliau, Monita ad Astronomos . . ., *Paris, 1667 (British Museum)*

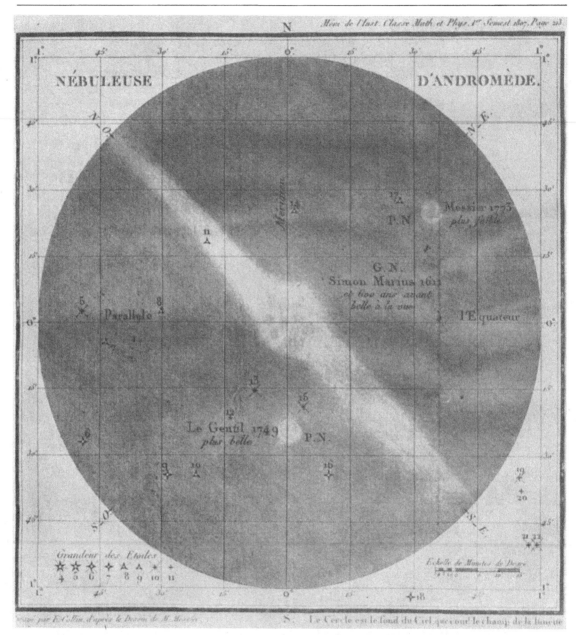

Messier's drawing of the Andromeda Nebula showing the two elliptical companion nebolae, M 32, below centre, and M 110, upper right.

flame of a candle seen through horn' or 'like a cloud consisting of three rays; whitish, irregular and faint; brighter toward the centre'.

G. B. HODIERNA,   before 1654, made an independent re-discovery. 'A very admirable [nebula] never seen (to my knowledge) by anyone. It is the unresolvable star over the right thigh of Andromeda . . . No multitude of close stars can be distinguished in it but [has a glow] similar to a comet.'

BULLIALDUS   rediscovered it when his attention was drawn to it by the passage of the comet of 1664 across that part of the sky.

HALLEY   considered that it appeared 'to emit a radiant beam.'

DERHAM   (No. 1, 1733) 'It is as considerable as any I have seen.'

LE GENTIL,   who discovered the companion nebula, M 32, in 1749, considered M 31 as round at first but later described it as oval and 'always of uniform light in all its parts'. It is included in de Chéseaux's list of 1746 but not described (No. 15).

MESSIER   (Aug. 3rd, 1764) 'The beautiful nebula, shaped like a spindle. . . . No star recognised. Resembles two cones or pyramids, base to base with axes NW–SE. The two points of light at the apices are about 40' apart. The common base of the pyramids, about 15'. Discovered by S. Marius. Diam. 40'. Reported on the *English Atlas*.'

BODE   (No. 3) 'Visible to the naked eye; a nebulous patch 15' in diameter.'

W. H.   Thought it the nearest of all nebulae. 'The brightest part of it approaches resolvable nebulosity and begins to show a faint, red colour, from which I believe its distance in the coloured part not to exceed 2000 times the distance of Sirius.'

M 31 (Trouvelot) Drawn at Harvard College Observatory in 1874 (B.A.A.)

J. H.	Called it 'this most magnificent object' and made a detailed drawing of it which was widely published.
SMYTH	'An overpowering nebula (with a companion, M 32, about 25' in the S vertical). It is of an oval shape, light, broadening toward S.f. edge of the general mass and of a milky, irresolvable nebulosity. Numerous telescopic stars around and 3 minute ones involved in the glow but can have no connection with it. Axis of direction trends S.p. & N.f. and may be caught by a good eye on a fine night.'
WEBB	'One of the grandest in the heavens, long, oval or irregularly triangular; ill-bounded and brightening to the centre. Plain to the naked eye.'
D'ARREST	'The preceding margin very clearly terminated on the eastern side. 12 mag. star precedes nucleus by $11^s.26$.'
G. P. BOND	Made many observations with the 15-inch refractor of Harvard College Observatory, tracing the length of M 31 to 4° and its breadth to $2\frac{1}{4}°$. In 1847 he also produced a very detailed drawing showing the dark streaks among the nebulosity.
ROBERTS	On Oct. 1st, 1888 he made a very successful 3-hour exposure photograph which plainly showed its spiral structure.
ROSSE	Considered it could be resolved into stars.
HUGGINS	Found the spectrum to be a continuous one and unlike the gaseous nebulae such as M 42.
LICK XIII	'This wonderful object, showing an almost stellar nucleus, with traces of spiral structure in the surrounding nebular matter.'
FLAMMARION	reports a note which Messier added in manuscript to his copy of the catalogue. 'I have employed different instruments, especially an excellent Gregorian telescope of 30 feet F.L., the large mirror 6 inches in diameter, magnification 104×. The centre of this nebula (M 31) appears fairly clear in this instrument without any stars appearing. The light gradually diminishes until it becomes extinguished. The former measurements were made with a Newtonian telescope of $4\frac{1}{2}$-feet F.L., provided with a silk thread micrometer. Diameter, 40'. Aug. 3rd, 1764.'

The first supernova ever to be observed outside our own Galaxy was detected in M 31 on Aug. 20th, 1885 by Hartwig at Dorpat Observatory. This stellar super-explosion, named *S Andromedae*, attained a brightness above mag. 6 between the 17th and 20th Aug. and was seen by several other observers including Baroness Podmaniczky in Hungary on Aug. 22nd and 23rd, Max Wolf at Heidelberg on Aug. 25th and 26th and Prof Ludovic Gulley at Rouen on Aug. 17th. However, it was only Hartwig who realised the true significance of this event. By Feb. 7th, 1890, the supernova had dimmed to mag. 16.

Two ordinary novae were observed in 1917 as faint specks on photographic plates and in the last 25 years, many more have been seen. Hubble found a total of 82 and Dr Arp at Mount Wilson found 30 more in 290 nights between June 1953 and January 1955. The number of novae appears to fall off with increasing distance from the centre of the galaxy. They are similar to novae in our own system, the shortest having a duration of 5 days and abs. mag. −8.5 and the longest, 150 days and abs. mag. −6.1.

In M 31, we are able to see as a whole a galaxy similar to our own, whereas,

from our position inside the disc of the galactic plane much of our neighbourhood, including the nucleus, is largely hidden from us by obscuring gas and
dust.

In 1944, Baade's great feat of resolving the central region of M 31 into stars
gave us much insight into the conditions prevailing in the nucleus of a spiral
galaxy. The nucleus of M 31 is spheroidal, having dimensions about $2\frac{1}{2}'' \times 1\frac{1}{2}''$
with the major axis aligned with the spiral's disc. Its actual size is not very
much greater than one of the larger globular clusters, such as M 92, but its
mass is nearly one hundred times as great and luminosity about twenty times
that of the globular.

The nucleus has been found to be in very rapid rotation; making one complete rotation in a little more than half a million years. The outer portions of
the spiral are moving more slowly and the direction of rotation has been
observed to be with the spiral arms convex to the spin, confirming similar
observations in our own Galaxy.

The spiral arms of M 31 are most conspicuous on blue-sensitive photographic plates and most of the gas, dust and blue stars are found at the junction of the spiral arms with the nucleus. These are regions of young,
Population I stars and recent photography with the new 79-inch reflector at
Trautenberg near Jena shows that different segments of the spiral arms have
different ages. Nevertheless, possibly more than 90 per cent of the mass of
M 31 consists of the older stars of Population II.

At least 300 globular clusters, which are mainly composed of Population II
stars, have been found surrounding M 31 and in 1955, Baade discovered 5
planetary nebulae – very faint objects of apparent magnitude 22 – more than
$1\frac{1}{2}°$ away from the centre.

In the best photographs, M 31 covers an area of about $3\frac{1}{2}$ square degrees
but, using a micro-densitometer to measure the faintest darkening of the

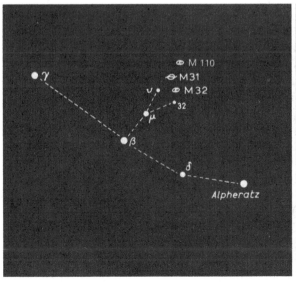

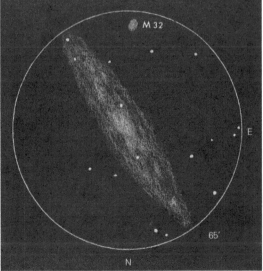

M 31 John Herschel's detailed drawing of the Andromeda Nebula

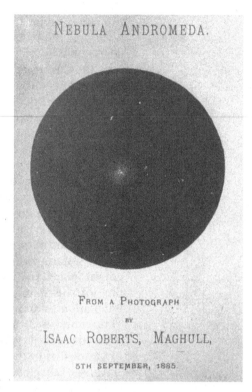

NEBULA ANDROMEDA.

FROM A PHOTOGRAPH
BY
ISAAC ROBERTS, MAGHULL,
5TH SEPTEMBER, 1885.

M 31 An engraving from a photograph taken by Isaac Roberts in 1885 showing the supernova S And. at the centre of the nebula

negative, this area is multiplied almost 10 times, revealing a total volume that is almost spherical.

Baade has estimated that the angle between our line of sight and the central plane of M 31 is 11°.7, which makes it very nearly edge-on.

The dimensions of M 31 are thought to be more than $1\frac{1}{2}$ times our own and its total mass is between 3 and 4 times ours, thus giving a similar density for the two systems.

In Erik Holmberg's *Catalogue of External Galaxies*, M 31 is given a total mass of 320 000 million Suns, an absolute magnitude of −21.4, a spectral class of G5 with a colour index of +0.50 and a true diameter of 38.9 kpc.

Intensive radio mapping of M 31 was undertaken in 1964 at the Mullard Radio Astronomy Observatory in Cambridge which showed a radio spectrum very similar to the Galaxy except that the brightness temperature seemed to be less than one-fifth of that displayed by our own system. Another survey (5C3) with the one-mile radio array in 1969 revealed that the spiral arms in M 31 have a comparatively low surface brightness, and although the nuclear region is of similar radio luminosity to that in the Galaxy, the compact source

Sagittarius A in our system is more than twenty times as intense as any counterpart in M 31. Observations with the 100-metre radio telescope of the Max-Planck Institute have also found that the electron halo of M 31 is considerably weaker than that which is known to surround the Galaxy.

In 1985 Van den Berg and Pritchet at the Dominion Astrophysical Observatory in British Columbia discovered *30 RR Lyrae* variables in the halo of M 31, and from their mean magnitudes obtained a distance to the Great Nebula of 740 kpc or 2.41 million light-years, a figure which agrees well with the results of other recent measurements. (For observing details, see M 32.)

Numerous attempts to trace any remnant of the historic 1885 supernova, S And. proved fruitless until November 1988, when Robert A. Fesen and colleagues at Colorado University detected a dark absorption feature close to the original location of the outburst. Using a charge coupled divice (CCD) with the 4-metre reflector at Kitt Peak Observatory, they measured a faint iron line at 3860 angstroms over an area of about 0.3 arc second, which corresponds to about 1 light-year at the distance of M 31. This indicates an expansion rate for the remnant of 4000 to 5000 km/sec since 1885. (For observing details see p. 135.)

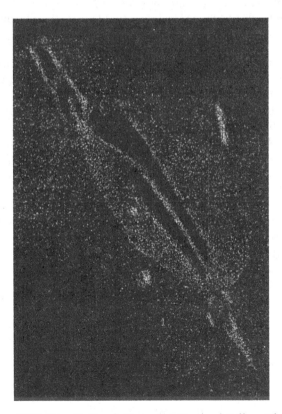

M 31 (Geo P. Bond) Drawn at Harvard College Observatory in 1847.

M 32 NGC 221
A dwarf elliptical galaxy, type E2 in ANDROMEDA
R.A. 0 h 40.0 m.   Dec. N 40° 36' (1950).
Size 3'×2'   Mag. 8.7   Distance 680 kpc, 2.22 million lt-yrs.

Discovered by Le Gentil on Oct. 29th, 1749.

LE GENTIL   'While observing the Andromeda Nebula with a fine 18-foot telescope . . . I saw another small nebula, about one minute in diameter which appeared to throw out two small rays; one to the right and the other to the left.'

MESSIER   (Aug. 3rd, 1764) 'Small nebula without a star. Round, much fainter than M 31. Messier saw it first in 1757 and has found no change in it. Diam. 2'.'

W. H.   (*Phil. Trans.* 1785) 'A pretty large, round nebula, much condensed in the middle.'

D'ARREST   'Le Gentil's companion of the great Andromeda nebula; its size and brightness are well known. Almost perfectly circular; gradually brighter towards the centre. The nucleus like a 10-11 mag. star.'

WEBB   'In the same field as M 31 on L.P. Small but bright. Rosse resolved into stars. Spectrum like M 31.'

LICK XIII   'The well-known companion, south of M 31. Exceedingly bright. On long exposures it appears as a "burnt out" oval, 2'.6×1'.8 with no trace of spiral character in the outer portions. It lies in P.A. 150° approx. The shortest exposures show a nucleus surrounded by bright nebular matter far brighter than the brightest part of M 31.'

M 32, like the other companion to M 31, NGC 205, is an elliptical galaxy. It was resolved into stars by Baade at the same time as he resolved the central portion of M 31.

The nucleus of M 32 has been found to be in comparatively rapid rotation. M. F. Walker, using an electronic camera attached to the Coudé spectrograph of the 120-inch Lick telescope found that the rotational rate appears to increase from the centre to a value of 60 km/s at a distance of $2\frac{1}{2}''$ and then declines to zero at 9" from the centre.

The mass of the nucleus has been calculated at about 100 million solar masses with a rotational period of about 600 000 years and has a mean density of about 5000 suns per cubic parsec. These properties make the nucleus of M 32 very similar to the nucleus of M 31 despite their very different appearances overall.

The *total* mass of M 32 is about 3000 million solar masses and its diameter is about 8000 light-years: its absolute magnitude is −16.5.

Although M 32 may well have been formed at the same time as M 31, it is considered to be at a much later stage of development, and, having probably passed more than once through the disc of the larger galaxy, has lost most, if not all, of its interstellar gas and dust.

Dr Whitford, of the University of Wisconsin, has adopted M 32 as a standard elliptical galaxy by which to measure the colour and reddening effects in other elliptical galaxies.

M 32 has a spectrum of type G3 and a colour index of +0.75.

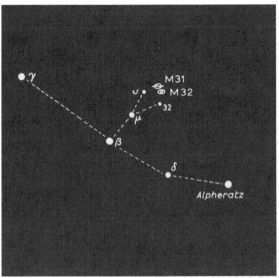
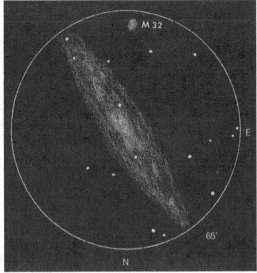

To observe   **(M 31 and M 32)**

On a clear night, M 31 can be seen clearly with the naked eye as a small hazy patch just over 1° W and slightly N of 4½ mag. *v Andromeda*. In binoculars, this area appears as a bright, featureless oval but the first sight of this famous nebula in the telescope is often disappointing to the amateur observer.

It would be better if he were first to try the nearby M 33 in Triangulum which, though diffuse, reveals much detail on close inspection. If now the telescope is turned on to M 31, its brighter, luminous glow can be seen to extend right across the field of view of even the widest-angle eyepiece and some idea of its great size begins to be apparent. The lowest available power should, of course, be used. With 12-inches aperture or greater, some slight trace of one or two dark rifts may be made out in the very best conditions.

M 32 can be seen as a typical elliptical nebula about 25′ S of the centre of M 31 but no detail can be seen other than the P.A. of its major axis, which is about 150°–330°.

An indication of the 'real' extent of M 31 can be gauged from the fact that its area extends outwards so that M 32 is included in the edge of the larger galaxy.

The other elliptical companion of M 31, NGC 205, can be seen about 45′ to the NW of the centre of M 31. It appears as a fairly bright oval, about 8′×4′ in P.A. about 320°–140° among a fairly rich field of faint stars, mostly in pairs. (See M 110.)

M 33 NGC 598 (17 HV)
A spiral galaxy, type Sc in TRIANGULUM
R.A. 01 h 31.0 m.   Dec. N. 30° 24' (1950).
Size 55'×40'   Mag. 6.7   Distance 720 kpc, 2.35 million lt-yrs.

Discovered by Messier in 1764.

MESSIER   (Aug. 25th, 1764) 'The nebula is a whitish light of almost even brightness. However, along two-thirds of its diameter it is a little brighter. Contains no star. Seen with difficulty in a 1-foot telescope. Found from α Trianguli. (Diam. 15').'

BODE   (No. 5, Aug. 18th, 1775) 'A faintly nebulous patch of disorderly shape.'

W. H.   'Has a mottled aspect.'

J. H.   'Only fit for low powers, being actually imperceptible from want of contrast with 144×.'

SMYTH   'A large, distinct but faint and ill-defined, pale white nebula with a bright star a little N.p. and five others following at a distance; between them and the nebula there is an indistinct gleam of mere nebulous matter.'

WEBB   'Very large, faint and ill-defined. Visible from its great size.'

ROSSE   'Full of knots. Spiral arrangement. Two similar curves like an "S" cross in the centre.'

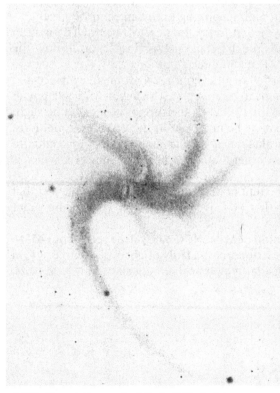

M 33 (Rosse) Drawn by Mitchell (B.A.A.)

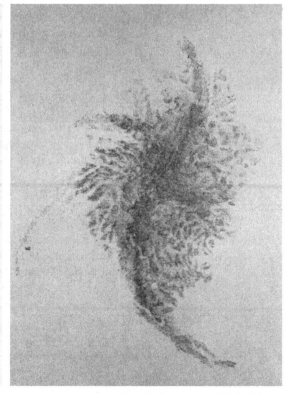

M 33 (Rosse) Drawn by Bindon Stoney (B.A.A.)

136

LICK XIII   'A close rival to M 31 as the most beautiful spiral known. With its faintest extensions it covers an area of at least 55'×40'. It is uncertain whether there is an actual stellar nucleus. A multitude of stellar condensations in the whorls. Best example of resolution into stars.'

This galaxy is a member of our own Local Group and is the most distant known member of it. Its diameter is about 17 kpc compared with about 25 kpc for our Galaxy. Its mass has been estimated at about 8000 million suns which is only about one twenty-fifth of ours.

It is a fine example of a 'normal' spiral, i.e. without a 'bar' and the spiral arms are very open like an 'S' as Lord Rosse noticed. A great amount of stars and gas lies in the spiral arms and comparatively little in the nucleus, showing that it is probably in a fairly early stage of development.

Baade detected Population II stars between the spiral arms and its rotational period has been measured as about 200 million years in the region of the rim.

Globular clusters, similar to those in our Galaxy, have been detected in M 31 and M 33 but those in M 33 appear to be bluer and fainter than those in M 31.

No supernova has yet been detected in this galaxy but four 'ordinary' novae have been seen.

NGC 604, which is one of the most conspicuous of the separate 'knots' in M 33, has been found to have a spectrum similar to the Orion Nebula, M 42. Altogether about 80 diffuse emission nebulae have been detected by photography, mostly in the spiral arms.

In Holmberg's catalogue, M 33 has a mass of 7900 million Suns, a diameter of 17.4 kpc and an absolute magnitude of −19.0. It has a spectrum of type A7 and colour index +0.20.

M 33, being an open, face-on system, is eminently suitable for observing the processes of star-formation. In 1978 a co-operative project by Oxford, Cambridge and Toronto astronomers found that the luminous blue stars which are the sites of recent star formation are not exclusively confined to spiral arm structures. The supergiant Cepheid variables also conform to the distribution of star-forming regions in areas outside the spiral arms.

Several supernova remnants have been discovered optically in M 33 and mapped at radio frequencies with high resolution. Among the 112 or so variables found, some 25 are Cepheids, the remainder being *RR Lyrae* short-period types of mag. 16. M 33 also contains at least one strong X-ray emitter which seems to have many of the characteristics of a 'black hole'.

The numerous variables which are observable in M 33 provide an important 'measuring rod' for establishing an extragalactic distance scale. In a detailed analysis of the data available up to 1988, M. W. Feast of the South African Astronomical Observatory has produced a 'best estimate' of the true distance modulus for this galaxy, a value of 24.4 magnitudes, which is equivalent to 759 kpc or 2.47 million lt-yrs. (For observing details, see next page.)

137

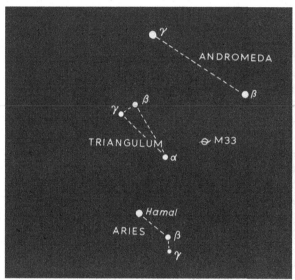
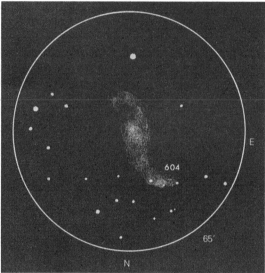

To observe   The visual magnitude of M 33 is given as 5.8 for the whole object and mag. 6.7 for the central portion of it only. It has been seen by some observers with the naked eye in exceptional conditions. It can, however, be quite difficult to pick up even with a good telescope unless the sky is quite dark and clear. Sir John Herschel's remark about it being 'fit only for low powers' is very apt.

First locate α *Trianguli* which is just N of the line *Hamal*-β *Andromedae*. M 33 is just S of the line α *Trianguli*-β *Andromedae* and about one-third of the distance from α *Trianguli*.

The central nucleus will be detected first being a round, hazy patch of light about 10′ diam. with a central, brighter area about 3′ to 4′ diam. It lies to the E of the centre of the base of an almost equilateral triangle of 7–8 mag. stars.

In an 8-inch telescope or larger, a small, bright patch may be detected 10′ NE of the centre: this was given a separate number, 604 in the NGC. Other fainter patches to the SE and SW, NGC 588, 592, 595 and 603, may be seen in a 12-inch. They are all parts of M 33 being 'knots' of stars and gas in the spiral arms.

Even with small telescopes this fine galaxy responds well to 'scanning' observation. Using a fairly high power, the object should be set at the upper limit of the visible image, allowed to drift slowly through the field and sketched carefully *en route*. The process is then repeated through the next declination 'strip', and so on. By this means an accumulative map can be built up which should reveal much more detail than a single, low-power overall view.

M 34 NGC 1039
A galactic cluster, type 'd' in PERSEUS
R.A. 02 h 38.8 m.   Dec. N 42° 34' (1950).
Size 35' diam.   Mag. 5.5   Distance 440 p, 1450 lt-yrs.

Discovered by Messier in 1764.

MESSIER (Aug. 25th, 1764) 'A cluster of small stars a little below the parallel of γ (Andromedae). In an ordinary telescope of 3 feet one can distinguish the stars. Its position was determined by β (Persei). Diam. 15'.'

BODE (No. 7, Sept. 2nd, 1774) 'A star cluster, visible to the naked eye.'

W. H. Resolved it in 1783 into 'A coarse cluster of large stars of different sizes.'

SMYTH 'A double star in a cluster. A & B 8 mag. both white. P.A. 249°, d. 20″ (1880). A scattered but elegant group, 8 to 13 mag. on a dark ground; several of them form coarse pairs.'

WEBB 'Just perceptible to the naked eye; a very grand low-power field: one of the finest objects of its class. Contains a little 8 mag. pair 20″ apart.'

ROBERTS Photo. Dec. 1892. 'A loose cluster of stars down to 15 mag. but not very rich and easily counted.'

Trumpler's classification for M 34 is I,3,m and the brightest star in the cluster is of spectral class B8 and magnitude 8.6.

In H. S. Hogg's catalogue the apparent diameter is given as 30', the integrated magnitude as 5.8 and the number of stars as 100 plus.

Dr Wallenquist counted 81 stars brighter than magnitude 12.7 in a diameter of 42'. He estimates the density of the centre of the cluster to be 20.9 stars per cubic parsec and, for the cluster as a whole, 0.83 stars per cubic parsec. Wallenquist's figure for the true diameter of the cluster is 5.7 parsecs or about 18½ light-years but Becvar's estimate in the *Atlas Coeli Katalog* is only 2.3 parsecs or about 7½ light-years.

In 1953, Harold L. Johnson of Yerkes Observatory made three-colour observations of 57 stars in the area of the cluster which included 42 members and 15 non-members. From this it was found that the cluster is reddened by 0.09 mag., giving a true distance modulus of 8.6 mag. This is equivalent to a distance of 525 parsecs. Wallenquist, however, gives a distance of 468 parsecs while Becvar in the *Atlas Coeli Katalog*, and Johnson, in a later estimate in 1957, give 440 parsecs.

Van Hoerner, in 1957, estimated the age of M 34 to be about 110 million years which makes it an intermediate-age cluster. (For observing details see next page.)

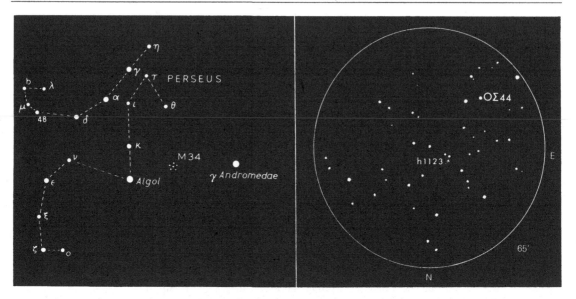

To observe M 34 can be found just N of the line *Algol-γ Andromedae* and due S of θ *Persei*.

It is a large, bright open cluster about 35′ in diameter. Radiating from the centre, the brighter stars form three distinctive curved arms. Many of the stars are distributed in pairs, and the neat double close to the centre of the cluster is h 1123, both components being white A0 stars of mag. 8.0, separation 20″ in P.A. 248°. On the SE edge of the cluster is another double, OΣ 44, A 8.4 mag., B 9.1 mag., separation 1″.4 in P.A. 55°.

M 34 seems to lack the background of faint stars which is common in most galactic clusters, and it does not respond well to high magnification. The best views are obtained with a wide field, and Walter Scott Houston has remarked 'I feel 15×65 binoculars give the best impression'.

The cluster can readily be picked up with the naked eye in good conditions, appearing as a faint blur of misty light.

M 35 NGC 2168

A galactic cluster, type 'e' in GEMINI

R.A. 06 h 05.8 m. Dec. N 24° 21' (1950).

Size 30' diam. Mag. 5.3 Distance 870 pc, 2850 lt-yrs.

Probably discovered by de Chéseaux in 1746.

DE CHÉSEAUX (No. 12) 'A star cluster above the northern feet of Gemini.'

MESSIER (Aug. 30th, 1764) 'A cluster of very small stars near the left foot of Castor; a little distance from the stars μ and η of that constellation (Gemini). Reported on the chart of the comet of 1770. Reported on the *English Atlas*. Diam. 20'.'

BODE (No. 14) 'A nebulous patch among small stars, close above the star η in Castor's foot.'

J. H. 'A very large, rich cluster with stars 9 to 16 mag.'

SMYTH 'A gorgeous field of stars from 9 to 16 mag. but with the centre of the mass less rich than the rest. From the small stars being inclined to form curves of 3 or 4 and often with a large one at the root of the curve, it reminds one of a bursting sky-rocket. In favourable circumstances, naked eye.'

WEBB 'Beautiful and extensive region of small stars, a nebula to the naked eye. Elegant festoon near the centre, starting with a reddish star.'

ROSSE 'Magnificent! – 300 stars in field 26' diam.'

LASSELL (In his 24-inch reflector) 'A marvellously striking object: no one can see it for the first time without exclamation . . . the field of view, 19' diam. is perfectly full of brilliant stars unusually equal in magnitude . . . of exquisite beauty. . . .'

Trumpler's classification for M 35 is III,3,r. In Helen Sawyer Hogg's catalogue of clusters in the *Handbuch der Physik*, its apparent diameter is given as 29', the integrated magnitude of the whole cluster, 5.6 mag., the magnitude of the brightest star is 9.0 and of spectral class B4.

Dr Åke Wallenquist counted 119 stars in a diameter of 30' and estimated the density of stars in the centre of the cluster to be 6.21 stars per cubic parsec. The overall density is about 0.7 stars per cubic parsec.

Wallenquist makes the true diameter of the cluster about 6.9 parsecs or $22\frac{1}{2}$ light-years but in the *Atlas Coeli Katalog* it is given as 9.4 parsecs or about $30\frac{1}{2}$ light-years. (For observing details see next page.)

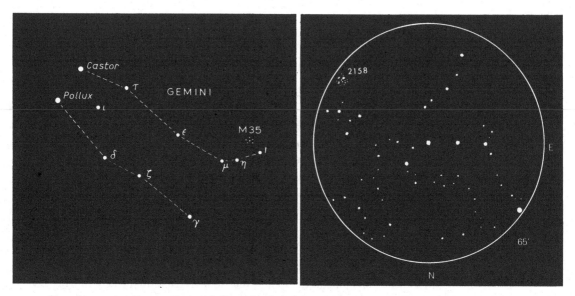

To observe   From Castor, follow the line of stars which forms the northern twin down to the foot, where three stars, μ, η and *1 Geminorum* are almost in line. M 35 is 1° N and 1° E of *1 Geminorum* or 2° N and 1½° W of η *Geminorum*.

M 35 justly deserves the praise bestowed upon it by the early observers. It contains more than 20 8–9 mag. stars, forming delicate patterns of loops and curls, with numerous fainter stars in the background. The brighter stars can be imagined to form an elaborate figure 1 with the base to the E and the tip ending in a bright pair. More stars extend to the NW and a curve to the NE ends in the bright, 6 mag. star *5 Geminorum*.

M 35 is altogether a superb cluster in which the imaginative eye can wander freely and with pleasure.

Objects near by   On the SW edge, just beyond a little cruciform group, and about 15' from the centre of the cluster, is a small, dense cluster of faint stars spreading fan-wise from an 11 mag. star close to the SE of it. This is NGC 2158.

55' to the WSW of M 35 is yet another cluster, IC 2157. This is actually brighter than 2158, but as it contains only three 10 mag. stars in a small triangle and a few others which are fainter, it is hardly recognisable as a cluster.

The star, *1 Geminorum* (2° SW of M 35), is almost exactly at the point of the summer solstice.

M 36 NGC 1960
A galactic cluster, type 'f' in AURIGA
R.A. 05 h 32.0 m.   Dec. N 34° 07' (1950).
Size 20' diam.   Mag. 6.3   Distance 1130 pc, 3700 lt-yrs.

Discovered by Hodierna before 1654 as 'a nebulous patch' in Auriga.

LE GENTIL 'An improperly-called nebula ... above the northern horn of Taurus. It requires a telescope of 12 feet to show that it consists only of stars.'

MESSIER (Sept. 2nd, 1764) 'A cluster of stars in Auriga, near the star φ. With an ordinary telescope of 3½ feet it is difficult to distinguish the stars. The cluster contains no nebulosity. Position determined from φ. (Aurigae) Diam. 9'.'

BODE (No. 12) 'A cluster of small stars below No. 9.' (M 38)

J. H. 'Bright, very large and very rich; stars of 9 and 11 mag.'

D'ARREST 'Extremely rich and elegant cluster. From the centre many swirls are arranged in three tight spirals.'

SMYTH 'A neat double star, Σ 737. A 8, B 9, both white, P.A. 309°, d. 12" (1836) in a splendid cluster. A rich, though open splash of stars 8–14 mag. with numerous outliers; like the device of a star whose rays are formed of small stars.'

WEBB 'Beautiful assemblage of stars, 8–14 mag. Very regularly arranged.'

ROSSE 'A coarse cluster.'

ROBERTS Photo in Feb. 1893 shows 'stars few in number; do not much exceed 100'.

J. E. GORE 'Poor compared with M 37.'

M 36 was classified by Trumpler as I,3,m and he assigns a spectral class of B3 to the brightest star which has a magnitude of 8.7.

H. S. Hogg gives the apparent diameter of the cluster as 16' and its integrated magnitude as 6.5.

Wallenquist counted 61 stars as probable members in a diameter of 19' and gives the density of the cluster as 14.6 stars per cubic parsec in the central region and 0.46 stars per cubic parsec overall. He includes M 36 among the intermediate age clusters.

The true diameter of the cluster is about 6.3 parsecs or about 20½ light-years, according to Wallenquist but Becvar, in the *Atlas Coeli Katalog*, gives only 4.0 parsecs or 13 light-years. (For observing details see next page.)

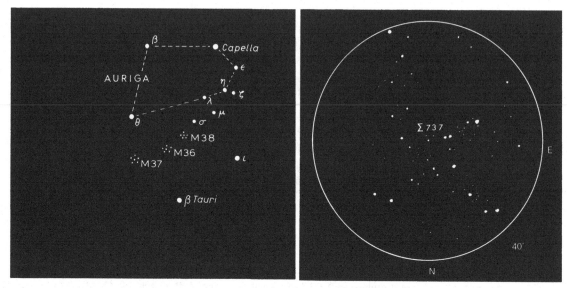

To observe   From M 38 move 2° S and 1½° E, *or*, from θ *Aurigae* (2nd mag.) move 3° S and 5° W.

M 36 consists of a pretty grouping of stars somewhat like a rocking chair in shape or similar to the constellation of Perseus in miniature. Two curved arms are distinguishable, the longer one extending to the SW. The field is densest in the centre and includes two close pairs, the more southerly one being Σ 737, described by Smyth. The cluster is brighter than M 38 but is slightly smaller, although stragglers to the SW extend its area to a total of about 30′ diam.

Objects near by   About 1° W and slightly N is NGC 1931, a diffuse nebula, about 3′ diam. with a small cluster of stars involved. It is very faint and needs a good aperture and the best conditions to pick up.

M 37 NGC 2099
A galactic cluster, type 'f' in AURIGA
R.A. 0.5 h 49.1 m.   Dec N 32° 32' (1950).
Size 25' diam.   Mag. 6.2   Distance 1105 p, 3600 lt-yrs.

Discovered by Hodierna before 1654 as 'a nebulous patch' in Auriga.

MESSIER (Sept. 2nd, 1764) 'A cluster of small stars a little removed from the preceding (M 36), the stars are smaller, more close together and enclosing some of the nebulosity. With an ordinary telescope of 3½ feet, it is difficult to see the stars. Reported on chart of comet of 1771. Diam. 9'.'

BODE (No. 13) 'A nebulous patch in Auriga, below the stars θ and ν.'

J. H. 'A rich cluster with large and small stars.'

SMYTH 'A double star; A and B 10 mag. both yellow. P.A. 346°, d. 17".4 (1879). A magnificent object; the whole field being strewn, as it were, with sparkling gold dust and the group is resolvable into about 500 stars from 10–14 mag. besides outliers.'

WEBB 'Even in small instruments, extremely beautiful; one of the finest of its class. Gaze at it well and long!'

D'ARREST 'Wonderful loops and curved lines of stars.'

KNOTT Observed a brighter star near the centre, which Burton described as 'ruby', mag. 10 and Espin as pale red, mag. 9.

This very rich galactic cluster contains 170 stars brighter than mag. 13 and 570 brighter than mag. 16. The density of stars of absolute magnitude greater than +4.5 in the central region has been estimated to be about 18 stars per cubic parsec which is about 50 per cent denser than in M 36.

Frederick R. West found the distance of the cluster to be about 1400 parsecs and its age to be about 220 million years.

Dr Upgren has used photographs of this cluster taken with the 40-inch Yerkes refractor at different times from 1903 to 1963 to obtain accurate proper motion comparisons for 212 cluster members down to mag. 12. From 11 photographs, 74 stars of almost certain membership were carefully plotted to show their drift over 60 years. These measurements indicated that the cluster is probably contracting slowly in a direction parallel to the plane of the galactic equator.

Similar motions were found for all the 212 stars, and studies by Swartz and Meuriers in Germany in 1960 and W. H. Jeffreys III at Van Vleck Observatory in 1962 lead to the same conclusion.

In Trumpler's classification, M 37 is designated I,1,r. The brightest star of spectral class B9 has a magnitude of 11.0. Wallenquist gives 507 stars in a diameter of 34', a central density of 43.6 stars per cubic parsec and overall density of 0.74 stars per cubic parsec.

He quotes 10.9 parsecs as the true diameter and distance as 1105 parsecs, while Becvar, in the *Atlas Coeli Katalog* gives 8.4 parsecs for the true diameter and 1450 parsecs for the distance.

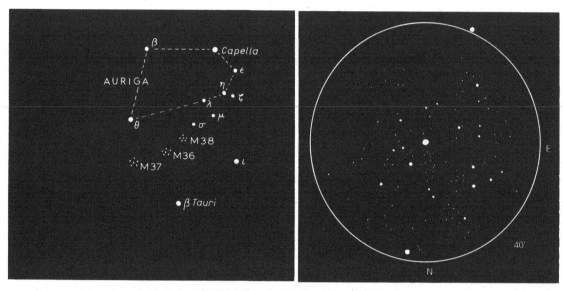

To observe   From M 36 move $1\frac{1}{2}°$ S and 3° E, *or*, from θ *Aurigae* move $4\frac{1}{2}°$ S and 1° W.

This is a fine cluster, more condensed than M 36 and containing hundreds of faint stars visible against a fairly dark background. The brighter stars appear to form the outline of a trapezium with the shorter side to the W, and in the centre is the brightest star of the cluster which, as noted by Burton and Espin, has a distinctly red tinge to it.

The closest concentrations of fainter stars are toward the N and W. The brighter portion of the cluster is about 20' in diameter but the total field is a good deal larger than this.

M 37 can almost be compared with M 11 in richness and it is almost twice as large.

Objects near by   50' to the SW of M 37, beyond a triangle of 7–8 mag. stars is a double star, Σ 796; A, 7.1, B, 8.0, P.A. 62°, d. 3".8. It is quite easy to split although the *comes* looks fainter than the 8 mag. quoted.

M 38 NGC 1912
A galactic cluster, type 'e' in AURIGA
R.A. 05 h 25.3 m.   Dec. N 35° 48' (1950).
Size 25'×10'   Mag. 7.4   Distance 850 pc, 2750 lt-yrs.

Discovered by Hodierna before 1654 as 'a nebulous patch' in Auriga.

LE GENTIL 'Above the former (M 36) and about 2¾° distant from it: seen in a telescope of 18 feet it appears as a star cluster.'

MESSIER (Sept. 25th, 1764) 'A cluster of small stars in Auriga, near the star σ, a little distance from the two preceding clusters; this one is of square shape and contains no nebulosity if examined carefully with a good telescope. It extends to 15' of arc.'

BODE (No. 9, Nov. 2nd, 1774) 'A lively nebula in which some stars can be found.'

SMYTH 'An oblique cross with a pair of large stars on each arm and a conspicuous single one in the centre, the whole followed by a bright individual of 7 mag.'

WEBB 'Noble cluster arranged as an oblique cross; pair of larger stars in each arm, brighter star in centre. Larger stars dot it prettily with open doubles. Glorious neighbourhood.'

R. J. Trumpler classed M 38 as II,2,r and gave the spectrum of the brightest star, which has a magnitude of 9.7, as B5.

In Helen Sawyer Hogg's list of clusters in the *Handbuch der Physik*, M 38 is given an apparent diameter of 18' and an integrated magnitude of 7.0.

Åke Wallenquist counted a total of 154 stars as probable members of the cluster in a diameter of 26'. According to the same authority, the density of stars in the central portion of the cluster is 7.76 stars per cubic parsec and the average density for the whole cluster is 1.11 stars per cubic parsec. He also includes M 38 among the older galactic clusters.

The true diameter of the cluster is about 6.4 parsecs or approximately 21 light-years in both Wallenquist's and Antonin Becvar's catalogues.

Becvar gives the distance as 1100 parsecs; Hogg quotes 800 parsecs and Wallenquist 849 parsecs. (For observing details see next page.)

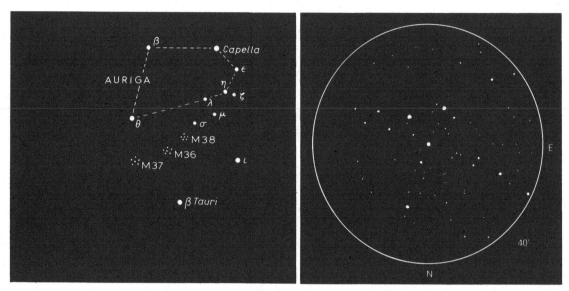

To observe M 38 lies very slightly N of the mid-point of the line joining ι and θ *Aurigae, or* 7° due N of β *Tauri*.

Although M 38 is the least bright of the Messier clusters in Auriga, it can be recognised instantly by its distinctive cruciform shape.

The longer arm of the cross lies in P.A. about 65°–245° and the shorter arm cuts it at almost a right angle; the intersection being well marked by a bright star. There is a considerable concentration of faint stars at the centre and the whole field is very rich altogether.

Even in small telescopes, its neat, bright outline makes it a favourite of many amateur observers.

Objects near by 30′ to the SSW of M 38 is a small, dusty cluster, NGC 1907, about 5′ diam. and mag. 10. It can be seen quite clearly about 15′ SE of a 6 mag. star.

M 39 NGC 7092
A galactic cluster, type 'e' in CYGNUS
R.A. 21 h 30.4 m.   Dec. N 48° 13' (1950).
Size 32' diam.   Mag. 5.2   Distance 250 pc, 825 lt-yrs.

Probably discovered by Le Gentil in 1750. (Bigourdan.)*

LE GENTIL   'At the tip of the tail of Cygnus ... It is a large cloud, bigger at one end than the other; the small end facing south-east ... It seems opaque and very dim ... (but) can be seen without the telescope.'

MESSIER   (Oct. 24th, 1764) 'A cluster of stars near the tail of the Swan, they can be seen with an ordinary telescope of 3½ feet. Diam. 1°.'

BODE   (No. 75, Oct. 27th, 1774) 'A small star cluster west of the star π in the tail of Cygnus.'

SMYTH   'A loose cluster or rather, splashy field of stars in a very rich vicinity. Several pairs included.'

ROSSE   'Very large but poor, little concentration. In a region of the Milky Way rich in stars.'

FLAMMARION   'Unusual curved runners of stars, with a compressed cluster of 20 stars, difficult to separate from the rest.'

Trumpler classed M 39 as a Class 1a object containing only main-sequence stars and none of spectral types earlier than A.

It is not included in H. S. Hogg's catalogue of clusters in the *Handbuch der Physik* nor in Wallenquist's 1959 catalogue of galactic clusters.

It was examined by Harold Johnson of Yerkes Observatory in 1952 for magnitudes, colours and spectral types of 28 stars in and around the cluster. He found that the cluster was essentially unreddened by interstellar absorption. He obtained a true distance modulus of 7.2±0.2 mag. which is equivalent to a distance of 275 parsecs ±30 parsecs.

A later estimate of distance in 1957 gave 250 parsecs. The *Atlas Coeli Katalog* gives the number of stars as 25 and the true diameter of the cluster as 2.2 parsecs or about 7 light-years.

The age of M 39 has been estimated as 300 million years according to Lohmann and 230 million years according to Van Hoerner. (For observing details see next page.)

---

*   Bigourdan's identification of Le Gentil's observation with M 39 is extremely doubtful, especially as he gave no position. The discovery is more properly allowed to Messier whose observation is supported by good co-ordinates.

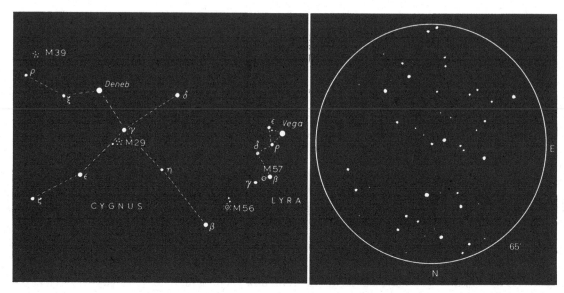

To observe   The whole of this area is very rich in the faint stars of the Milky Way and care must be taken to arrive at the correct position.

First pick up 4th mag. ϱ *Cygni* which is 9° E of *Deneb*. From ϱ *Cygni* move 3° N and ¼° W.

The position can be checked from π² *Cygni* (4½ mag.), from which move 2½° W and 1° S.

It is strange that M 39 seems to have received little attention from early observers for it is large, bright and distinctive. Even Smyth's description of 'large and splashy' does it little justice for it has a distinct geometrical shape.

Three bright 8 mag. stars lie at the corners of a large, almost equilateral triangle with the southerly side aligned E–W. Many stars are grouped in pairs, including a bright pair near the centre.

The cluster is very similar in form to M 48 in Hydra only here the triangle is inverted.

M 40   (Not included in NGC)
A pair of stars in URSA MAJOR
R.A. 12 h 21 m. Dec. N 58° 20′ (1950).

HEVELIUS   (1660) No. 1496 in his *Prodromus Astronomiae* was reported as 'Supra tergum nebulosa' (a nebula above the back (of Ursa Major)).

DERHAM   Included it as No. 14 in his list of 1733 in *Phil. Trans.*

MAUPERTUIS   Included it in his *Figure des Astres* 2nd Edn., 1742.

MESSIER   (1771 Catalogue) 'On the night of 24–25 Oct. (1764) I looked for the nebula which is above the back of the Great Bear, noted in the book, *Figure des Astres*, 2nd Edn. Its 1660 position should be R.A. 182° 32′ 41″ and Dec. North 60° 20′ 33″. I found by means of this position, two stars, very close together and of equal brightness, about 9th mag., situated at the root of the tail of the Great Bear. They are difficult to distinguish with an ordinary telescope of 6 feet. Their position is R.A. 182° 45′ 30″ Dec. N 59° 23′ 50″. It is presumed that Hevelius mistook these two stars for a nebula.'

(In his later catalogue of 1781 Messier described the stars merely as 'very close together and very small . . .')

BODE   (No. 23) 'Two small nebulous stars near each other, close above the star δ in Ursa Major.'

SMYTH   (Under δ *Ursae Majoris*) 'This being the reported site of Hevelius' nebula of 1660 and Messier's No. 40, I searched for them by "fishing" but found only a couple of small stars lying N.p. and S.f. with gleams of others. This group, however, resolved by my telescope, may have been the one seen by those astronomers.'

FLAMMARION   'We checked this position vainly at Juvisy. There is no nebula here, no cluster, and M 40 corresponds with nothing in particular . . . It is difficult to understand what this "nebulose" of Hevelius signifies. No nebula can be seen there only 2 stars, of 8th mag. 4′ apart, accompanied to the S by another little one of 9th mag. at 3′.4 to S. The 3 stars might, in an old, imperfect instrument, give a vague impression of a faint nebulosity.'

(In the Bulletin of the Astronomical Society of France for 1919 in which these remarks appear, Flammarion also included a sketch of the area, selecting two stars about 50′ N of 71 *Ursae Majoris*. These appear to be of unequal magnitude which is contrary to Messier's 1771 description; the position, too, is about 30′ S of that of Messier's.)

While M 40 consists only of a pair of stars and is of little astronomical importance, it is an actual Messier Object and the observer may well be tempted to look for it – if only to complete his 'set'. Naturally, it is difficult to identify with much confidence an object described merely as two close 9th mag. stars and neither Smyth nor Flammarion were particularly successful. In Becvar's *Atlas Coeli Katalog* (Messier's list) M 40 is given a position of R.A. 12 h 33m, Dec. N 58° 30′ (1950) but without any explanation. In this position, which is about 1½° f. Messier's position, there is a pair of stars but they, like Flammarion's are of unequal brightness.

Almost exactly in Messier's position (R.A. 12 h 20 m, Dec. N 58° 22′ (1950)) however, there is a distinct pair of 9th mag. stars in P.A. approx. 85°,

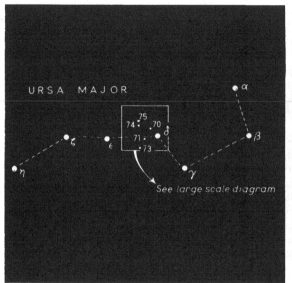

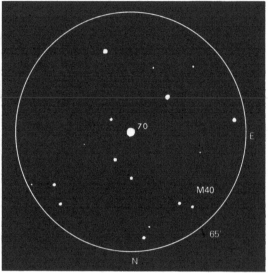

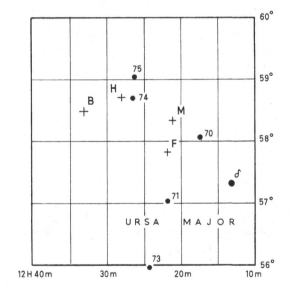

The various positions for M 40 (epoch 1950) are given below.

Hevelius—(1660) R.A. 12 h 29 m Dec. N 58° 44'

Messier—(1764) R.A. 12 h 20 m Dec. N 58° 22'

Flammarion—(1919) R.A. 12 h 22 m Dec. N 57° 50' (approx.)

Becvar—(1964) R.A. 12 h 33 m Dec. N 58° 30'.

d. about 50″ which is just separable in a small telescope similar to that used by Messier.

This identification is confirmed by John H. Mallas of Scottsdale, Arizona in a letter to *Sky and Telescope* (Aug. 1966, p.83), and he has further identified the pair as the double, Winnecke 4 (mags. 9.0 and 9.3, d. 49″). This was discovered by A. Winnecke at Pulkovo in 1863, according to Mallas.

M 41 NGC 2287
A galactic cluster, type 'e' in CANIS MAJOR
R.A. 06 h 44.9 m.   Dec. S 20° 41' (1950).
Size 30' diam.   Mag. 4.6   Distance 500 pc, 1600 lt-yrs.

Discovered by Hodierna before 1654 as 'a nebula near Sirius'.
Rediscovered by Flamsteed, Feb. 16, 1702, and by Le Gentil in 1749.

FLAMSTEED   'Near this star (12 CMa) there is a cluster.'

LE GENTIL   'In the collar of Canis Major: it is only by using a telescope of 8 feet that the cluster of stars of which it is composed can be distinguished: it appears to be only a simple nebula in a telescope of two or three feet.'

MESSIER   (Jan. 16th, 1765) 'A cluster of stars below Sirius; this cluster appears nebulous in an ordinary telescope of one foot; it is nothing more than a cluster of small stars.'

SMYTH   'A double star, A 9, lucid white, B 10, pale white, P.A. 85° d. 45" (1836) in a scattered cluster. Divided into five groups of which the central one is the richest and marked by 3 bright stars forming a crescent. In N.p. is an open double star (above).'

WEBB   'Larger stars in curves with ruddy star near centre. Superb group, visible to naked eye.'

J. E. GORE   'Referred to by Aristotle in his "Meteorologies" as a star with a tail.'

LICK XIII   'A large, very coarse and sparse cluster about 25' diam.'

This galactic cluster has received some attention in recent years: the magnitudes, colours and proper motions of 106 stars were measured in order to determine its size and age.

Its linear diameter, assuming its distance to be 725 parsecs, works out to be about 6.3 parsecs or about 20 light-years. The cluster is unusual for the fact that, among its members are some K-type giant stars which are not usually found in galactic clusters. Helfer, Wallerstein and Greenstein found from investigations of one of these K-type giants that it contained roughly the same metal abundancies as the Sun.

M 41 is a comparatively young cluster having an age of about 100 million years.

In Trumpler's classification, M 41 is designated I,3,r, and its brightest star is of spectral type A0.

Wallenquist estimated a total of 96 probable member stars in a diameter of 51' and obtained a value for the density of stars of 6.25 stars per cubic parsec in the central portion and an average of 0.05 stars per cubic parsec for the cluster as a whole.

Wallenquist's true diameter is based on a distance of 501 parsecs and is given as 7.4 parsecs or about 24 light-years.

In Becvar's *Atlas Coeli Katalog* the true diameter is taken to be 6.6 parsecs or about 21½ light-years using a distance of 760 parsecs. (For observing details see next page.)

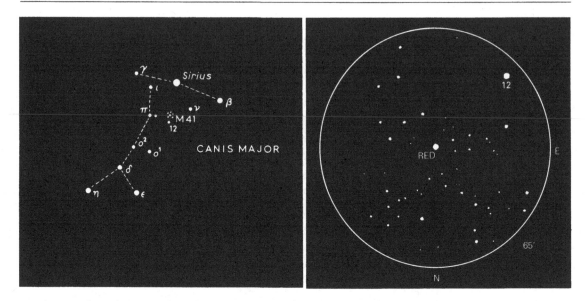

To observe   From Sirius move 4° S and ½° E. The cluster has a 6 mag. star, *12 Canis Majoris*, on its SE edge.

It contains about 50 stars and, although bright, Webb's remark that it is visible to the naked eye is a little optimistic for observers in the higher latitudes. For southern observers it is a brilliant object.

The cluster is roughly circular in outline but with irregular extensions, mostly to the north. The most striking features are the distinct curved arms radiating from the central, slightly orange star, toward the north and east.

Objects near by   There is a double star on the western edge which may be the one noted by Smyth.

M 42 NGC 1976
The 'Great' Nebula. A gaseous emission nebula in ORION
R.A. 05 h 32.9 m.   Dec. S 05° 25′ (1950).
Size 66′×60′   Mag. 4.0   Distance 460 pc, 1500 lt-yrs.

M 42 is one of the best known and is certainly the most spectacular of the Messier
Objects. As a star in Orion's sword, it has been known since the beginnings of
recorded astronomy; Ptolemy, Tycho Brahe and Bayer all noted it as of the 3rd
magnitude and the last named gave it the designation of θ *Orionis* in his
Uranometria in 1603.

Galileo selected the stars of Orion's belt and sword as one of the first targets for
his telescope and drew a diagram in *The Sidereal Messenger,* in which he said, 'I
have added eighty other stars recently discovered in their vicinity . . .' He made no
mention, however, of any nebulosity around θ *Orionis.*

Its discovery as a nebula is generally attributed to Nicholas Peiresc in 1610, a
year after Galileo's first use of the telescope. The Jesuit astronomer, Cysatus
(1588–1657) also made an independent discovery of the same nebula, comparing it
with the comet he had seen in the telescope in 1618.

Some time before 1654 the nebula was observed by G. B. Hodierna in Sicily,
describing it merely as containing 22 stars and some nebulosity. He also made a
sketch depicting three of the stars of the 'trapezium' surrounded by a hatched
ellipse representing the nebulous outline. This is probably the first 'drawing' (as
opposed to a symbolic indicator) ever made of a nebula. (See below.)

The first detailed description of the Orion Nebula was given by Huyghens in
1656, apparently unaware of the previous observations of Peiresc, Cysatus or
Hodierna.

HUYGHENS   'There is one phenomenon among the fixed stars worthy of mention, which as far
as I know, has hitherto been noticed by no one and indeed, cannot be well
observed except with large telescopes. In the sword of Orion are three stars quite
close together. In 1656 as I chanced to be viewing the middle one of these with the
telescope (a 23-foot F.L. refractor), twelve showed themselves – not an uncommon

*M 42 Drawing by G. B. Hodierna some time before 1654. The first known
representative sketch of a nebula. (Palermo Astronomical Observatory)*

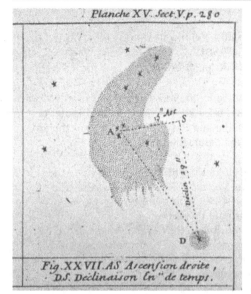

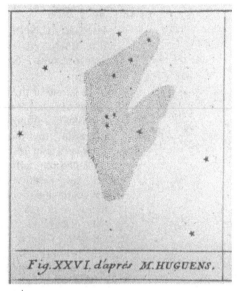

M 42 and M 43 (De Mairan) Drawn
in 1731. De Mairan's nebula, M43, is
shown at 'D' (British Museum)

M 42 (Huyghens) Drawn in about
1656 (British Museum)

circumstance. Three of these almost touched each other and, with four others, shone through a nebula so that the space around them seemed brighter than the rest of the heavens which was entirely clear and appeared quite black, the effect being that of an opening in the sky through which a brighter region was visible.'

Later, in 1684, using a 34-foot telescope, Huyghens found that the three stars in the centre were accompanied by a fourth and in 1695 drew a sketch of the nebula with the 'trapezium'.*

HALLEY   (No. 1, 1715) 'Two very contiguous stars environed with a very large transparent bright spot, through which they appear with several others.'

DE CHÉSEAUX   (No. 16) 'The orion nebula discovered by M. Huyghens.'

LE GENTIL   (Mar. 10th, 1758) 'It appeared to have the shape of the open jaws of some animal. I observed toward the west, an extension of light forming a rectangle: this light is very diffuse. The three stars in a straight line in the "lower jaw" appear completely detached from the nebula.'

(Apr. 3rd, 1758) 'The four stars in the centre appear extraordinarily brilliant!'

MESSIER   (Mar. 4th, 1769) 'Position of the fine nebula in Orion's sword, around the star theta which is included there with three other stars, smaller than can be seen in some good instruments. Messier has prepared a drawing in which he has included

*   According to J-J de Mairan in his *Treatise on the Aurora Borealis* of 1773, the fourth star, 'D', was first seen by Picard in 1673. Surprisingly, the *fifth* star, 'E', was claimed as a discovery by Robert Hooke using a $3\frac{1}{2}$-inch refractor in 1664. In his *Micrographia* of 1665, Hooke declared, 'Whereas Huyghens in 1665 saw only 3 little stars in a cluster, I discovered 5, and the twinkling of divers others ...' As with divers other claims by Hooke, this failed to get acceptance by his contemporaries, and star 'E' is usually credited to F. G. W. Struve, who found it on Nov. 11th 1826. Star 'F' was discovered by John Herschel on Feb. 13th 1830.

156

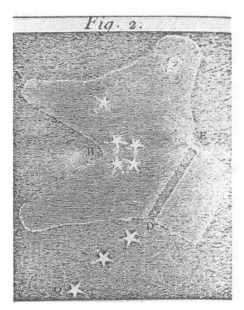

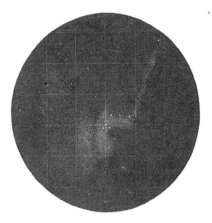

M 42 Drawing by Messier in 1771

M 42 Drawing by Le Gentil in 1758

the details of this Great Nebula with the greatest care, which can be seen in the *Memoirs* of the Academy for 1771. Huyghens discovered it in 1656. It has been observed since by many astronomers. Reported on the *English Atlas*.'

In the introduction to his 1781 catalogue, Messier refers again to his drawing of the Orion nebula. 'This drawing will help to recognise it again, provided that in the future, it is not subject to change. Comparing the present drawing with those of Ms Huyghens, Picard, de Mairan and le Gentil, it is surprising to find such a change in it that, considering its shape alone, one would have difficulty in recognising it as the same nebula. M. leGentil's drawing can be seen on page 470, Plate XXI, of the volume of the Academy (of Sciences) for 1759.'

BODE (No. 10) 'The most remarkable nebula in the heavens. θ shines four-fold through a good telescope; there are three very small stars next to it. There are altogether seven stars to be seen, all involved in a lively nebula or light-shimmer.'

W. H. observed the Orion Nebula in 1774 as one of the first objects on which to test his newly constructed 6-foot F.L. reflector. When, in August 1789, he saw it again in his huge 48-inch aperture, 40-foot F.L. telescope, he described it as 'an unformed fiery mist, the chaotic material of future suns'. This was a typical example of his almost prophetic vision.

J. H. with the 18-inch aperture, 20-foot F.L. telescope in 1820. 'A curdling liquid or surface strewn with flocks of wool – or like the breaking up of a mackerel sky. Not very unlike the mottling of the Sun's disc only coarser and the flocculi not round but wisps. No appearance of being composed of stars – altogether different from resolvable nebulae.'

And at a later date: 'In its more prominent details may be traced some slight resemblance to the wings of a bird . . .'.

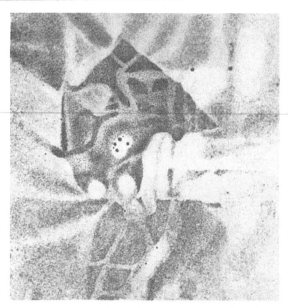

M 42   (Rosse) The Huyghenian region of the Orion Nebula as drawn by Bindon Stoney in 1852 (B.A.A.)

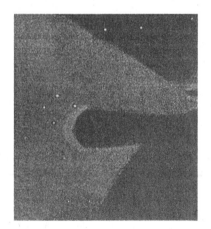

M 42 (Smyth) Showing the 'Fishes Mouth' and 'Trapezium' (B.A.A.)

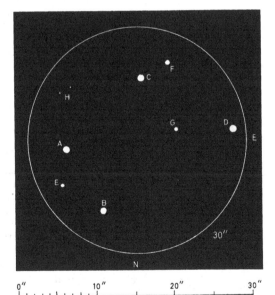

M 42 The stars of the Trapezium in M 42. Stars A and B are both Algol-type eclipsing variables: A, period 65.4325 days, range 6.73 to 7.53 mag.; B, period 6.4705 days, range 7.95 to 8.52 mag.

158

LORD ROSSE	noticed that many faint stars embedded in the central parts are strikingly red when observed with the eye.
G. P. BOND	(Feb. 26th, 1861) 'It is now impossible to see in it any other aspect than as a maze of radiating, spiral-like wreaths of nebulosity or filamentous tentacles; the centre of the vortex being about the trapezium.'
LASSELL	(Also in 1861) 'Large masses of cotton-wool packed one behind another, the edges pulled out so as to be very filmy.' 'Pea-green colour.'
MITCHEL	(*Stellar Worlds* (1869) pp. 256–61, Wm Collins, Glasgow) 'In exhibiting this object ... I have frequently heard the remark that a part of the nebula was hid by a *black* cloud.'
SMYTH	(Describing the trapezium) 'A multiple star, the trapezium, in the Fish's mouth of the Great Nebula. 'A' 6 mag. pale white, 'B' 7 mag. faint lilac, 'C' 7½ mag. garnet, 'D' 8 mag. reddish, 'E' 10½ mag. blue.'

(The fifth star in the trapezium, 'E' was discovered by Wilhelm Struve, using the 9½-inch refractor at Dorpat in 1826; the sixth, 'F' by John Herschel on Feb. 13th, 1830 with Sir James South's 11¾-inch refractor; the seventh, 'G' by Alvan G. Clark in 1888, when testing his 36-inch objective of the Lick Observatory; the eighth, 'H', by E. E. Barnard found later in the same year, using the same telescope. This last star Barnard found later to be itself double, both of magnitude 16 and at the limit of visibility and resolution.

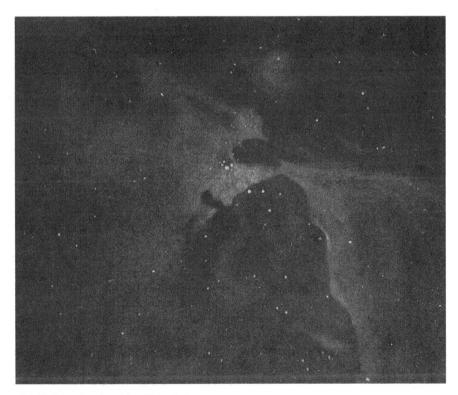

M 42 Drawing by John Herschel

159

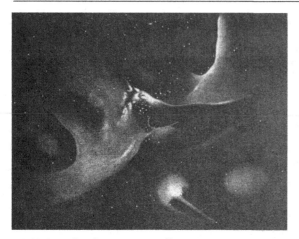

M 42 Drawing by Wm. Lassell

M 42 A sketch by Philip Braham, RCS (From English Mechanic, *1883)*

All these members of the trapezium have been photographed in recent years, using infra-red plates taken at Yerkes Observatory and at Flagstaff, Arizona but another star, fainter than any previously seen, and which was glimpsed by Barnard in 1889, has not been recorded since.

In 1879, William Huggins first detected that the star 'A' was in fact a spectroscopic binary and this was confirmed by the Yerkes astronomers, Frost and Adams, in 1904.)

HUGGINS   found the spectrum of M 42 to be gaseous, containing the same green lines he had discovered in other nebulae. It was here that the hypothetical element 'nebulium' was supposed to exist.

WEBB   described M 42 as 'One of the most wonderful objects in the heavens, readily visible to the naked eye, yet strangely missed, as Humboldt says, by Galileo who paid great attention to Orion. An irregular, branching mass of greenish haze; in some directions moderately well defined where the dark sky penetrates it in deep openings.'

M 42 was first photographed by Henry Draper on Sept. 30th, 1880, by Janssen in France in 1881, followed by Dr Common in 1883 and by Isaac Roberts with his 20-inch reflector at Crowborough in 1886.

In about 1888, William Huggins showed by photographing the spectra of both the trapezium and the nebula that the stars were actually involved in the clouds of glowing gas. Radial velocities in the nebula were measured by Campbell and Moore, using the 36-inch refractor of the Lick Observatory with the Brashear grating spectrograph from 1915 to 1918. The different values of radial velocity found in different parts of the nebula appeared to give evidence of turbulent motion within it. Recently, large turbulence effects have been confirmed by Georges Courtes of Marseilles using interference techniques.

The American astronomer R. J. Trumpler in 1931 discovered that the trapezium is associated with a rich cluster of red stars which can be detected only by infra-red photography. Within a circle of 1' radius around the trapezium centre he counted 41 red stars and 62 stars inside a 2' circle.

Trumpler's unique infra-red cluster was next investigated by W. Baade and R. Minkowski in 1938: they confirmed its existence and concluded that the cluster was probably larger than could be observed, the outlying areas being almost certainly obscured by nebulous matter. Later still, in 1952, from photographs taken with the Schmidt telescope of Tonanzintla Observatory, Dr Haro has shown that the cluster is a good deal larger than Trumpler had observed and that the outer parts suffer heavy absorption.

Also associated with the nebula are several hundred irregularly variable stars including some 'flare stars'. Ten of these have been detected by Haro and Morgan in one square degree centred on the trapezium: they are very faint, of magnitudes between 16.4 and 17.6 but are liable to sudden 'flares' of $\frac{1}{2}$ to 2 magnitudes in as little as 5 to 30 minutes.

In addition, the compact cluster of faint stars around the trapezium appears to be expanding at the rate of about one part in 300 000 each year. Dr Strand of Dearborn Observatory has compared 1957 photographs with the 40-inch Yerkes refractor with similar plates taken 50 years before by Ritchey and Parkhurst to arrive at this result. These observations would appear to make the age of the cluster about 300 000 years which would put it among the youngest yet known.

V. C. Fessenkov has made calculations of the critical density of gas required for star formation in our part of the Galaxy and has found that in some parts of M 42 the gas is above this critical value.

That new stars are being formed in M 42 is at least, highly probable, for at a distance of about $1\frac{1}{2}°$ from the centre of the nebula, a group of hazy stars which had five members in 1947 was seen to contain seven in 1954.

The luminous portion of the nebula is only a part of the much more extensive dark nebulosity and W. Baade was of the opinion that the hot, new stars of the trapezium had, comparatively recently, burned a 'hole' in the dark nebula.

Several stars in other parts of the sky have been found to have common motions directly *away* from M 42. Two of these are *AE Aurigae* with its associated nebulosity (IC 405), and μ *Columbae*, each having the same speed of 127 km/s in opposite directions away from the Great Nebula. Blaauw and Morgan, of Yerkes Observatory, have calculated that these two stars, both of which are about 337 parsecs distant from the nebula, could have originated there about $2\frac{1}{2}$-million years ago. That these are young, blue stars of about this age is a supporting factor of this remarkable conjecture.

The Orion Nebula is composed mainly of hydrogen but spectroscopic evaluations show that for every 1000 atoms of oxygen present, there are 1600 of neon, 130 of nitrogen and sulphur, 40 of argon and 2 of chlorine.

In 1979 a curious 'bubble' was detected just south of θ^1 *Ori* (star A of the trapezium). It has the 'appearance of a nearly perfect circle with filaments radiating from the centre like the spokes of a cartwheel' according to Dr K. Taylor of Wise Observatory. This object may indicate the presence of a stellar wind 'snow plough' effect which can be typical of the early stages in star formation in H II regions.

Infra-red observations have revealed that behind the visible region of M 42 lies the huge 'Orion Molecular Cloud' which has an angular diameter of more than half a degree. In 1985 this cloud was seen to contain a compact, flattened 'doughnut' of gas in which the carbon monoxide emission indicated a temperature of more than 80 K. The heating may be caused by radiation from the stars in the trapezium region. (For observing details see p. 164.)

M 43 NGC 1982
A gaseous emission nebula in ORION (part of the 'Great' Nebula)
R.A. 05 h 33.1 m.   Dec. S 05° 18' (1950).
Size 2' diam.   Mag. 9.0   Distance 460 pc, 1500 lt-yrs.

Discovered by de Mairan before 1750. (Bigourdan.)

DE MAIRAN (1733) 'Close to the luminous space of Orion, one sees one of the stars . . . surrounded by a brilliance very similar to that which produces, as I believe, the atmosphere of our Sun, if it were dense enough and extensive enough to be visible in telescopes at a similar distance.'

LE GENTIL (April 3rd, 1758) 'The light of this star is pallid and feeble and does not twinkle at all.'

MESSIER (Mar. 4th, 1769) 'Position of a little star surrounded by nebulosity and which is below the nebula of Orion's sword. M. Messier has included it in the drawing of the Great Nebula.'

J.H. (In 1864 catalogue) 'A remarkable and very bright object, very large, round, with tail much brighter in the middle where there is a star of 9th mag.'
(In *Memoirs* of the Royal Astronomical Society) 'The Nebula Minima – a rounded mass which appears as if just drawing together into a star.'

LICK XIII 'No reason to doubt it is part of the Great Nebula. It surrounds Bond's star.'

NGC (Additional note by J. H.) 'There are two observations by W. H. of 1 H. III, but they differ enormously. One agrees with M 43.'
(Note by Dreyer) 'It is the so-called Mairan's nebula . . .'

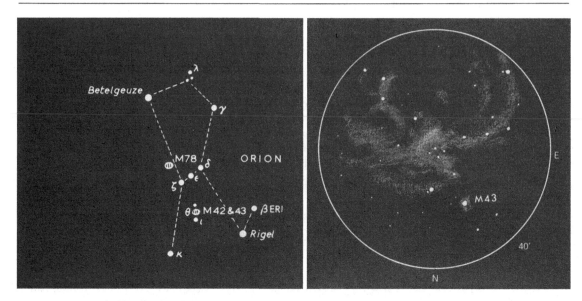

To observe   **(M 42 and M 43)**

Surrounding θ *Orionis*, in the sword of Orion.

This complex of bright and dark nebulae and embedded stars is a magnificent and inspiring sight in almost any kind of telescope. A low power and wide field of view shows the wide-spreading wings and the nebula as a whole. The stars of the trapezium and the mottled aspect of the bright, central 'Huyghenian' region, require a higher magnification and a large aperture will reveal more fine detail.

In many ways, the Orion Nebula is a finer object, even in a small telescope, than can be seen in the familiar photographs taken by large instruments.

In photographs, the central area is generally 'burnt out' by the long exposures necessary to show the faint, outer portions of the nebula. The line of three bright stars to the SE of the trapezium is often completely obscured as is the trapezium itself. The bright, mottled, triangular region to the south of the trapezium is also seen more easily by visual means than by normal photography and the delicacy of the extended 'wings' to the NW and SE has to be seen directly at the telescope to be properly appreciated.

The dark lane of the 'Fish's mouth', extending east from the trapezium can be very clearly seen and there is a distinct bifurcation in the SE portion of the diffuse nebula. The whole of the SW sector of the nebula can be observed to glow with a soft, milky light, but here the eye cannot compete with the camera and only vague and fleeting detail reveals itself visually.

M 43 is about 10' N and slightly E from the trapezium in M 42. It appears as a separate nebulous star at first sight, as Messier described it, but closer examination shows wisps of faint nebulosity extending toward the main body of the larger nebula.

M 44 NGC 2632
'Praesepe.' A galactic cluster, type 'd' in CANCER
R.A. 08 h 37.4 m.   Dec. N 20° 00' (1950).
Size 70' diam.   Mag. 3.7   Distance 182 pc, 600 lt-yrs.

'Praesepe' (The Manger), also known as the 'Beehive', although not as immediately conspicuous as the Pleiades, was well known to the earlier astronomers. Aratus and Theophrastus, who wrote between 300–250 BC, considered it a portent of rain if Praesepe looked dim or was invisible in an apparently cloudless sky. Hipparchus included Praesepe in his star catalogue of 130 BC but only as a 'cloudy star' and although some later observers described it as containing 'nebulous stars emitting a strange light', its true identity as a star cluster had to await the use of the telescope. It was catalogued by Ptolemy as No. 449, by Ulugh Begh as No. 446, by Tycho Brahe as No. 576 and by Hevelius as No. 291.

GALILEO (In 1609) 'The nebula called Praesepe contains not one star only but a mass of more than 40 small stars. We have noted 36 besides the Aselli.'
(The Aselli (The Asses) are two naked-eye stars, Asellus Borealis (γ *Cancri*) 5th mag. Simon Marius observed the cluster in 1612.)

HODIERNA (before 1654) counted 38 stars.

DE CHÉSEAUX (No. 11) 'Praesepe, the well known cluster.'

MESSIER (March 4th, 1760) 'A cluster of stars known by the name of the nebula of Cancer. The position given is that of star c.'

BODE (No. 20) 'The familiar cluster, Praesepe.'

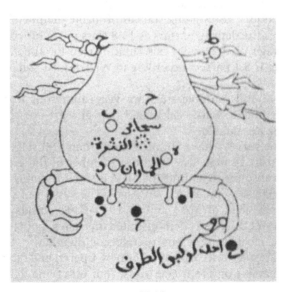

M 44

The constellation of Cancer as shown on an early Persian MS. The cluster 'Praesepe' is denoted by a dotted circle (British Museum)

165

SMYTH   'Long known as a nebula, components not being separately distinguished by the naked eye.'

WEBB   'Two triangles will be noted. Too large for visual fields: full of fine combinations.'

LICK XIII   'Stars not nebulous with an exposure of 2 hours.'

M 44 is a cluster very different in many ways from M 45 (the Pleiades). As the Lick XIII observation shows, Praesepe contains no nebulosity at all, even in a long exposure photograph. In addition, many of its stars are yellow or orange in colour and have spectra similar to that of the Sun while the Pleiades are all, very distinctly, whiter, hotter stars. One point of similarity to the Pleiades, however, is the fact that some of the stars show spectra exhibiting a broadening effect due to fairly rapid rotation.

M 44 is considered to be a much older cluster then the Pleiades and Baade has estimated its age at about 400 million years.

Although it is a very open cluster – its mean density is about 1.5 stars per cubic parsec – and so is likely to be easily disrupted by galactic gravitational forces, its position, being nearly 100 parsecs above the galactic plane, may help to mitigate this effect.

The most closely related cluster to Praesepe is the Hyades, which, while not included in Messier's list, is a true cluster of associated stars. There are, in fact, many pointers to the possibility that Praesepe and the Hyades had, at one time, a common origin.

According to Trumpler, the total diameter of the cluster, including all the fainter stars, is nearly 6°. His classification for the cluster is I,2,r and the spectrum of the brightest star is B9.

Wallenquist counted 62 stars as probable members in a diameter of 110' and gives the density of stars as 13.8 stars per cubic parsec in the centre and an average of 0.92 stars per cubic parsec for the whole cluster. Its true diameter is 5.1 parsecs according to Wallenquist and 3.9 parsecs in the Atlas Coeli Katalog.

Double and multiple stars are very numerous in this cluster: in 1940 Haffner and Heckman calculated that at least one-fifth of all the stars in Praesepe are double.

Recent surveys have counted more than 200 stars as members, ranging from 3.3 to 14 mag. The brightest member of the group is ε Cancri, which with an absolute magnitude of +0.2, is some 70 times brighter than the Sun.

The eclipsing binary RY, an Algol-type system of period 1.093 days, lies within the field of M 44, but in 1980 N. S. Awadalla and E. Budding of Manchester University demonstrated that it lies about four times further away than the cluster and is not, therefore, a member.

The latest distance determination by Upgren and others in 1979 works out at 182 parsecs or about 600 lt-yrs. (For observing details see next page.)

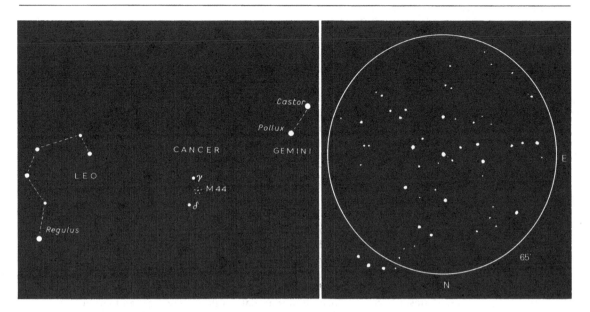

To observe Praesepe is easily recognised and is visible to the naked eye. It lies roughly half way between Regulus and Pollux.

On a dark, clear night it seems as if individual stars can be picked out with unaided vison. This is almost certainly an illusion as the stars are much too close to separate without the use of the telescope.

The two 4th mag. stars, γ and δ *Cancri* (the Aselli) bracket the cluster to the east.

In the telescope, Praesepe is a brilliant, glittering and lavish feast for the eye. In anything but a rich-field instrument, however, it is likely to be too large for the field of view and its beauty has to be appreciated by instalments.

Naturally, the lowest possible power should be used and the best views may well be obtained with firmly mounted field glasses.

(Perhaps the oddest thing concerning this well-known and beautiful cluster is the name by which it was known among the Chinese. This, according to Richard Hinckley Allen's classic *Star Names and their Meanings*, was 'Tseih She Ke' or 'The Exhalation of piled-up Corpses'.)

167

M 45 (Not included in NGC)

'The Pleiades.' A galactic cluster, type 'c' in TAURUS

R.A. 03 h 44.1 m.   Dec. N 23° 58' (1950).

Size 120' diam.   Mag. 1.6   Distance 126 pc, 410 lt-yrs.

The Pleiades have been known from the earliest times; nearly 1000 years B C. Hesiod referred to their disappearance behind the Sun for 40 days in the summer. Their first appearance in the morning sky before the Sun – the heliacal rising – was an important astronomical event of the ancient world and Julius Caesar made this date in May the beginning of summer in the Julian calendar.

Homer, in the *Odyssey*, Book 5, describes how Ulysses 'sat at the helm and never slept; keeping his eyes upon the Pleiads . . .', holding the Plough on his left hand, according to the sailing instructions of the 'divine Calypso' whose island he had just left.

There are three references in the Bible:

Job, Ch. 9, v. 9: 'Which maketh Arcturus, Orion and Pleiades and the chambers of the South.'

Job, Ch. 38, v. 31: 'Canst thou bind the sweet influences of Pleiades or loose the bands of Orion?'

Amos, Ch. 5, v. 8: 'Seek him that maketh the Seven Stars and Orion.'

They were frequently known as the Seven Sisters or Seven Virgins and are generally taken to be the seven daughters of Atlas and Pleione: Alcyone, Asterope, Celaeno, Electra, Maia, Merope and Taygeta, all half-sisters to the Hyades. According to one version of the myth, these beautiful nymphs were pursued by the hunter Orion through the woods of Boeotia for five years until Zeus translated them all, including Orion's dog, into the stars.

One of the seven is always faint and this has been taken to be either *Electra*, mourning for the fall of Troy or *Merope*, ashamed of having married the mortal, Sisyphus; but in fact, the six visible to the naked eye are: *Alcyone, Atlas, Electra, Maia, Merope* and *Taygeta*.

Ovid referred to the fact that only six of the seven sisters were visible, in the well known line, 'Quae septem dici, sex tamen esse solent.'

Both Ptolemy and Al-Sufi gave positions for only four stars but strangely these do not include *Alcyone* which is the brightest of the group and it is possible that this star has become much brighter in comparatively recent times.

Other names have been given to the group at various times: some Greek writers referred to them as 'A bunch of grapes'; the Arabs knew them as 'Ath-thurayya' (The Crowd) or 'The Little Ones'. They are called in French 'Poussinière', in German, 'Gluckhenne', in Italian, 'Le Gallinelle' and in Spanish, 'Las Gabrillas' or little nanny goats. In English, they have also been known as 'The Hen with Chicks'. The Pythagoreans referred to them as 'The Lyre of the Muses'.

Although six is the number generally seen by the naked eye alone, the other three named members, *Pleione, Asterope* and *Calaeno* are all brighter than mag. 6 and can be seen without optical aid. *Pleione*, however, is only about five minutes of arc away from her bright consort, *Atlas* and they may be difficult to separate.

Various claims for super-perception among the Pleiades have been made from time to time. Miss Airy, daughter of one of Britain's Astronomers Royal, could pick out 12, while Möstlin, a contemporary of Kepler, could see 14 and this number was claimed in more recent times by Carrington and Denning. The record in this competition, however, is attributed to Carl von Littrow, who claimed to have seen 16.

With the invention of the telescope, many more stars could be counted; Galileo made out 36; Hodierna, 37; De la Hire, 64; Hooke, 78; Jeurat, 103; de Rheita, 188 and Wolf 499 down to mag. 14.

The Pleiades were included as the last item, No. 46, in Messier's first list of nebulae and clusters, published in 1771.

In about 1846, Mädler, a German astronomer working at Dorpat Observatory, noticed that the stars of the Pleiades had no motion relative to one another and rather rashly assumed that this represented the motionless centre of the whole stellar system which he fixed upon the star Alcyone. This hypothesis had a considerable vogue for a while but Wilhelm Struve considered the idea 'much too hazardous' and most of the astronomers of the time abandoned it.

The common proper motion of the Pleiades stars did, however, lead to the notion that they all belonged to a group and were moving together through space.

On Oct. 19th, 1859, Tempel, a German astronomer working in Venice, detected in a 4-inch telescope, 'A faint stain, like breath on a mirror, about 35′×20′, extending south from the star *Merope*. He first thought it was a new comet but on the next night it appeared in the same position. It was observed by Tempel and Pape on Dec. 31st, 1860 and later confirmed by Auwers who considered it to be triangular and 15′ in extent.

Lewis Swift, using only 2 inches aperture and a magnification of 25× claimed to have seen the nebulosity in 1874 when he, also, mistook it for a comet. At the same time, Burnham at Chicago, with an 18-inch was unable to see it and asserted that it was non-existent. Schiapatrelli in Milan, testing a new telescope on Feb. 25th, 1875, saw it clearly and was surprised at its size as it appeared to extend from *Merope* beyond *Electra* and as far as *Calaeno*. Maxwell Hall estimated its size as 45′×30′ when using a 4-inch refractor in Jamaica.

As the nebulosity was often invisible in quite large telescopes, it was thought that it was variable in brightness but photographs, first taken by the brothers Paul and Prosper Henry in Paris in 1885, showed that it enveloped most of the cluster and was a permanent feature. Isaac Roberts also photographed the nebula on one of his first plates taken in 1886.

MESSIER   (Mar. 4th, 1769) 'A cluster of stars, known by the name of the Pleiades. The position reported is that of the star Alcyone.'

WEBB   'Nebula seen in a 2-inch but invisible in 11-inch.'

D'ARREST   'Here are nebulae, invisible or barely seen in great telescopes which can easily be seen in their finders.'

The Pleiades are known to compose a comparatively young cluster of very white stars; 300 to 500 of them contained in a sphere of about 50 light-years diameter and embedded in nebulous matter which shines by the reflected light of the stars within it – a reflection nebula. (Portions of the nebulosity include NGC 1435 and IC 349.)

The density of the stars in the centre of the cluster is estimated at about 2.8 stars per cubic parsec: this is very low compared with M 11 with 83 per cubic parsec or even M 36 with 12 per cubic parsec. With this low concentration it is not expected that it will have a very long existence as a cluster. According to Baade, it has been in existence for about 20 million years only and has a total life-expectancy of perhaps no more than 1000 million years.

Many of the stars in the Pleiades have been found to be in rapid rotation and *Pleione* itself is one of the most rapidly rotating stars known, spinning on its axis about a hundred times as fast as does the Sun.

This rotation can be detected by spectroscopic observations when the absorption lines appear broad and diffuse. Theory would show that a star like *Pleione* in so rapid rotation should become unstable and would tend to shed some of its material, producing an equatorial ring or partial shell.

This was, in fact, discovered in 1938 by D. B. McLaughlin of Michigan and O. C. Mohler of Cook Observatory, Philadelphia and came to be known as *Pleione's shell episode.*

The intensity of this spectroscopic phenomenon gradually grew until the shell reached its maximum density in 1945 and 1946. The effect then gradually declined and by the end of 1951, it had disappeared: this was believed to be due to the supply of atoms which formed the shell, decreasing from below, while radiation pressure accelerated the shell material out into space until it was no longer detectable.

It is possible that such rapidly rotating stars may lose some appreciable part of their material by successive shedding from their equatorial zones.

The Pleiades also contains a number of 'flare' stars similar to those found in M 8 and M 42 and radio emission on the 21-cm hydrogen band seems to show some interaction between the stars and the interstellar gas.

Trumpler's classification for M 45 is II,3,r and the spectrum of the brightest star, B5. Wallenquist calculates the linear diameter to be 6.8 parsecs or about 22 light-years while Becvar gives 8.4 parsecs which is equivalent to about $27\frac{1}{2}$ light-years.

The nebulosity surrounding the star Merope was first discovered by Ernst Tempel with a 4-inch refractor at Vienna Observatory in 1859, when he compared it to 'a stain of breath upon a mirror'. The nebulosity was seen to extend to *Alcyone* in 1875 and to *Electra, Calaeno* and *Taygeta* by 1880. The full extent and complexity of the nebulous matter, however, was revealed by the first astro-cameras, notably those of the brothers Henry in Paris and Isaac Roberts in England between 1885 and 1888.

Pleione is variable, ranging from 4.77 to 5.50 mag. (For observing details see next page.)

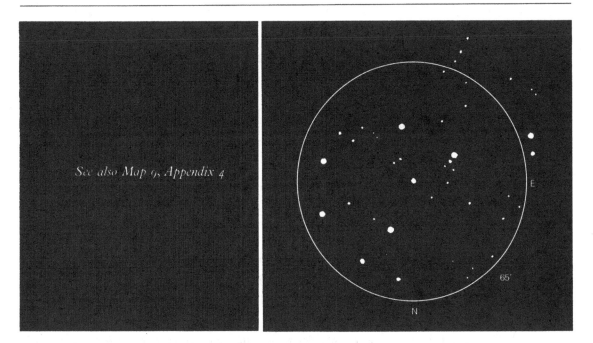

See also Map 9, Appendix 4

To observe   The Pleiades are best observed with a field of view of not less than 2°: this requires a telescope of about f.4 or less and a rich-field instrument of this type would very probably show the nebulosity in a dark, transparent sky. Failing this, a good pair of field glasses on a steady mounting would provide excellent views of this splendid cluster.

In the diagram of the Pleiades (Appendix 4, Map 9) an area of 2° square, centred on *Alcyone* is shown, containing 30 of the brightest stars, although two of them – marked with an X – have a different proper motion and are not true members of the cluster. Another eight members in the same range of brightness as the ones illustrated, lie outside the area. (The total number of stars in the cluster is not exactly known but, in 1958, Johnson and Mitchell at Lowell Observatory determined the colours and magnitudes for 262 members brighter than visual magnitude 16 with another 27 probable members.)

The double stars illustrated are:

A.D.S. 2748 with 11 mag. companion, P.A. 339° distance 3".7
A.D.S. 2767 with 10 mag. companion, P.A. 266° distance 5".6
A.D.S. 2795 with 10 mag. companion, P.A. 239° distance 3".1 ⎤
    and 11 mag. companion, P.A. 236° distance 10".0 ⎦ Triple

The nebulosity surrounding *Merope* can often be elusive, especially in Newtonians of long focus. However, a good Rich Field Telescope (RFT) in dark skies should show it clearly. Walter Scott Houston, using an 8-inch reflector in southern Arizona found the field 'laced edge to edge with bright wreaths of delicately structured nebulosity'.

171

M 46 NGC 2437

A galactic cluster, type 'f' in ARGO PUPPIS
R.A. 07 h 39.5 m.   Dec. S 14° 42' (1950).
Size 30' diam.   Mag. 6.0   Distance 980 pc, 3200 lt-yrs.

Discovered by Messier in 1771.

MESSIER   (Feb. 19th, 1771) 'A cluster of very small stars . . . The stars cannot be distinguished except in a good telescope. The cluster contains a little nebulosity.'

J. H.   'A superb cluster of stars 12–15 mag.'

D'ARREST   'Very large, many multiple stars in groups. Planetary nebula near the northern border.'

SMYTH   'A double star, A, 8½ mag., B, 11 mag. Both pale white, P.A. 90° d. 15" (1836). A noble though loose assemblage of stars 8 to 13 mag. Most compressed trending S.f. & N.p.'

WEBB   'Beautiful circular cloud of small stars about ½° diam.'

LICK XIII   'A very large, bright, sparse cluster about 25' diam. in which is involved the planetary nebula 2438.'

M 46 has been found by James Cuffey of Kirkwood Observatory, to contain a number of extremely blue stars of type A0. The cluster shows no evidence of absorption due to obscuring matter but a few of the stars, when examined spectroscopically, show absorption lines which are broadened due to rotation, a phenomenon which is well marked in the Pleiades.

The radial velocity of the member stars of M 46 was measured by Baade and found to be +41.4 km/s. The planetary nebula (NGC 2438) which can be seen in the cluster has also been subjected to radial velocity measurement and the result for this object was +77 km/s. This is so different from that of the cluster itself that it makes it unlikely that the planetary nebula is an actual member.

In Trumpler's classification the cluster is coded II,2,r and he gives the brightest star a spectral type of B9.

Wallenquist counted 197 stars as probable members in a diameter of 28' and estimates the density of stars to be 9.25 stars per cubic parsec in the centre and 0.74 stars per cubic parsec for the cluster as a whole. Wallenquist gives the cluster's linear diameter as 8.0 parsecs or about 26 light-years and the distance as 980 parsecs, while Antonin Becvar quotes the diameter as 12.8 parsecs or about 41½ light-years with the distance as 1820 parsecs. (For observing details see next page.)

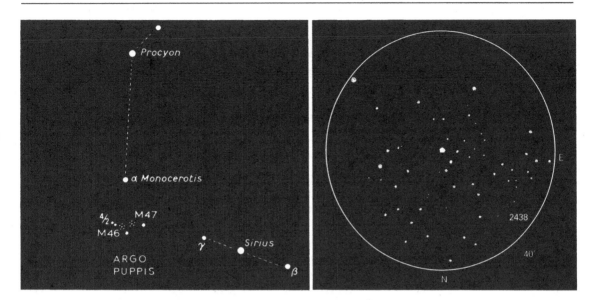

To observe   From Sirius move N about $2\frac{1}{2}°$ and then sweep E for 14° or until due S of *Procyon* and α *Monocerotis*.

If the sky is not completely clear or dark, M 46 may be so faint as to be almost invisible: the position may then be checked by moving back to the W a little less than 2° to pick up M 47 which is brighter and much more easily seen.

When atmospheric conditions are suitable, M 46 is a fine sight, being at least 30' in diameter and very rich in faint stars which are condensed in the central area of the cluster. The brighter stars stretch away to the N and E and there is a solitary 8 mag. red star about 20' away to the SW.

The planetary nebula, NGC 2438 (39 H. IV) lies within the boundary of the cluster about 10' NE of the centre and being fairly large – about 1' diam. – and diffuse, it needs the very best conditions and a good aperture to see well. Its annular form can just be made out in an 8-inch as it darkens slightly in the centre. On the SE edge is a faint but distinct star of about mag. 12. The nebula has a central star but this is mag. 16 and would need a large aperture to see at all.

173

M 47   NGC 2422 (38 H. VIII)
A galactic cluster, type 'd' in ARGO PUPPIS
R.A. 07 h 34.3 m.   Dec. S 14° 22' (1950).
Size 25' diam.   Mag. 5.2   Distance 548 pc, 1750 lt-yrs.

One of the 'missing' Messier Objects and here identified as NGC 2422.
Discovered by Hodierna before 1654. 'A Nebulosa between the two dogs.'

MESSIER   (Feb. 19th, 1771) 'A cluster of stars a little removed from the preceding (M 46), the stars larger. The cluster contains no nebulosity.'

SMYTH   (under 38 H. VIII) 'A double star, A, 7½ mag.; B, 8 mag. both bluish-white, P.A. 304°, d. 7".3 (1878) in a loose cluster of the Milky Way. A very splendid field of large and small stars disposed somewhat in a lozenge shape and preceded by a 7 mag. with a companion about 20" N.f. of it. 'A' is known as *34 Officinae Typographicae* of Bode.'*

WEBB   (under NGC 2422) 'Grand broad group visible to naked eye: too large even for 64×. Some brilliant 5 or 6 mag. stars including Σ 1121. About 2½° f. is a group around 4–5 mag. – a fiery 5 mag. leads region.' (This latter group obviously refers to NGC 2437 (M 46).)

M 47 is classed by R. J. Trumpler as his type II,3,m and he also gives the spectral class of the brightest star as B3.

In Helen Sawyer Hogg's catalogue of clusters in the *Handbuch der Physik*, it is quoted as having an apparent diameter of 30', an integrated magnitude of 4.3 and a magnitude of the brightest star of 9.8.

Dr Åke Wallenquist, in his survey of the *Structural Properties of Galactic Clusters* of 1959, reports 43 stars as probable cluster members in a diameter of 32' and the mean magnitude of the five brightest stars as 7.2.

In his estimate of the density of the cluster he gives 16.0 stars per cubic parsec for the central portion and 0.62 stars per cubic parsec for the whole cluster.

Wallenquist's figure for the linear diameter of the cluster is 5.1 parsecs or about 17 light-years: the distance of the cluster he calculates as 548 parsecs.

In Antonin Becvar's *Atlas Coeli Katalog*, the linear diameter is put at 8.4 parsecs or more than 27 light-years. Becvar's figure for the distance of M 47 is also larger, being as great as 1150 parsecs. (For observing details see next page.)

---

*   This constellation, 'The Printing Office', was one of J. E. Bode's introductions and was formed out of stars in Canis Major and Puppis. It was still in use on some charts until 1878.

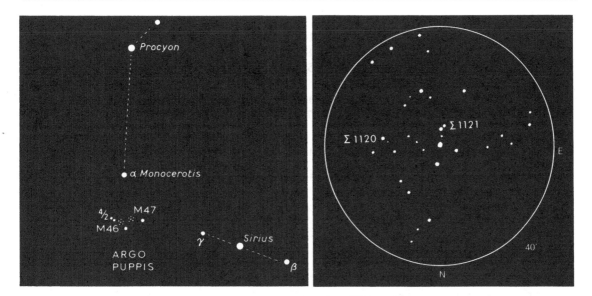

To observe   M 47 is a bright, open cluster, much more easily seen than its neighbour, M 46. In *Norton's Star Atlas* M 47 is not marked but the position is shown for 38⁸, together with the double star Σ 1121 and this is the cluster we are looking for.

From *Sirius* move N a little more than 2° and sweep to the E for 12° or 50 min. of R.A. until almost due S of *Procyon*. M 47 will be found about ½° f. a bright 5 mag. star.

M 47 contains relatively few bright stars but they are mostly very bright and form a pattern which somewhat resembles the constellation of Orion, including the stars of the 'Belt', here twisted into a N–S alignment. Two distinct arcs of three stars lie to the SW and near the centre of the cluster is the double star mentioned by Smyth, Σ 1121, P.A. about 30° and d. about 8″, both white.

On the western edge is another pair, Σ 1120, A, 5.6 mag., B, 9.5 mag., P.A. 35°, d. 20″ (1925). This is not quite so easy to distinguish as the *comes* is rather faint, although well separated.

M 47 is a brilliant and impressive cluster and full of interest. It is visible to the naked eye.

Objects near by   About 40′ to the north and slightly east, and connected to M 47 by a string of five 8 mag. stars, is a small cluster, NGC 2423. It is rather sparse and poor, containing about 20 9–10 mag. stars and many fainter still, but is well worth looking for.

M 48   NGC 2548 (22 H. VI)
A galactic cluster, type 'f' in HYDRA
R.A. 8 h 11.2 m.   Dec. S 5° 38' (1950).
Size 35' diam.   Mag. 5.5   Distance 480 pc, 1500 lt-yrs.

One of the 'missing' Messier Objects and here identified with NGC 2548.
Discovered by Messier in 1771.

MESSIER   (Feb. 19th, 1771) 'Cluster of very small stars without nebulosity. This cluster is a short distance from the 3 stars which form the beginning of the Unicorn's tail.'

SMYTH   (Description of 22 H. VI) 'A neat but minute double star in a tolerably compressed cluster. A, mag. 9½ and B, mag. 10: both white; P.A. 206°, dist. 7" (1880). A splendid group in a splashy region of stragglers which fills the field of view and has several small pairs chiefly of 9th mag.'

WEBB   (Description of NGC 2548) 'Group of pretty uniform 9th mag. stars with a profusion of lesser ones.'

ROSSE   'Riddled with dark lanes and openings.'

NGC   'A cluster, very large, pretty rich, pretty much compressed, stars 9 to 13 mag.' Observed also by Caroline Herschel in 1783.

In Trumpler's classification, M 48 is designated I,2,r and he also lists the spectral class of the brightest star at A0.

Helen Sawyer Hogg lists the apparent diameter as 30', the integrated magnitude of the cluster as 5.5 mag. and the magnitude of the brightest star as 9.7.

Wallenquist, in his survey of the properties of galactic clusters, reports a total of 56 probable member stars in a diameter of 34' and gives the density of the cluster as 16.9 stars per cubic parsec in the central area and 0.99 stars per cubic parsec for the cluster as a whole.

Wallenquist gives 4.8 parsecs for the linear diameter of the cluster; Doig, in 1926, made it 6.4 parsecs while Becvar gives 8.3 parsecs.

Hogg quotes 500 parsecs for the distance of M 48, Wallenquist gives 481 parsecs while Becvar's figure is much greater at 950 parsecs. (For observing details see next page.)

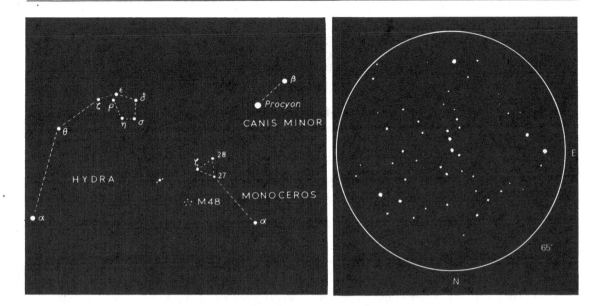

To observe   This cluster is a little difficult to find as there are many small aggregations of stars in the neighbourhood which look like clusters.

The position given in *Norton's Star Atlas* for 48^M is Messier's original position corrected for precession to 1950 but there is no recognisable cluster in this position. The position of NGC 2548 is marked in *Norton* as 26^6 (Wm Herschel's No.) and it is this with which we are now dealing.

Extend the line β *Canis Minoris-Procyon* to the SE for 3 times that interval and pick up the easily recognisable group of 3 close-in-line stars *1, 2* and *c Hydrae*.

M 48 is 3° W and 2° S of this group.

When found, the pattern of M 48 is easily recognised as it forms a distinct, almost equilateral triangle with the three brightest stars at the corners. Many of the fainter stars are grouped in small pairs. The surrounding background is fairly dark compared with the neighbouring star-groups and the cluster looks quite separate and self-contained.

It is visible to the naked eye in favourable conditions.

M 49 NGC 4472
An elliptical galaxy, type E4 in VIRGO
R.A. 12 h 27.2 m.   Dec. N 08° 16' (1950).
Size $4\frac{1}{2}'\times4'$ Mag. 8.6   Distance 12.5 Mpc, 41 million lt-yrs.

Discovered by Messier in 1771.

MESSIER — (Feb. 19th, 1771). 'Nebula discovered near ϱ Virginis. Seen only with difficulty in a $3\frac{1}{2}$-foot telescope. The comet of 1779 was compared by M. Messier with this nebula on 22nd and 23rd April. The comet and the nebula had the same light. M. Messier reported this on the chart of the track of the comet. Reviewed 10th April, 1781.'

ORIANI — (April 22nd, 1779) 'Very pale and looking exactly like the comet.'

SMYTH — 'A bright, round and well-defined nebula. Nebula has a pearly aspect.'

D'ARREST — 'Innumerable groups of stars: on 147× resolved into stars of 13–14 mag. near the edges.'

WEBB — 'Bright: faint haze in beautiful position between two 6-mag. stars. Bright open pair S 3".7.'

LICK XIII — 'The very bright nucleus is not stellar; shows well in a 3 min. exposure. Nearly round, 2' in diam. fading out rapidly towards the edges. No structure discernible, though spiral character is suspected near the centre in the short exposures.'

M 49 is one of the giants among the elliptical galaxies; it is larger and more massive than M 87 and comparable with M 60 in mass but larger in linear dimensions. Its mass, given in Erik Holmberg's *Catalogue of External Galaxies*, is put at 1 000 000 million Suns and its diameter about 42.7 kpc or about 140 000 light-years.

M 49 is not quite as intrinsically luminous as M 87, its integrated absolute photographic magnitude being −21.4, about the same as the Andromeda Nebula, M 31.

This galaxy seems to be low in high-luminosity blue giant stars. It has an integrated spectrum of Type G7 and colour index of +0.76, which makes it yellower than most ellipticals. It is a member of the Virgo Group.

A probable (unconfirmed) supernova (1969 Q) of magnitude 13.0 was reported in this galaxy in June 1969. (For observing details see next page.)

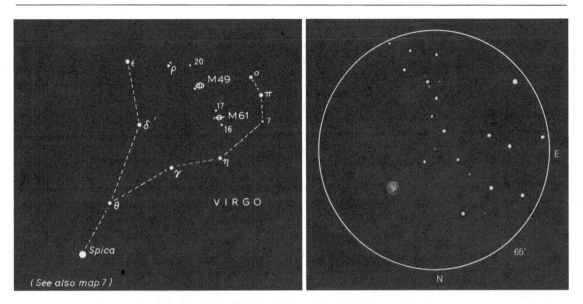

To find   See *Guide to Virgo Group* (p. 387).

M 49 is one of the brighter of the Virgo Group galaxies and is easily picked out.

To observe   It appears as a bright, round, glowing patch about 5′ in diameter and has a clear outline. It has a round, distinctly brighter central area, giving it the distinctive 'pearly' appearance noted by Smyth and this is typical of many of the elliptical galaxies of types E0 to E2.

There is a zig-zag chain of stars extending to the south and a bright 6th mag. star dominates the field about 40′ to the SE.

Objects near by   A little over one degree to the east, and slightly south, is NGC 4526, an E7 elliptical galaxy, bracketed by two 7 mag. stars.

179

M 50 NGC 2323

A galactic cluster, type 'e' in MONOCEROS
R.A. 07 h 00.6' m.   Dec. S 08° 16' (1950).
Size 15'×20'   Mag. 6.3   Distance 910 pc, 2950 lt-yrs.

Possibly discovered by G. D. Cassini before 1711.

MESSIER   (April 5th, 1772) 'Cluster of small stars, more or less brilliant. . . . It was while observing the comet of 1772 that M. Messier observed this cluster. He reported it on the chart of that comet.'

BODE   (No. 16, Dec. 2nd, 1774) 'While looking for the nebula which Cassini reported between Canis Major and Minor, I found finally, under the belly of Monoceros, a small star cluster against a nebulous background with four small stars to the west.'

J. H.   'A remarkable cluster, very large and rich, with stars 12–16 mag.'

D'ARREST   'Cluster of small stars, very large, contains a red star close to its edge.'

SMYTH   'A delicate and close double, A, 8 mag. and B, 13 mag, both pale white; P.A. 170°, d. 5" (1833). An irregularly round and very rich mass of stars 8 to 16 mag. A red star on S verge and a pretty little equilateral triangle of 10' just N of it.'

WEBB   'Brilliant cluster, straggling and containing a red star. In a superb neighbourhood.'

J. E. GORE   'Roberts' photo of March 1893 shows it to be not very rich; probably not more than 200 stars in main body.'

In J. R. Trumpler's classification, M 50 is coded as I,2,m. The brightest star, of magnitude 9.0, has a spectral class of B8.

In Helen Sawyer Hogg's catalogue of clusters, the apparent diameter is given as 16' and the integrated magnitude of the cluster is estimated to be 7.2.

Wallenquist counted 50 stars as probable members in a diameter of 14' and calculates the density of stars in the cluster to be 5.13 stars per cubic parsec in the centre and 1.03 stars per cubic parsec overall.

The linear diameter of the cluster is given as 4.5 parsecs by Wallenquist and 3.7 parsecs by Becvar.

Hogg quotes the distance of the cluster as 7000 parsecs; Becvar, 800 parsecs; Dr R. H. Garstang, in his list of the Messier Objects in the 1964 *Handbook* of the British Astronomical Association, gives 910 parsecs and Wallenquist gives 1110 parsecs.

M 50 is considered to be a cluster of intermediate age, and recent observations show that many of the member stars show some degree of reddening which may be due to a small, localised obscuration seen near the cluster. (For observing details see next page.)

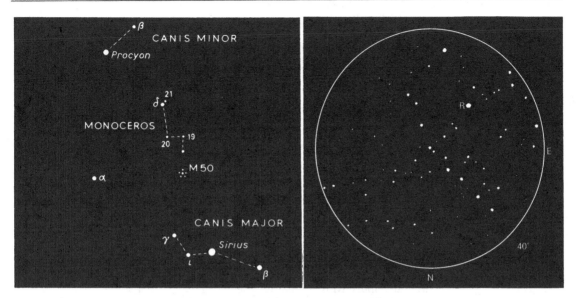

To observe From *Sirius* move E to ι and γ *Canis Majoris* (4th mag.) M 50 is 7° N of γ *Canis Majoris, or,* on the line from *Sirius* to β *Canis Minoris,* M 50 lies a little more than one-third of the distance from *Sirius.*

It is almost exactly on the galactic equator.

M 50 appears as a bright, dense concentration in the Milky Way. Although this area, like much of the Milky Way, is rich in groups of faint stars, M 50 stands out well as a separate cluster by virtue of having a ring of slightly darker sky surrounding it.

Stragglers from the cluster form a branching 'Y' shape, one of the branches being terminated by the red star noted by both Smyth and Webb.

Smyth's double is a bit uncertain.

(It is possible that M 50 is the object described by Jacques Cassini in his *Elements of Astronomy* as having been discovered by his father, G. D. Cassini, 'in the area between *Canis Major* and *Canis Minor*'. Messier stated at the end of his catalogue that he had failed to find the nebula in this region and supposed it to have been a faint, passing comet. That, however, was written before 1771 and Messier did not observe M 50 until 1772. Bode's remarks seem to imply this identification.)

M 51 NGC 5194 and 5195
The 'Whirlpool' Nebula. A spiral galaxy, type Sc in CANES VENATICI
R.A. 13 h 27.8 m.   Dec. N 47° 27' (1950).
Size 12'×6'   Mag. 8.1   Distance 11.5 Mpc, 37 million lt-yrs.

Messier discovered NGC 5194 in 1773 and Méchain saw the secondary nebula, 5195, in 1781.

MESSIER   (Jan. 11th, 1774) 'Very faint nebula without stars ... M. Messier discovered this nebula on the 13th of Oct. 1773, while observing the comet which appeared in that year. Seen only with difficulty in a 3½-foot telescope. Reported on the chart of the comet of 1773–74. It is double, each having a brilliant centre, 4' 33" apart. The two atmospheres touching; one fainter than the other. Reviewed several times.'

BODE   (No. 25, Jan. 15th, 1774) 'A small, faint, luminous nebula, possibly of an oblong shape.'

J. H.   Thought it a system similar to our own, the halo representing the Galaxy.

SMYTH   'A pair of lucid white nebulae, each with an apparent nucleus with their nebulosity running into each other. The southern object is truly singular, having a bright centre surrounded with luminosity, resembling the ghost of Saturn with his rings in a vertical position. A stellar universe, similar to that to which we belong, whose vast amplitudes are in no doubt peopled with countless numbers of percipient beings.'

ROSSE   (Spring 1845) 'Spiral convolutions; ... with successive increase of optical power, the structure has become more complicated ...' 'The connection of the companion with the greater nebula is not to be doubted ... the most conspicuous of the spiral class.' (1861) 'The outer nucleus unquestionably spiral with a twist to the left.'

MITCHEL   (*Stellar Worlds* (1869), pp. 256–61, Wm Collins, Glasgow) 'The curious spiral form

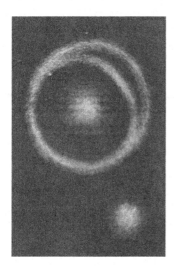  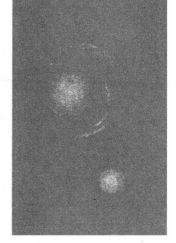

M 51 (John Herschel) From G. F. Chambers' Handbook, *1889 (B.A.A.)*   *M 51 (Rosse) The first represen- tation of a spiral nebula (B.A.A.)*   *M 51 (Smyth) From* A Cycle of Celestial Objects *(B.A.A.)*

is exhibited with great beauty, and seems to indicate the action of some powerful and controlling law.'

D'ARREST   'Noted nebula in Canes Venatici. Nucleus "A" very large and bright with a double annulus which is brighter in P.A. 37° and 98°. Nucleus "B" large, brighter and denser towards the centre.'

WEBB   'One of the Earl of Rosse's wonderful spirals: its wreaths of stars are beyond all but the finest telescopes: common ones will show two very unequal nebulae in contact, both brightening in centre. Traces of the halo encompassing the larger may, perhaps, be caught. 9.3-inch speculum showed plainly outer end of spiral and junction with smaller nebula. A misty spot in finder.'

ROBERTS   Photos in April 1889 and May 1896. 'Both nuclei stellar, surrounded by dense nebulosity and the convolutions broken up into star-like condensations with nebulosity around them.'

FLAMMARION   'In the margin of Messier's manuscript of his catalogue, there is a note in his handwriting with a little sketch. "M. Méchain saw this nebula double – March 20th, 1781, saw this nebula; effectively it is double. The centre of each is brilliant and clear; distinct and the light of each touches each other."'

LICK XIII   'Including very faint matter to the N of 5194, it covers an area about 12'×6' in P.A. approx. 30°. A sharp stellar nucleus in 5194 and the whorls show a multitude of stellar condensations. The satellite nebula, 5195, has a bright, elongated nucleus, its nebulosity is of a more diffuse type, without discernible spiral structure and with several rifts which suggest absorption effects.'

M 51 is a member of a small group of galaxies and is thought to be about one-quarter to one-half the size of our own Galaxy.

The spiral arms, of which four concentric coils can be traced, contain many super-giant stars and the irregular companion, NGC 5195 is connected to the main galaxy by an extension of one of the spiral arms, indicating some kind of tidal action between the two.

A recession velocity of 485 km/s has been measured for this galaxy, of which about 160 km/s is due to our own movement away from M 51; giving a true recession of about 325 km/s.

Holmberg, in his *Catalogue of External Galaxies*, gives the mass of M 51 as 160 000 million Suns, its diameter as 38.0 kpc or 124 000 light-years. Its absolute magnitude is −21.6, spectral class F8 and colour index +0.36. The mass-density works out to be about 0.025 suns per cubic parsec or one solar mass per 40 cubic parsecs.

The companion galaxy, NGC 5195, has an absolute magnitude of −20.1 and although smaller than its neighbour, appears to be more massive. Its linear diameter is estimated to be 23.4 kpc, and its mass, about 316 000 million suns.

The distance of the two galaxies, based on photometric methods, is about 11.5 Mpc but by measurements of the observed red shift, the distance is rather less at about 7¼ Mpc.

M 51 has been observed by radio astronomers to be surrounded by a large halo of hydrogen which emits radiation in the 21-cm band. A similar halo has been detected in M 81.

183

In 1971, using the Westerbork aperture–synthesis array at Leiden Observatory, radio-astronomers were able to detect two distinct 'radio' spiral arms in the structure of M 51. These were not coincident with the bright visual features but positioned along their inside edges. These arms produced more than half the total radio emission from the galaxy, the remainder appearing to emanate from the nucleus.

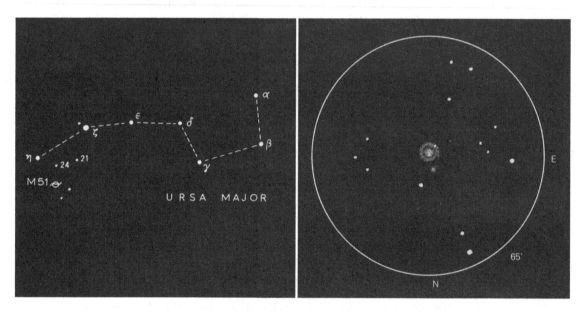

To observe   M 51, although not particularly faint can often be elusive. It lies about $3\frac{1}{2}°$ S.p. *Benetnash* (η *Ursae Majoris*) at the end of the tail of the Great Bear. The safest approach is first to move about 2° p. from *Benetnash* to 5 mag. *24 Canem Venaticorum* and then S 2° and a little p. When found, it is quite unmistakable and a unique sight.

Low power should be used at first as one needs all the light possible for this object which appears as two glowing patches, almost touching. The larger, to the south, is about 8′×6′ while the smaller is about 10′ to the north and only about $1\frac{1}{2}$′ diam. Both have very small, star-like centres.

In an 8-inch there is a faint trace of the connecting arm between the two which can be seen on the east side, and the 'halo' around the larger nebula is just visible. With averted vision, some of the brighter knots in the spiral arms may be made out in glimpses.

M 51 does not stand high magnification very well for, although one may obtain a better separation of the two portions of the nebula, detail may not necessarily be improved unless the seeing and the optics are perfect.

The 'Whirlpool' is a good description of M 51 which shows something of its spiral structure even in apertures as small as 8 inches.

Objects near by   There is a very faint elliptical galaxy, NGC 5198, about 30″ in diameter about 30′ to the south of M 51.

184

M 52 NGC 7654

A galactic cluster, type 'e' in CASSIOPEIA
R.A. 23 h 22.0 m.   Dec. N 61° 19' (1950).
Size 20'×12'   Mag. 7.3   Distance 925 pc, 3000 lt-yrs.

Discovered by Messier in 1774.

MESSIER   (Sept. 7th, 1774) 'A cluster of very small stars mingled with nebulosity which can be seen only with an achromatic telescope. M. Messier saw this cluster close to the comet of 1774 on the 7th of Sept. of that year. It was below the star δ Cassiopeiae and that star was used to determine the position of both comet and cluster.'

J. H.   'Large, rich, round and much compressed, with stars 9–13 mag.'

SMYTH   'An irregular cluster of stars. This object assumes somewhat of a triangular form with an orange-tinted 8 mag. star at the vertex, giving the resemblance of a bird with out-stretched wings. It is preceded by two stars of 7–8 mag. followed by another of similar mag. and the field is one of singular beauty on M.P.'

WEBB   'Irregular, with orange star, as is frequently the case.'

ROSSE   'Contains about 200 stars.'

LICK XIII   'A very sparse, open cluster, 16' diam. of stars 12–16 mag.'

From observations of the types of stars in this cluster and their evolutionary sequence, its age has been estimated to be approximately the same as that of the Pleiades. This means that it is a comparatively young cluster, about 20 million years old.

The distance of the cluster has been measured by several different observers and they are all in fairly close agreement. The latest figure is that obtained by Petch of Yerkes Observatory in 1960 who gives a true distance modulus (i.e. corrected for space absorption) of 11.1 mag. which is equivalent to 1660 parsecs or about 5400 light-years.

M 52, according to R. J. Trumpler's classification, is coded II,2,r. The brightest star, with a magnitude of 11.0, has a spectrum of B7.

Hogg gives the apparent diameter as 13' and the integrated magnitude of the cluster as 8.2.

Wallenquist counted 193 probable member stars in a diameter of 18' and gives the density of the cluster as 55.7 stars per cubic parsec near the centre and an average density for the cluster as a whole, 3.24 stars per cubic parsec. This high overall density is only exceeded, among the galactic clusters in Messier's list, by M 67 and M 11.

The linear diameter of the cluster is 4.8 parsecs or 16 light-years, according to Wallenquist and 4.1 parsecs or 13½ light-years in Becvar's *Katalog*.

Hogg makes the distance 700 parsecs; Wallenquist, 924 parsecs; Becvar, 1170 parsecs; while Garstang gives 2100 parsecs in the B.A.A. 1964 *Handbook*. (For observing details, see next page.)

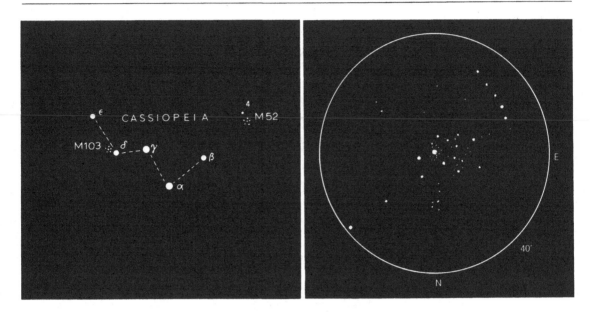

To observe   Extend the line α–β *Cassiopeiae* for about $6\frac{1}{2}°$ to the NW to the 5 mag. star *4 Cassiopeiae*. M 52 is about 1° S and slightly W of *4 Cassiopeiae*.

It is a small cluster, containing a few concentrations of faint stars with several brighter members of 9–10 mag. and one slightly orange star of about mag. 8.

As Smyth noted, it has a definite triangular shape and, including the addition of two bright stars in line to the NW, a little imagination reveals a figure quite like a slim arrow-head pointing to the SE. The 'bird with out-stretched wings' eluded me.

It is not a conspicuous object but there are several pleasing groups of stars in the near vicinity. It is almost on the galactic equator.

Objects near by   About 35' to the SW is a planetary nebula, NGC 7635. This consists of a large, faint, diffuse oval, about $3\frac{1}{2}' \times 3'$ surrounding an $8\frac{1}{2}$ mag. star. The nebula, popularly known as the 'Bubble', is quite difficult to see because of its low surface brightness.

M 53 NGC 5024
A globular cluster, class V in COMA BERENICES
R.A. 13 h 10.5 m.   Dec. N 18° 26' (1950).
Size 10' diam.   Mag. 7.6   Distance 21 kpc, 69 000 lt-yrs.

Discovered by Bode in 1775.

BODE (No. 26) 3rd Feb. 1775. '1° E of star 42 Comae, a new nebula, appearing through the telescope as round and pretty lively.'

MESSIER (Feb. 26th, 1777) 'Nebula without stars discovered in Coma Berenices, a little distance from the star 42 in that constellation, after Flamsteed. This nebula is round and conspicuous. The comet of 1779 was compared directly with this nebula and M. Messier has reported it on the chart of that comet. Reviewed the 13th April, 1781: it resembles the nebula which is below Lepus.' (M 79)

W. H. 'One of the most beautiful sights I remember to have seen in the heavens. The cluster appears under the form of a solid ball, consisting of small stars quite compressed into a blaze of light with a great number of loose ones surrounding it and distinctly visible in the general mass. Similar in appearance to M 10.'

J. H. 'Curved appendages of stars, like the short claws of a crab, running out from the main body.'

SMYTH 'A mass of minute stars 11–15 mag. and from thence to gleams of star-dust, with stragglers to N.p. and pretty diffuse edges.'

ROSSE 'Not compressed to one point but apparently to four or five different points within a small area, diameter 3'.'

WEBB 'Brilliant mass of minute stars, blazing in centre.'

LICK XIII 'A fine bright globular cluster 10' in diameter.'

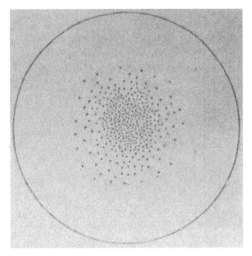

M 53 (William Herschel)
From Phil. Trans. *of the Royal*
Society (R. A. S.)

Although this globular cluster looks perfectly round to the eye, Shapley found from photographs that it has an ellipticity of 9 in P.A. 170°.

M 53 was also studied by Baade, who estimated that it contained about 50 cluster-type variable stars. More recently, in 1958, Dr James Cuffey of Goethe Link Observatory, found 46 variables in the cluster and obtained a distance modulus of 16.9 mag. which is equivalent to a distance of 24 kpc. He also observed that, in its composition of different star-types, it resembled the globular cluster, M 3 in Canes Venatici.

In Helen Sawyer Hogg's catalogue of clusters in the *Handbuch der Physik*, M 53 is given an apparent diameter of 14'.4, an integrated magnitude of 8.68, a spectrum of A8 and colour index − 0.08. The average magnitude of the 25 brightest stars is 15.07.

The linear diameter of the cluster is 17 parsecs or about 55 light-years, according to Becvar's *Atlas Coeli Katalog*.

Shapley gives the distance of M 53 as 18.2 kpc; Becvar and Garstang both quote 20 kpc and Kinman's figures work out at an average of about 21 kpc.

In 1964 R. Margoni of Asiago Observatory made studies of the *RR Lyrae* variables in this cluster. Some of them were found to exhibit secular variations of period. He also found one new variable, bringing the total to 47. (For observing details see next page.)

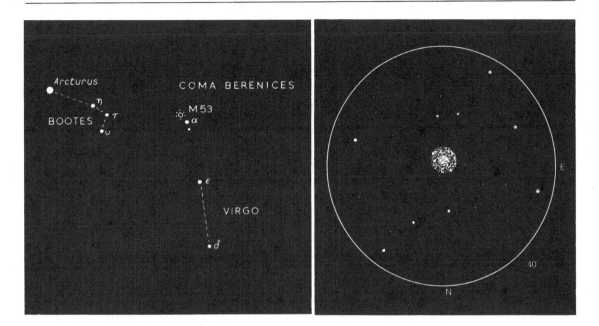

To observe M 53 is easy to find as it is only 1° NE of the 4th mag. star, α *Comae Berenicis*. α *Comae Berenicis* itself can be found by extending the line *Arcturus*-η *Bootis* for 11° to the W or by following the curve of γ–ζ–ε *Virginis* another 7° to the NNE.

Although it is one of the smaller globular clusters, M 53 is certainly one of the most beautiful. It consists of a brilliant nucleus about 2' in diameter, surrounded by a dusty halo of light that gives it a glittering, gem-like appearance. The background is dark but glimpses of numerous, faint stars can be obtained in clear conditions. The cluster is bright enough to stand a moderate magnification but it is very dense in the central region and requires a large aperture and high magnification to resolve completely.

Wm Herschel compared it to M 10 and Messier himself likened it to M 79. In Shapley's classification, it has the same concentration as the latter and the visual appearance also suggests this. From a purely aesthetic point of view, however, it surpasses both.

Objects near by 1° to the SE is another globular cluster of about the same size but, at mag. 10, a good deal fainter. This is NGC 5053.

α *Comae Berenicis* – 1° to the SW is double (Σ 1728). Both components are 5.2 mag. but with a maximum d. of only 0".7 it needs a good 8-inch aperture to split. It is a binary with a period of 25.9 years.

M 54 NGC 6715
A globular cluster, class III in SAGITTARIUS
R.A. 18 h 52.0 m.   Dec. S 30° 32′ (1950).
Size 6′ diam.   Mag. 7.3   Distance 15.1 kpc, 49 000 lt-yrs.

MESSIER (July 24th, 1778) 'Very bright nebula, discovered in Sagittarius. It is bright in the centre and contains no star, seen with an achromatic telescope of 3½ feet (F.L.). Its position has been determined from ζ Sagittarii, 3rd mag.'

J. H. 'A globular cluster, very bright, large, round, at first gradually, then suddenly much brighter in the middle; well resolved, clearly seen to consist of stars which are chiefly of 15 mag. with a few outliers of 14 mag. 2½′ diam.'

LICK XIII 'Remarkably condensed globular cluster, 2′ in diameter.'

FLAMMARION (Quoting J. H. at Cape of Good Hope, July 31st, 1834) 'Globular cluster, bright, round, gradually much brighter in the middle. Dimensions in R.A. − 9 s. Clearly resolved with the left eye.' (Flammarion's note: 'It is known that J. H.'s right eye is stronger than his left.')

Shapley observed this globular cluster to be perfectly round in outline: he detected no variable stars in it. Its diameter he took to be 2′.1 and its integrated magnitude 7.1 but Mrs H. S. Hogg, in her catalogue of clusters, gives it an apparent diameter of 5′.5 and an integrated magnitude of 8.7. Its spectrum is of type F7 and its colour index +0.01.

The linear diameter of the cluster, according to Becvar, is 17 parsecs or about 55 light-years: this is based upon a distance of 15.1 kpc which is in close agreement with T. D. Kinman's figure in his survey of the distance moduli of globular clusters.

In recent years some 82 variable stars have been found in the cluster, 55 of them being of the *RR Lyrae* type. (For observing details see next page.)

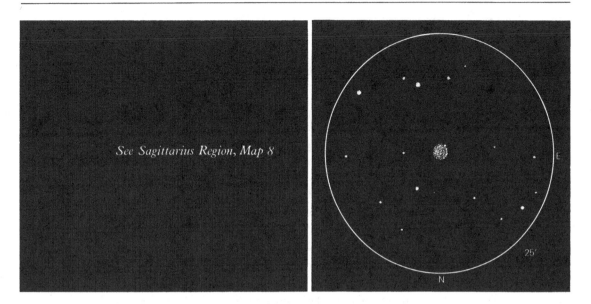

See Sagittarius Region, Map 8

To observe   M 54 is very easy to find being $\frac{1}{2}°$ S and $1\frac{1}{2}°$ W of ζ *Sagittarii* (the most southerly star in the Sagittarius 'dipper').

M 54 is bright but very small and looks almost like a planetary nebula at first sight. It is quite round and its diameter, although given as 6′, seemed only about half that. One bright, small star could be seen distinctly in the SE border but this may not be a member of the cluster.

It will stand H.P. well but even a magnification of 250× failed to resolve it although it appeared to break up into several imperfect rings.

It is too far south to be seen easily from Great Britain but in all parts of the U.S.A. except the extreme north, it is accessible at culmination.

M 55 NGC 6809
A globular cluster, class XI in SAGITTARIUS
R.A. 19h 36.9 m.   Dec. S 31° 03' (1950).
Size 15' diam.   Mag. 7.6   Distance 6.6 kpc, 20 000 lt-yrs.

Discovered by Lacaille at Cape of Good Hope 1751–52.

LACAILLE 'It resembles the shadowy nucleus of a large comet.' (Lac. I 14.)

MESSIER (July 24th, 1778) 'A nebula which is a whitish spot. Extending for 6' around the light is even and does not appear to contain a star. Found from ζ Sagittarii. Discovered by M. l'Abbé de LaCaille in 1775. Messier looked for it in vain on July 29th 1764 as reported in his memoir.'

J. H. 'A globular cluster, pretty bright, large, round, very rich, very gradually brighter in the middle; all clearly resolved into stars, 11–13 mag. 6' in diameter.'

J. E. GORE 'Observing with a 3-inch refractor in the Punjab . . . saw glimpses of stars in it with power 40×; it will not bear high powers with this aperture.'

LICK XIII 'A fine globular cluster, 10' diam. The background of faint stars is less dense near the centre than in most clusters of this type.'

FLAMMARION 'Its appearance is plainly stellar: it is a huge agglomeration of stars uniformly distributed and immersed in a pale nebulosity. Diam. about 6' but a little elongated N–S. This cluster should be admirable in the southern hemisphere; for us it is a little pale.'

Harlow Shapley did not include M 55 in his catalogue of globular clusters in 1930. In the report on variable stars in the Harvard *Annals* for 1902 it is described as 'a very open cluster' while a detailed star-count at that time revealed a total of 1195 stars, including two variables.

Helen Sawyer Hogg, in her *Handbuch der Physik* catalogue of clusters, gives the apparent diameter as 14.8', integrated magnitude 7.08, the average magnitude of the 25 brightest stars 13.58 and the colour index −0.09.

The linear diameter of the cluster given by Becvar in the *Atlas Coeli Katalog*, is 25 parsecs or about 81 light-years. This is based on a distance for the cluster of 5.8 kpc: T. D. Kinman's distance modulus of 13.6±0.6 mag. is equivalent to a mean value for distance of about 6.0 kpc which is in reasonable agreement.

Up to 1968 six *RR Lyrae* type short-period variables have been found.

In 1988 Schade and others investigated the cluster and arrived at an estimate of its age as 14 000 million years. Its distance modulus of 14.1 is equivalent to about 6.6 kpc or 20 000 lt-yrs. (For observing details see next page.)

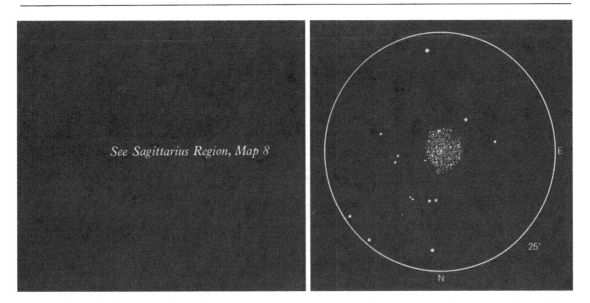

See Sagittarius Region, Map 8

To observe   M 55 is a little isolated and the best approach is from ζ *Sagittarii* in the little 'dipper' of that constellation.

From ζ *Sagittarii* move 1° S and 7° E.

This globular cluster makes a striking contrast with the three Messier globulars to the west. M 55 was later included by Shapley in his class XI and is one of the most open of globular clusters; the other three are very concentrated.

M 55 is bright and large, having a diameter of at least 10'. It is, of course, easily resolved on M.P. Higher magnification shows it to have a very irregular outline and to be elongated N–S. At times it appears to have a small semi-circular 'bite' out of the SE edge. Many faint stars of magnitude 12 or less make up the glow.

Being more diffuse than M 54, it is very difficult to see at all from Great Britain and latitudes lower than about 40° N are required to see it well.

M 56 NGC 6779
A globular cluster, class X in LYRA
R.A. 19 h 14.6 m.   Dec. N 30° 05′ (1950).
Size 5′ diam.   Mag. 8.2   Distance 14 kpc, 45 600 lt-yrs.

Discovered by Messier in 1779.

MESSIER	(Jan. 23rd, 1779) 'Nebula without star, having little light; M. Messier discovered it on the same day as he found the comet of 1779, Jan. 19th. On the 23rd, he determined its position by comparing it with the star 2 Cygni, following Flamsteed. It is near the Milky Way and close to it is a 6 mag. star. M. Messier reported it on the chart of the comet of 1779.'
W. H.	Resolved it into stars of 11–14 mag. in 1784.
D'ARREST	'Cluster containing a very large number of stars, none greater than 12–13 mag. With magnification 356× resolved into minute stars.'
SMYTH	'A globular cluster in a splendid field.'
WEBB	'Faintish, perhaps resolvable in 3.7-inch. In a fine field and rich region.'
J. E. GORE	'A photograph by Dr Roberts shows it to be a globular cluster with rays of stars projecting from it.'
LICK XIII	'Rather bright, condensed cluster, 3′ in diameter, probably globular.'

Shapley gives the ellipticity of M 56 as 8 in P.A. 45° and reports only one variable star among its members. He gives the apparent diameter as 1′.8 and the integrated magnitude as 8.8. Hogg, however, makes the diameter 5′.0 and the magnitude 9.55. Other details of the cluster given by Mrs Hogg are: the average magnitude of the 25 brightest stars 15.31; spectrum of type F5 and colour index −0.04.

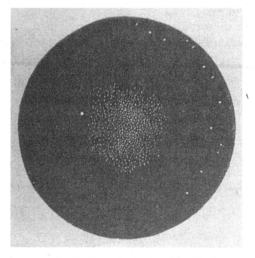

M 56 (Smyth) From A Cycle of Celestial
Objects *(B.A.A.)*

Shapley's early estimate of the distance of M 56 was 2.3 kpc but Becvar gives 13.8 kpc and T. D. Kinman's revision puts the distance near to 14.0 kpc which is in good agreement.

Based on this latter figure, the linear diameter of the cluster is about 10 parsecs or about 33 light-years.

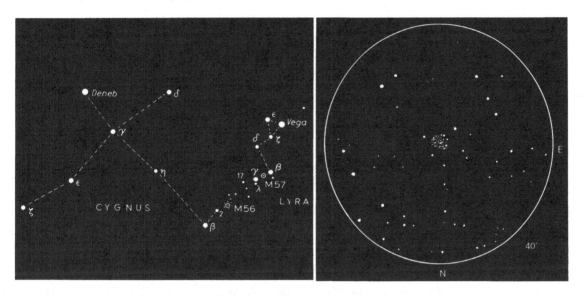

To observe From M 57 move 2½° S and 4½° E, *or,* from *Albireo* (β *Cygni*) move 2° N and 3° W. It lies 20' SE from a 5½ mag. star.

M 56 appears as a small, moderately bright cluster of even density and no very marked central condensation. The central portion of about 4' diameter gives the impression of being slightly three-lobed and this area is surrounded by a faint rim of stars. The outer edges are easily resolved in a 6-inch but even in an 8-inch, it was too faint to stand enough magnification to resolve the centre. The background consists of the Milky Way and provides a rich field of view.

M 57 NGC 6720
The 'Ring' Nebula. A planetary nebula, type IV in LYRA
R.A. 18 h 51.7 m.   Dec. N 32° 58' (1950).
Size 80"×60"   Mag. 9.3   Distance 430 pc, 1410 lt-yrs.

Discovered by Darquier in 1779 in a telescope of about 3 inches aperture.

DARQUIER   'A very dull nebula, but perfectly outlined; as large as Jupiter and looks like a
fading planet.'

MESSIER   (Jan. 31st, 1779) 'A patch of light between γ and β Lyrae, found when looking for
comet of 1779 which passed close. It seems that this patch of light which is
rounded, must be composed of very small stars but with the best telescope it is
impossible to distinguish them; they are merely suspected. Reported on chart of
comet of 1779. M. Darquier of Toulouse discovered it when observing the same
comet.'

W. H.   (*Phil. Trans.* 1785) 'A perforated Nebula, or Ring of Stars.'
'Among the curiosities of the heavens should be placed a nebula that has a
regular, concentric, dark spot in the middle, and is probably a ring of stars. It is of
an oval shape, the shorter axis being to the longer as about 83 to 100; so that, if
the stars form a circle, its inclination to a line drawn from the sun to the centre of
this nebula must be about 65 degrees. The light is of the resolvable kind, and in the
northern side three very faint stars may be seen, as also one or two in the southern
part. The vertices of the longer axis seem less bright and not so well defined as the
rest.'

J. H.   'Interior filled with a feeble but very evident nebulous light: like gauze stretched
over a hoop.'

ROSSE   'Evident symptoms of resolvability at its minor axis: strips or wisps inside and
regularity of outline broken by appendages branching into space, especially in
minor axis.'

MITCHEL   (*Stellar Worlds* (1869), pp. 256–61, Wm Collins, Glasgow) 'Appears as a ring of
misty light hung in the heavens, with a diameter as large as the full moon, when
seen with the naked eye.'

D'ARREST   'Annular nebula, very luminous, elliptical, flocculent appearance. Within is full of
fainter light.'

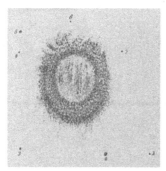

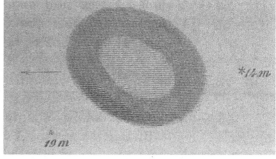

M 57 (Rosse) From Collected
Scientific Papers *(B.A.A.)*

M 57 (D'Arrest) Drawing from Siderum
Nebulosorum *1867 (B.A.A.)*

J. E. GORE   'Photographs by Schaeberle show that it is really a spiral nebula.'

SMYTH   'Only annular nebula within reach of small telescopes: an elliptic ring, the major axis of which trends S.p. & N.f. Central vacuity black. A solid ring of light in the vacuity of space.'

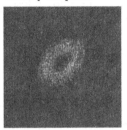

M 57 (Smyth) From A Cycle of Celestial Objects
(B.A.A.)

SECCHI   Resolved into minute stars 'glittering like silver dust.'

WEBB   'Light imagined unsteady like other planetary nebulae – probably an illusion.'

VON HAHN   first detected the central 15 mag. star in 1800 in Germany with a 20-foot F.L. reflector.

HOLDEN   in 1874–75 observed with the 26-inch refractor of the U.S. Naval Observatory. 'Minor axis more sharply terminated on N side: near N end of minor axis is a bright patch. The star inside the ring is seen occasionally. The whole of the interior glistening in points. At S. end of minor axis, filaments of nebulosity extend 15″ to 30″ off. These are very faint. Two brighter spots are seen in N.p. & S.f. parts of ring and possibly a third S.p.'

BARNARD   in 1893–94 took detailed measurements of the ring as follows:

Outer diameter of major axis	80″.89
Inner diameter of major axis	36″.52
Outer diameter of minor axis	58″.81
Inner diameter of minor axis	29″.36

LICK XIII   'Central star mag. 15.4 visual and about mag. 13 photo. The outside dimensions are about 83″×59″ in P.A. 66°.'

The ring shape is due to the fact that we see the surrounding, spherical gaseous shell of the nebula from a long distance away, and as the eye observes more glowing matter at the periphery than through the centre, it appears to us as a luminous ring.

The 12 mag. star is about 1′ E from the centre of the ring: there is a 14 mag. star 1′ W from the centre and another just inside the SW edge. The P.A. of the major axis is now considered to be 63°. Many early observers thought the central star to be variable but it is not now considered to be so.

Recent observations by astronomers at Harvard Observatory and also in the Soviet Union show that the ring is expanding at a rate very close to 1″ of arc per century.

Colour photographs taken in 1959 with the Palomar 48-inch Schmidt telescope show that M 57 has a crimson rim, shading inward through yellow to a greenish-blue interior.

Long exposure photographs by J. C. Duncan in 1937 had shown another fragmentary ring outside the bright one and having a 'curdled' appearance. In 1987 Mexican astronomers using the 2.12 metre telescope in Baja California, found a new, filamentary halo surrounding the nebula, extending its diameter nearly three times. It is believed to be the remains of a shell of expanding matter thrown off at a date much earlier than the event which caused the familiar 'shell'.

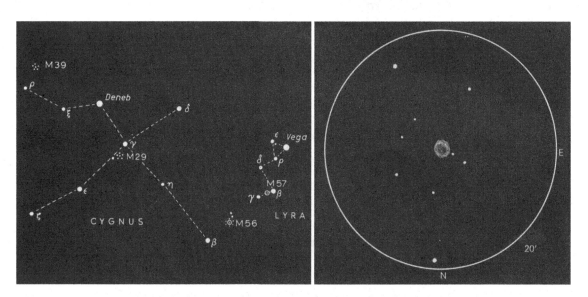

To observe   M 57 though small is quite easy to find. It is about 7° SE from *Vega* and lies almost half-way between the two 3rd mag. stars β and γ *Lyrae*.

When in position between β and γ a power of about 60× should be used, for on lower magnification it will be hardly distinguishable from a star. When found, still higher magnification can easily be used as it is a bright object.

It appears as a distinct ring, very small but clear. It looks circular at first but on closer examination it can be seen to be distinctly longer in P.A. about 60°–240°, the longer extremities being slightly less distinct than the shorter. The centre of the annulus can be seen to be not quite dark but the central 15.4 mag. star needs about 18-inch aperture for detection.

The observer should 'practise' carefully on M 57. Even in a 3-inch telescope it is possible to estimate the P.A. of the major axis with some accuracy. The 12 mag. star close to the east can be detected in a 4-inch. The training acquired will be valuable when looking for smaller and fainter planetary nebulae later on.

Objects near by   β and γ *Lyrae* are both double: β (Σ I 39) – A, mag. 4 (var.), B, mag. 7.8, P.A. 149°, d. 46".6 (1925); γ – A, mag. 3.3, B, mag. 12, P.A. 300°, d. 13".8 (1898).

M 58 NGC 4579
A barred spiral galaxy, type SBc in VIRGO
R.A. 12 h 35.0 m.   Dec. N 12° 05′ (1950).
Size 8′×6′   Mag. 8.2   Distance 12.5 Mpc, 41 million lt-yrs.

Discovered by Messier in 1779.

MESSIER (April 15th, 1779) 'Very faint nebula discovered in Virgo, almost on the same parallel as ε, 3rd mag. With the slightest illumination of the micrometer wires, it disappears. M. Messier reported it on the chart of the comet of 1779.'

J. H. (Catalogue of 1864) 'Bright, large, irregularly round, very much brighter in the middle – hardly resolvable; rather mottled as if with stars.'

ROSSE (May 3rd, 1851) 'Gradually much brighter in the middle: a little extended S.p. N.f. Edges fade off very gradually.'

D'ARREST 'A large and luminous nebula; the centre elliptical: major axis 100″; minor axis 80″.'

FLAMMARION 'A little disc, almost circular, fairly strongly condensed toward the centre; about 2′ diam. flanked by two stars, the brighter 7 or 8 mag. p. a little N by 23 s.: the second f. to NE 10 mag.'

Erik Holmberg, in his 1964 *Catalogue of External Galaxies*, gives the mass of M 58 as about 160 000 million solar masses: this is about one half the mass of M 31 and comparable with that of our own Galaxy. The diameter of M 58, according to Holmberg, is about 35.5 kpc which is greater than ours. The resulting mass-density for this galaxy works out to be about 0.03 solar masses per cubic parsec.

The absolute magnitude is given as -21.1 which is thought to be about that of our Galaxy. The spectral class for M 58 is G3 and the colour index $+0.44$.

From its distance modulus, found to be 30.5 mag. by both photometric and red-shift determinations, it is a true member of the Virgo Group of galaxies.

A Type II supernova (1988 A) was found in this galaxy by Ikeya on 1988 Jan. 18, in a position 40″ S of the nucleus, magnitude 13.5.

Another supernova, Type I (1989 M) was discovered by Kimeridze on 1989 June 28, at magnitude 12.2 pv. in a position 33″ N and 40″ W of the centre of M 58. (For observing details see next page.)

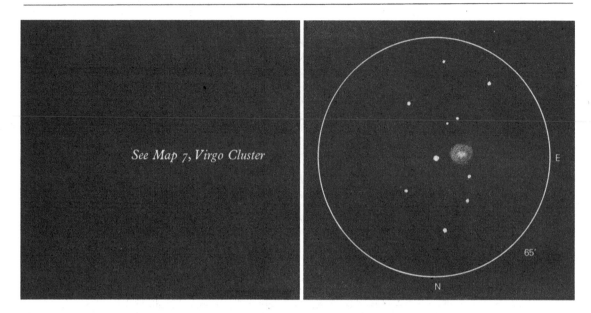

To find   See *Guide to Virgo Group* (p. 387).

To observe   M 58 is the brightest of the Messier galaxies in the Virgo Group and is an easy object to see.

Although it is an SB galaxy, it looks very much like the nearby elliptical galaxy, M 60 but it is larger, a little brighter and has a larger central condensation. The 'bar' can just be made out in an 8-inch as an extension of the central nucleus in an EW direction.

The ends of the minor axis – aligned approximately N–S – are more clearly terminated than those of the major axis.

The galaxy responds well to higher magnification in telescopes of moderate aperture.

Objects near by   About half a degree to the SW can be found NCG 4567 and NGC 4568, an interesting pair of 11th mag. galaxies known as 'The Siamese Twins'.

M 59 NGC 4621
An elliptical galaxy, type E5 in VIRGO
R.A. 12 h 39.5 m.   Dec. N 11° 55′ (1950).
Size 2′×1½′   Mag. 9.3   Distance 12.5 Mpc, 41 million lt-yrs.

KOEHLER   (April 11th and 13th, 1779) (M 59 and M 60) 'Two very small nebulae, hardly visible in a 3-foot telescope; one above the other.'

MESSIER   (April 15th, 1779) 'A nebula in Virgo, in the neighbourhood of the preceding (M 58) on the parallel of ε, which was used to determine its position. It is of the same light as that above and as faint. M. Messier reported it on the chart of the comet of 1779.'

J. H.   (Catalogue of 1864) 'Bright, pretty large, little extended, very suddenly much brighter in the middle. Two stars precede.'

SMYTH   (see M 60).

D'ARREST   'Rapidly denser toward the centre like most of the nebulae in this region.'

The mass of M 59 is only a quarter of that of the nearby galaxy, M 60. The figure for M 59 is estimated at about 250 000 million solar masses.

The spectrum is of Type G7 which is the same as that for M 60. Becvar, in the *Atlas Coeli Katalog*, gives the photographic magnitude as 11.0 and visual magnitude as 9.6. The difference between these two quantities would give a colour index of +1.4 mag. which is larger than that of any other Messier galaxy and shows a considerable reddening. This figure, however, cannot be considered very reliable.

M 59 is probably a true member of the Virgo Group but Holmberg does not include it in his catalogue and gives no distance modulus for it. Becvar's distance modulus is 30.3 mag. equivalent to a distance of about 11.0 Mpc.

A supernova was discovered in this galaxy by Zwicky: it reached a maximum brightness of 11.8 mag. in May 1939. (For observing details see M 60.)

M 60 NGC 4649
An elliptical galaxy, type E2 in VIRGO
R.A. 12 h 41.1 m.   Dec. N 11° 49′ (1950).
Size $2\frac{1}{2}′ \times 2′$ Mag. 9.2   Distance 12.5 Mpc, 41 million lt-yrs.

Discovered by Koehler, April 11–13th, 1779 while observing the comet of that year.

ORIANI   (April 12th, 1779) 'Very pale and looking exactly like the comet.'

MESSIER   (April 15th, 1779) 'A nebula in Virgo, a little more distinct than the two preceding, (M 58 and M 59) on the same parallel as ε (Virginis) which was used to determine its position. Reported on the chart of the comet of 1779. He discovered the three nebulae while observing the comet which passed very close to them. The latter passed so near on the 13th and 14th April that they were both in the same field of view and he could not see it. It was not until the 15th, while looking for the comet, that he perceived the nebula. None of the three nebulae appears to contain a star.'

SMYTH   'A double nebula, lying N.p. & S.f. about 2′ or 3′ centre to centre, the preceding one being extremely faint. The following, or brighter one, is that seen and imperfectly described by Messier in 1779 and is nearly between two telescopic stars N–S. A fine field is exhibited under 93× because the bright little nebula, M 59 is quitting the N.p. verge and another small one is seen in the upper part, H. 3171; in fact, 4 nebulae at once.'

WEBB   'Double, p. extremely faint, not seen in 3.7-inch. In large field with M 59 N.p. & H. II 70 S.p. like a hazy star.'

D'ARREST   'Double nebula. S.f. round, much brighter in centre, diam. 100″. Satellite nebula precedes, also round, not more than 3′ diam.'

LICK VIII   Gives a photograph of the two nebulae together and describes NGC 4649 (M 60) as a globular cluster about 1′ diam. and NGC 4647 as a spiral nebula about 80″ diam.

LICK XIII   'M 60 (NGC 4649) 2′ in diameter, growing rapidly brighter to a very bright central portion which shows no true nucleus in short exposures: no spiral whorls discernible.'
'NGC 4647. nearly round, 2′ in diameter, a faint, rather patchy spiral with an almost stellar nucleus.'

SHAPLEY   (*Galaxies*, 1943) 'There are many similar examples of an open spiral and an elliptical galaxy in close proximity. It is very difficult to account for as they are thought to be very different in age and also of development. There is the possibility of chance encounter, but this is also remote.'

M 60, like M 49 and M 87, is a giant elliptical galaxy. Its mass, as given by Erik Holmberg in his 1964 *Catalogue of External Galaxies*, is about 1 000 000 million solar masses which is the same as M 49. Compared with M 32, the elliptical companion of the Andromeda Galaxy, it is 200 times as massive.

Its diameter is about 36.3 kpc which makes it about 15 times as big in linear dimensions as M 32. Its absolute magnitude is −20.9; its spectrum is of type G7 and Holmberg gives the colour index as +0.83 which shows a large amount of reddening.

Both the red shift and photometric distance moduli give the same result of 30.5 mag., equivalent to a distance of about 12.5 Mpc which Holmberg has established as the distance of the Virgo Group of galaxies.

The faint Sc galaxy, NGC 4647, which is close to the NW of M 60 is also a member of the Virgo group: it has a mass of about 10 000 million Suns, a diameter of about 17.8 kpc, an absolute magnitude of −19.1 and a colour index of +0.23.

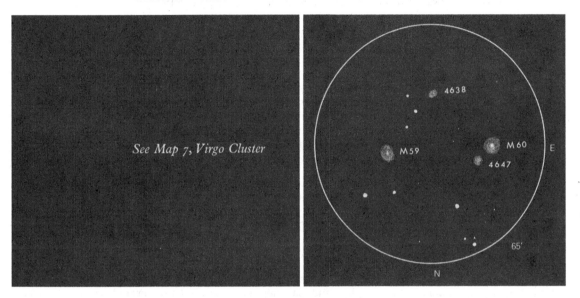

To find   See *Guide to Virgo Group* (M 59 and M 60) (p. 387).

M 59 and M 60 can both be seen in the same field on L.P. M 60 is about 25′ E and slightly S of M 59.

M 59 is a small oval patch which appears to have its major axis in P.A. about 10°. A brighter nucleus is apparent but it is not strongly marked. The whole nebula is quite bright and easily seen.

M 60 is distinctly brighter than M 59 and almost as bright as nearby M 58. It seems almost perfectly round in outline and has a very bright central nucleus which is small and almost star-like. Both M 59 and M 60 are elliptical galaxies and have the bright, pearly appearance typical of these objects.

Objects near by   Very close to the NW of M 60 is the Sc galaxy, NGC 4647: this is very faint (11.5 mag.) and it needs good conditions to see it well. It is almost round and although it brightens slightly towards the centre, it has no distinguishable nucleus. A large aperture is needed to make out any detail.

A fourth nebula, mentioned by both Smyth and Webb, lies to the S, almost equidistant from M 59 and M 60. It is very small and faint and appears elongated in P.A. about 120°. This is NGC 4638, an E5 galaxy of magnitude about 12.

M 61   NGC 4303 (139 H. I)
A spiral galaxy, type Sc in VIRGO
R.A. 12 h 19.4 m.   Dec. N 04° 45′ (1950).
Size 6′ diam.   Mag. 9.6   Distance 12.5 Mpc, 41 million lt-yrs.

Discovered by Oriani in 1779 while observing the comet of that year.

ORIANI   (May 5th, 1779) 'Very pale and looking exactly like the comet.'

MESSIER   (May 11th, 1779) 'A nebula, very faint and difficult to distinguish. M. Messier mistook this nebula for the comet of 1779 on the 5th, 6th and 11th of May. On the 11th he found it was not a comet but a nebula which was on its path and in the same part of the sky.'

SMYTH   'A large, pale-white nebula. This is a well-defined object but is so feeble as to excite surprise that Messier detected it with his 3½-foot telescope in 1779. Under the best action of my instrument it blazes toward the middle. In W.H.'s reflector it is faintly seen to be bi-central, the nuclei 90″ apart and lying S.p. and N.f. preceded by 4 telescopic stars and followed by another. This object is an outlier of a vast mass of discrete but neighbouring nebulae, the spherical forms of which are indicative of compression.'

ROSSE   'A spiral with a bright centre and 2 knots; another neb. 10′ N.f.'

D'ARREST   'Large and fairly bright; nucleus like a 13 mag. star, surrounded by a diffuse, luminous halo.'

WEBB   'Faint, bright centre.'

LICK XIII   'Nearly round, 6′ in diameter, very bright. A beautiful spiral with a very bright, almost stellar nucleus and many almost stellar condensations in its open, somewhat irregular whorls.'

ROBERTS   (Photo May 1899) 'A right hand spiral.'

M 61 is one of the larger of the Messier galaxies; Holmberg gives its linear diameter as 38.9 kpc. Its mass, however, according to the same authority, is only about 50 000 million solar masses – about a quarter of the mass of our own Galaxy.

Its mass-density is accordingly low; Holmberg quotes about 0.008 solar masses per cubic parsec or about one solar mass to every 125 cubic parsecs.

Its absolute magnitude is given as −21.2 which is about the same as the suggested luminosity of our own system. The spectrum is of type G1, and its colour index of +0.17 shows it to be the least reddened of the galaxies in Messier's list.

Its apparent dimensions, according to the *Atlas Coeli Katalog*, are 5′.6×5′.3. It is an established member of the Virgo Group.

In 1926 a supernova was discovered in M 61 by Wolf and Reinmuth: it reached a maximum of 12.8 mag. in May 1926. In June 1961 a second supernova reached mag. 13, and on June 11 1964, a third was found, of mag. 12. (For observing details, see next page.)

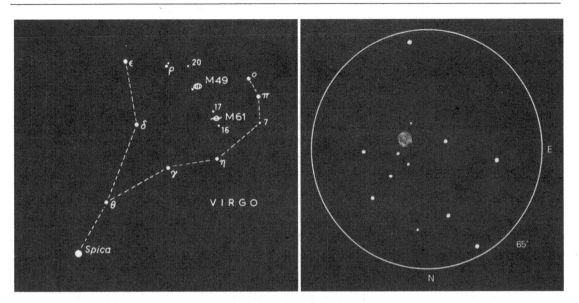

To find   See *Guide to Virgo Group* (p. 387).

To observe   M 61 is rather diffuse and pale and there is only slight evidence of central condensation. The general outline is extremely difficult to make out and appears irregular. With patient use of averted vision it appears to be faintly three-lobed: it is almost round with a diameter of about 5′. The background is dark and the field contains several 8–9 mag. stars.

Objects near by   About 50′ N lies a double star *17 Virginis*, A, 6.6 mag., B, 9.4 mag. Separation, 20″.0 in P.A. 337°. Close to the double is the faint (12th mag.) galaxy, NGC 4324, a difficult object to pick up.

M 62 NGC 6266
A globular cluster, class IV in OPHIUCHUS
R.A. 16 h 58.1 m.   Dec. S 30° 03' (1950).
Size 6' diam.   Mag. 6.6   Distance 6.9 kpc, 22 500 lt-yrs.

Discovered by Messier in 1771.

MESSIER   (June 4th, 1779) 'A very fine nebula; it resembles a little comet. It is bright in the centre and is surrounded by a faint glow. M. Messier had seen this nebula before on 7th June 1771 but obtained only its approximate position. Reviewed 22nd March, 1781.'

W. H.   First resolved it and pronounced it a miniature of M 3.

J. H.   (1847) 'A fine cluster but very low: asymmetrical.'

SMYTH   'A fine, large, resolvable nebula: an aggregated mass of small stars running up to a blaze in the centre.'

J. E. GORE   'The most condensed part is a perfect blaze but not quite in the centre.'

LICK XIII   'Bright globular cluster, greatly condensed at the centre: the central part is 1½' in diameter: main portion of cluster, 6'.'

Sir John Herschel, 1847, noticed that this globular cluster was asymmetrical and this was also observed by Bailey in 1915. Shapley referred to it as 'strikingly non-symmetrical; the most irregular globular cluster' and gives this analysis of star counts:

P.A. in deg.	15	45	75	105	135	165	195	225	255	285	315	345
No. of stars	67	56	67	45	35	42	49	56	68	79	72	72

When opposite sectors are combined, this gives the P.A. of the major axis about 75° and an ellipticity of 8.

Twenty-six cluster-type variables have been counted in M 62. It is the nearest of the Messier globular clusters to the galactic nucleus and the strong gravitational forces to which it is subjected in this part of our Galaxy may, conceivably, be a factor to account for its somewhat malformed appearance.

M 62 is not included in Helen Sawyer Hogg's catalogue but in the Harvard *Annals*, Vol. 38, it is given this description: 'A very condensed and unsymmetrical cluster. The lack of symmetry is marked not only in the distribution of stars but especially in the distribution of the variables, 19 being found N of the centre and 7, S.' The total count of stars was 1162, of which 24 were variable (two variables were thought to be non-members).

The linear diameter of the cluster, according to Becvar, is 14 parsecs or about 45 light-years. Shapley's early distance determination gave 18.6 kpc but later estimates put it between 6.0 and 9.0 kpc. Becvar gives 6.9 kpc.

Up to 1973 some 89 variables have been counted, mostly *RR Lyrae* type. (For observing details, see next page.)

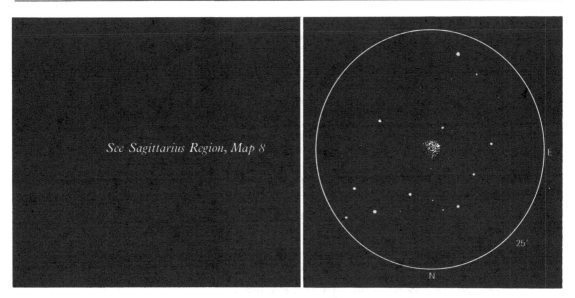

See Sagittarius Region, Map 8

25'

E

N

To observe   From τ *Scorpii* move 2° S and then sweep 5° E. Being bright, M 62 is easy to pick up.

M 62 is a very bright but small globular cluster which has a slightly bluish glow. Its most remarkable feature is its distinct lack of symmetry, the most condensed portion being well to the SE of the centre. It must be one of the most comet-like of all the Messier Objects (except perhaps M 78 in Orion) as it presents something of a rounded 'nose' to the SE and the fainter portions to the NW resemble the beginnings of a 'tail'.

Moderate magnification, however, resolves all but the denser portions into stars.

M 62 is difficult to see from Great Britain owing to its 30° southerly declination. From most of the U.S.A. it can be picked up at culmination without much difficulty.

Objects near by   1° to the WSW is the Mira-type variable RR *Scorpii* of period 279.74 days, ranging from 5.1 to 12.0 mag.

M 63 NGC 5055

A spiral galaxy, type Sb in CANES VENATICI
R.A. 13 h 13.6 m.   Dec. N 42° 18′ (1950).
Size 10′×5′   Mag. 10.1   Distance 7.3 Mpc, 23.8 million lt-yrs.

Discovered by Méchain in 1779.

MESSIER   (June 14th, 1779) 'Nebula discovered by M. Méchain. M. Messier looked for it; it is faint, it has nearly the same light as M 59. Contains no star and with the slightest illumination of the micrometer wires, it disappears. Near to it is an 8 mag. star, preceding the nebula. M. Messier reported its position on the chart of the comet of 1779.'

W. H.   'Very bright, extending from N.p. to S.f. 9′ or 10′ long and near 4′ broad, with a very brilliant nucleus.'

SMYTH   'An oval nebula, on the chest of Asterion, the northern dog. Of milk-white tint and brightens in the centre where the nucleus resembles a small star.'

ROSSE   Found 'indications of a spiral structure.'

D'ARREST   'Bright, fairly large, oval, 85″. Nucleus like an 11 mag. star in preceding part of the nebula.'

WEBB   'Oval, not bright, 8 mag. star p. A minute triplet, f.'

LICK XIII   'A bright, beautiful spiral, 8′×3′ in P.A. 98°. Has an almost stellar nucleus. The whorls are narrow, very compactly arranged and show numerous, almost stellar condensations.'

In this galaxy the spiral pattern is tightly wound; the two best spiral arms can be traced for about a quarter turn and dust clouds can be seen lying across the spiral arms themselves.

M 63 was used, with other galaxies, to try to determine their direction of tilt relative to our line of sight by Dr John B. Irwin of Indiana University, using photographs taken with the 36-inch telescope of Goethe Link Observatory. One feature which became apparent was that, on short exposures, the nearer edge of the galaxy appears sharper on the negative. This effect is clearly shown in M 63 in which the nearer edge is the southern one.

Danver, in 1942, estimated that the angle of tilt of the equatorial plane of M 63 was 31°.4 to our line of sight. However, in 1961, R. A. Fish of Yerkes Observatory corrected this value to 36°.8.

Calculations of the mass of M 63, made at Yerkes Observatory in 1959, gave a value of about 76 000 million Suns. Holmberg's figure is similar at about 79 000 million Suns.

The linear diameter of M 63 is about 33.1 kpc; the absolute magnitude −20.9, the spectral class F8 and the colour index +0.36. The mass-density is calculated to be about 0.02 solar masses per cubic parsec or one solar mass for every 50 cubic parsecs.

From rotation measures, E. M. & G. R. Burbidge and K. H. Prendergast obtained in 1960 an estimate of the mass of M 63 at about 150 billion times that of the Sun. The distance also derived to be 10.7 Mpc or nearly 35 million lt-yrs.

A type I supernova reached a maximum of 11.8 mag. in this galaxy on May 25th, 1971.

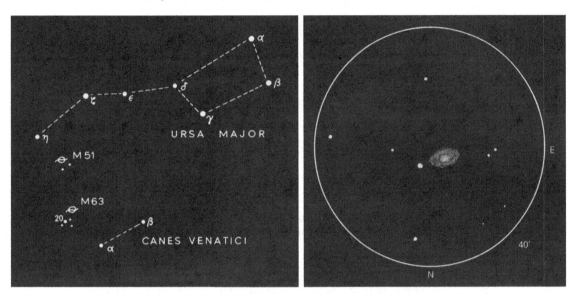

To observe From α *Canem Venaticorum* move 2° N and 3½° E to the little group of three stars which contains the 4th mag. *20 Canem Venaticorum*. M 63 is a little less than 2° N of this star.

M 63 is not difficult to recognise: it appears as a rather pale, glowing oval with a brighter and more solid-looking nucleus. This nucleus is about 2′×1′ but the size of the fainter, outer area is rather difficult to determine. One can perceive traces out to about 8′×4′ or more in good conditions. It stands M.P. quite well when the P.A. of the major axis can be fairly easily seen at about 100° but little further detail can be made out in an 8-inch.

The nucleus, which Smyth noted as star-like, in fact shows a small but bright measurable area.

The 8 mag. star close preceding on the major axis can be seen, in photographs, to be superimposed upon the outer part of the disc of the galaxy.

M 64 NGC 4826

The 'Black-eye' Nebula. A spiral galaxy, type Sb in COMA BERENICES
R.A. 12 h 54.3 m.   Dec. N 21° 57' (1950).
Size 8'×5'   Mag. 6.6   Distance 3.6 Mpc, 11.8 million lt-yrs.

Discovered by J. E. Bode in 1779.

BODE — (Apr. 4th, 1779) 'A small nebulous star about 1° NE 35 Com. Ber.'

MESSIER — (Mar. 1st, 1780) 'Nebula discovered in Coma Berenices which is about half as bright as that which is below the hair (M 53). M. Messier has reported its position on the chart of the comet of 1779. Reviewed 17th Mar. 1781.'

J. H. — 'I am much mistaken if the nucleus be not a double star in the general direction of the nebula: 320× much increases this suspicion; 240× shows well a vacuity below the nucleus.'

D'ARREST — 'Fine, very large, elliptical, 140"×95". Unexpectedly much brighter in the centre where it is probably partly resolvable.'

SMYTH — 'A conspicuous nebula, magnificent both in size and brightness, being elongated in a line N.p. & S.f. and blazing to a nucleus.'

ROSSE — 'Circular shaped with dark and light spot on one side around which is a close cluster of well-defined little stars.'

WEBB — 'Magnificent bright nebula, blazing to a nucleus. (Faint in 3.7-inch.)'

LICK XIII — 'The central portion of this fine nebula is very bright and there is a bright, almost stellar nucleus. It is 8'×4' in P.A. 110°. The whorls are rather compact and of very uniform texture, without irregularities or condensations. The most striking feature of this spiral is the somewhat irregular but very clear-cut absorption area in the north of the nucleus.'

It has been observed in this galaxy that, on short exposure photographs, the centre of the galaxy is distinctly placed off-centre relative to the surrounding bright regions. This, in addition to the sharper cut-off on one edge, enables the direction of tilt of the galaxy relative to our line of sight to be determined. In the case of M 64, the nearer edge is to the south.

Similar effects have been determined by Dr Irwin of Indiana University for other galaxies including M 63, M 65, M 66, M 81 and M 104. In all these galaxies, except M 104, the spiral arms have been observed to be 'trailing'.

Erik Holmberg, in his *Catalogue of External Galaxies*, gives the linear diameter of M 64 as 49.0 kpc which makes it the largest of all the Messier galaxies except M 77 in Cetus. Holmberg gives the mass of M 64 as 790 000 million Suns and the mass-density as about one solar mass per 16 cubic parsecs.

It is also a very luminous object with an absolute magnitude of −22.3, which again is only exceeded by M 77 among the galaxies in Messier's catalogue. The spectral class is G7 and the colour index, +0.52.

The distance of M 64 is given as 6.0 Mpc in Dr Garstang's list in the 1964 *Handbook* of the B.A.A. The distance modulus in Becvar's *Atlas Coeli Katalog* is 27.8 mag. which is equivalent to about 3.5 Mpc. Holmberg's modulus of 30.7

mag. by photometric methods only, gives a much larger value of about 13.5 Mpc.

Although M 64 lies in the same direction as the Virgo Group of galaxies, it is now known to be a foreground object, and not so distant.

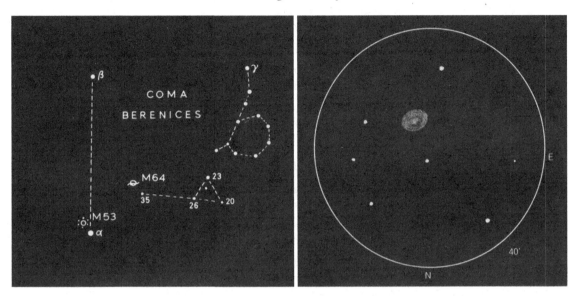

To observe Below the bright cluster of stars in the centre of Coma Berenices is a distinct equilateral triangle of 5 mag. stars, *23, 20* and *26 Comae Berenicis*. 3° E of the most easterly of these (*26*) is another 5 mag. star, *35 Comae Berenicis*. M 64 is just over 1° NE of this star, *or*, from M 53 move 3° N and 3° W.

M 64 is a very bright and conspicuous object and of large dimensions compared with most of the galaxies in Virgo. The outline is surprisingly clear and can be seen to be slightly oval. The central nucleus is small but decidedly not star-like. To the averted vision especially, the whole object glows strongly as if lit from within.

The dark, 'black-eye' region to the north of the centre needs larger than 8 inches aperture to distinguish with certainty.

Objects near by *35 Comae Berenicis*, 50' to the WSW is triple, Σ 1687: A, 5.1; B, 7.3, P.A. 134°, d. 1".0 (1957); C, 9.0, P.A. 126°, d. 28".8 (1921).

M 65 NGC 3623
A spiral galaxy, type Sa in LEO
R.A. 11 h 16.3 m.   Dec. N 13° 22' (1950).
Size 8'×2'   Mag. 9.5   Distance 9.0 Mpc, 29 million lt-yrs.

Discovered by Méchain in 1780.

MESSIER	(Mar. 1st, 1780) 'Nebula discovered in Leo: it is very faint and contains no star.'
W. H.	'A very brilliant nebula, extended in the meridian, about 12' long. It has a bright nucleus, the light of which suddenly diminishes on its border, and two opposite, faint branches.'
ROSSE	(Mar. 31st, 1848) 'Curious neb, with bright nucleus at left; a little above and towards the right is a streak; spiral; resolved very well about the nucleus but no other part.' (Feb. 25th, 1854) 'Suspect dark spaces on either side of nucleus.'
D'ARREST	'Very fine, very elongated, gradually denser towards the centre; resolved into faint stars with 147×: emitting arms on both sides, almost N–S.'
WEBB	(M 65 and M 66) 'Two rather faint objects, elongated in different directions, in a low power field with several stars. S (66) rather the larger and brighter.'
A. M. CLERKE	'Photo by von Gothard in 1888, M 65 showed bright centre with four appendages, resembling the sails of a windmill.'
ROBERTS	(M 65 and M 66) 'Photos show ovoid formations composed of closely winding luminous coils, thick inlaid – in the case of M 66, the chief of the pair – with nebulous condensations.'
LICK XIII	'A beautiful bright spiral, 8'×2' in P.A. 174°. Whorls rather indistinct, with one, almost stellar condensation; bright, almost stellar nucleus. Absorption lane on the east.'

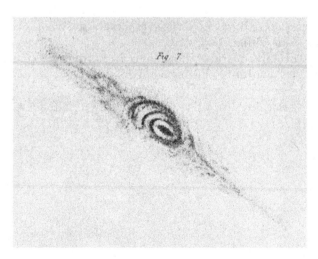

M 65 (Rosse)
From Collected Scientific Papers *(B.A.A.)*

M 65 is actually a member of a triple system of galaxies which includes M 66 and NGC 3638 – another Sb galaxy about 35' to the NE.

Hubble made observations of M 65 in an attempt to measure its direction of rotation. He found the inclination of the plane of its disc to our line of sight to be approximately 14°.

This latter measurement was complicated by the fact, peculiar to this galaxy, that the dark and bright spiral arms appear to have different major axes.

On photographs, the galaxy can be seen to extend up to 200" of arc from the centre.

Attempts to measure its mass in 1960 gave a figure of 140 000 million Suns but this was thought to be an under-estimate and in 1961, Margaret Burbidge obtained a figure of 200 000 to 300 000 million Suns.

The central nucleus of M 65 can be seen to have a dust arm emerging from it and the whole galaxy appears to contain a considerable amount of internal absorption material.

The linear diameter of M 65, according to Holmberg, is about 31.6 kpc and its mass-density is fairly high at about 0.08 solar masses per cubic parsec. Its spectral type of G0 and the colour index +0.51. The absolute magnitude is −21.2.

According to de Vaucouleurs, the plane of M 65 is inclined at about 15° to our line of sight. (For observing details see M 66.)

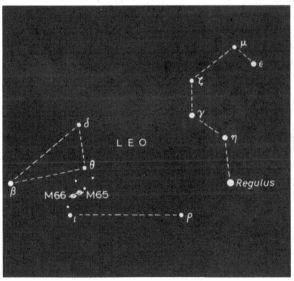

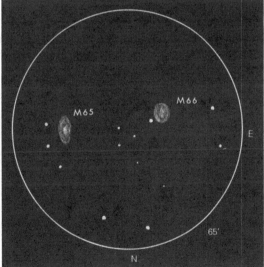

M 66 NGC 3627
A spiral galaxy, type Sb in LEO
R.A. 11 h 17.6 m. Dec. N 13° 16′ (1950).
Size 8′×3′ Mag. 8.8 Distance 9.0 Mpc, 29 million lt-yrs.

Discovered by Méchain in 1780.

MESSIER (Mar. 1st, 1780) 'Nebula discovered in Leo; the light is very faint and it is very close to the preceding (M 65): they both appear in the same telescopic field. The comet of 1773–74 passed between them on Nov. 2nd, 1773 and M. Messier no doubt missed them then because of the light of the comet.'

SMYTH (M 66) 'A large, elongated nebula with a bright nucleus trending N.p. & S.f. Preceded at about 73 s by another of similar shape which is M 65.'

ROSSE 'Suspected darkness on either side of nucleus; extended.'

D'ARREST 'Fine, large, much brighter in the centre; elongated, 6′×2′. A twin to M 65 which is conspicuous to the west.'

ROBERTS 'Spiral with well-defined stellar nucleus, forming the pole of the convolutions, in which I have counted 14 nebulous, star-like condensations.'

A. M. CLERKE 'A photo by von Gothard in 1888 shows a complex arrangement of envelopes, partially surrounding a nucleus somewhat like the paraboloidal veils flung round the head of a comet near perihelion.'

LICK XIII 'A very bright, beautiful spiral 8′×2½′ in P.A. 180°. Bright, slightly elongated nucleus; the whorls are somewhat irregular and show numerous condensations.'

M 66 is a good deal larger than its neighbouring galaxy, M 65, but it has only a little more than half the mass of the latter. Holmberg gives the mass as

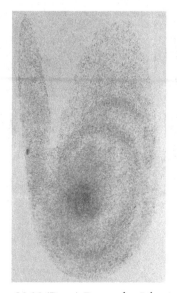

M 66 (Rosse) Drawn by Johnstone Stoney (B.A.A.)

214

about 130 000 million solar masses and the linear diameter as 36.3 kpc. The mass-density is accordingly only about 0.025 solar masses per cubic parsec or one solar mass per 40 cubic parsecs.

The absolute magnitude of M 66 is given as −21.4 which is a little greater than M 65. The spectrum is of type G2 which is close to that of M 65 but the colour indices are considerably different, M 66 having an index of +0.34 which is much less reddened than M 65.

Becvar, in the *Atlas Coeli Katalog*, gives a distance modulus of 28.9 mag. for both M 65 and M 66. This is equivalent to a distance of about 6.0 Mpc but Holmberg obtains a distance modulus of 29.8±0.13 mag. for all the Leo Group of galaxies (which consists of at least eight members) and this is equivalent to a distance of about 9.0 Mpc.

A Type II supernova of mag. 15 was found in M 66 on Dec. 12th, 1973.

Another, much brighter, supernova was discovered on Jan. 30th, 1989, which reached a maximum of 12.2 mag. on Feb. 1st, 1989.

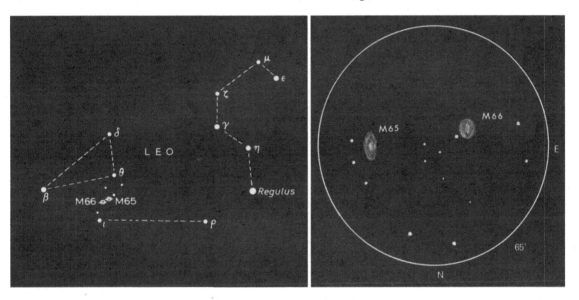

To observe   (M 65 and M 66)

From *Regulus* move 1° N and sweep to the E for about 18° to a point half-way between the bright stars θ and ι Leonis.

M 65 and M 66 are only about ½° apart and can be seen together in an L.P. eyepiece. M 66 is slightly S of E from M 65.

Both galaxies are bright and distinct and, although M 66 is actually the brighter, M 65 tends to appear more conspicuous owing to its more 'streaky' outline. Their major axes can be seen to be not quite parallel, M 66 being precisely in the meridian while M 65 has a P.A. about 5° less.

Both have clear central nuclei but in M 65, the nucleus is larger and more distinguishable. There is a distinct bright star about 5' to the NW of M 66.

The whole L.P. field is an impressive sight but each galaxy will stand appreciable magnification when more detail can be made out.

215

Objects near by   35' to the NE of M 65 is the Sb galaxy, NGC 3628 (not in *Norton's Atlas*). It is large, 12'×1½'; clearly elongated but quite pale and needs averted vision; mag. 10.9.

A little over 1° to the W and slightly S, is NGC 3593, another Sb galaxy, smaller and fainter than 3628.

M 67   NGC 2682
A galactic cluster, type 'f' in CANCER
R.A. 08 h 47.8 m.   Dec. N 12° 00' (1950).
Size 27' diam.   Mag. 6.1   Distance 830 pc, 2700 lt-yrs.

Discovered by Koehler between 1772 and 1779.

KOEHLER   'A fairly discernible nebula of oblong shape near α Cancri.'

MESSIER   (Apr. 6th, 1780) 'A cluster of small stars with nebulosity below the southern claw of the Crab. The position determined from the star α.'

W. H.   Resolved it in 1783. 'A very beautiful and pretty much compressed cluster of stars, easily to be seen by any good telescope and in which I have observed above 20 stars at once in the field of view of my great telescope with a power of 157.'

J. H.   'A remarkable object, very bright, large and rich, with stars 10 to 15 mag.'

D'ARREST   'A most elegant cluster of stars, 11 to 13 mag. Very large, 20' or 30' in diameter.'

SMYTH   'Rich but loose cluster, consisting principally of a mass of 9 and 10 mag. stars, gathered somewhat in the form of a Phrygian cap; followed by a crescent of stragglers.'

WEBB   'Visible in finder.'

FLAMMARION   'It resembles a sheaf of corn.'

M 67 has been more extensively studied than almost any other galactic cluster and it has revealed much important information regarding the age of clusters and the evolution of stars.

It is believed to be one of the oldest of all galactic clusters, most estimates being between 4000 and 5000 million years. It is very dense and being about 440 parsecs above the galactic plane, it is far from the Milky Way and so fairly free from much of the disrupting effects of the Galaxy's gravitational forces.

Its life expectancy has been estimated to be more than 5000 million years and, although as a galactic cluster, it contains only Population I stars, its age is more nearly that of the globular clusters.

Stellar evolution theory predicts that a very old cluster like M 67 should contain many white dwarfs, formed from hot, massive stars. At the distance of M 67, a normal white dwarf would be about magnitude 21. In 1963, professor Luyten examined Baade's 200-inch plates and found 79 objects which were thought to be possibly white dwarfs, but he could detect no concentration of these toward the centre as was expected. The validity of this theory is thus not yet confirmed.

M 67 is known to contain 500 stars between mag. 10 and mag. 16 and a very large number of stars which are fainter still.

V. C. Reddish, of London University, considers M 67 to be a peculiar galactic cluster, and a transitional type between Population I and Population II.

Trumpler's classification for M 67 is II,2,r and he gives the spectrum of the brightest star as B9. The linear diameter of the cluster, according to Wallenquist, is 4.2 parsecs: he also gives the central density as 27.3 stars per cubic parsec and 0.77 stars per cubic parsec overall.

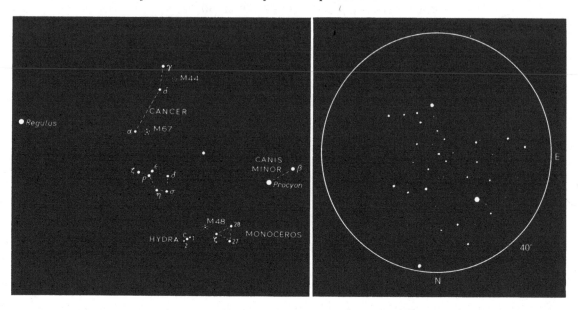

To observe   M 67 is rather isolated from the brighter stars but, being a bright cluster, it is easy to see and in good conditions can just be made out by the naked eye.

First pick up α *Cancri* (4th mag.) which is a little NE of the mid-point between *Regulus* and *Procyon*. M 67 is 2° W of α *Cancri*.

It is a bright, prettily-formed cluster in a fairly dark part of the sky. The central portion is almost circular and about 12′ in diameter. The brightest star in the group is on the NE edge and about mag. 8. There are about 20 stars of mag. 10 to 12 in the central area and more than 60 stars can be counted within a diameter of 15′.

The stars form a pleasing pattern but Smyth's description of it being like a Phrygian cap – a high, conical head-dress like a bishop's mitre, or the 'Cap of Liberty' worn by the French revolutionaries – does not exactly leap to the eye. Flammarion's description of it resembling a sheaf of corn is much more evident.

M 68 NGC 4590

A globular cluster, class X in HYDRA

R.A. 12 h 36.8 m. Dec. S 26° 29' (1950).

Size 4' Mag. 8.2 Distance 11.2 kpc, 36 000 lt-yrs.

Discovered by Méchain in 1780.

MESSIER (April 9th, 1780) 'Nebula without stars in Hydra; it is very faint, very difficult to see in the telescope; near to it is a 6 mag. star.'

W. H. (1786) 'Resolved into a rich cluster of small stars so compressed that most of the components are blended together. 3' to 4' diam.'

J. H. 'A globular cluster, irregularly round, gradually brighter in the middle . . . all clearly resolved into stars, 12 mag.; very loose and ragged at the borders.'

SMYTH 'A large, round nebula, nearly half-way between 2 small stars, one N.p. and the other S.f. It is very pale but so mottled that a patient scrutiny leads to the inference that it has assumed a spherical figure in obedience to attractive forces.'

WEBB 'Globular cluster of stars, well resolved, 12 mag. stars, one red.'

This globular cluster has not been greatly explored: Shapley found that it contained 28 cluster-type variable stars and, in 1930 gave its ellipticity as 9. In 1949, however, he described it as being accurately round for a count of 2000 of the brightest stars.

It contains about 250 giant stars of absolute magnitude greater than zero compared with twice that number for M 3 and M 13.

In Helen Sawyer Hogg's catalogue of clusters, the apparent diameter is given as 9'.8; its integrated magnitude as 9.12; the spectral type as A6 and the average magnitude of the 25 brightest stars as 14.8.

The linear diameter of the cluster, according to Becvar, is about 13 parsecs or about 42 light-years. Shapley's early determination of distance was about 15.5 kpc but Becvar gives 11.5 kpc and the mean of T. D. Kinman's moduli gives a distance of approximately 12.0 kpc. In 1937 McClure and others obtained 11.2 kpc (36 000 lt-yrs.).

Up to 1973, 38 variables had been counted, but one of Shapley's variables (No. 27) has since been shown to be not a member of the cluster. (For observing details see next page.)

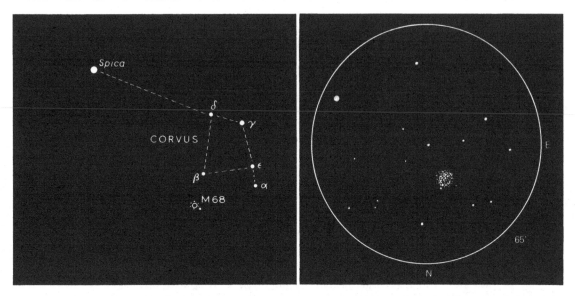

To observe   This is a rather difficult cluster to see from latitudes of 50° N or higher unless the observer's southern aspect is fairly unimpeded. Apart from this, it is easy to find as it is only about 45′ NE of a bright, 5½ mag. star which is itself 3° SSE of β *Corvi* (mag. 3).

If this 5½ mag. star is picked up first, M 68 will be in the same field of view in an L.P. eyepiece. Low power is required if its altitude is low as the amount of light will then be small. For observers in lower latitudes, for whom it rises higher in the sky, it is an easy object and very bright.

M 68 appears as a perfectly round disc, about 4′ diam; bright at the centre and gradually fading off towards the edges. It is situated at the lower apex of a large 'V' of stars with the western arm curving away to terminate in the 5½ mag. star mentioned above.

It is not difficult to resolve if it can be seen bright enough to stand a power of 150× to 200×. Several arcs of brighter stars can be seen in the S and W portions of the cluster and about 5′ NW of the centre is a single star, referred to by Webb as red but which appeared more orange to me.

Objects near by   The bright star to the SW, mentioned by Messier, is double. It is catalogued as A.D.S. 8612; A, 5.4 mag.; B, 12.2 mag.; P.A. 152°; d. 1″.6 (1926) and, for small telescopes it can be considered a very difficult test object.

About 25′ to the NE is *FI Hyd.*, a *Mira*-type variable with a period of 324 days which reaches ninth magnitude at maximum. It has no association with the globular.

M 69 NGC 6637
A globular cluster, class V in SAGITTARIUS
R.A. 18 h 28.1 m. Dec. S 32° 23' (1950).
Size 4' diam. Mag. 8.9 Distance 7.2 kpc, 23 500 lt-yrs.

Discovered by Lacaille at Cape of Good Hope 1751–52.

LACAILLE   'It resembles the small nucleus of a comet.' (Lac. I. 11.)

MESSIER   (Aug. 31st, 1780) 'Nebula without star in Sagittarius. Near to it is a 9 mag. star; the light is very faint; can be seen only in a good sky and the least illumination of the micrometer wires extinguishes it. The position was determined from ε Sagittarii. This nebula has been observed by M de La Caille and reported in his catalogue. It resembles the nucleus of a little comet. Diam. 2'.'

J. H.   'A globular cluster, bright, round, 3' diam. All clearly resolved into stars, 14–15 mag. A blaze of stars.' (An additional note by J. H. in his *General Catalogue* of 1864 reads: 'Piazzi, in a note in his catalogue, says that both M 69 and M 70 are 1° more to the south. But he is wrong.')

LICK XIII   'Bright globular cluster, 3' diam.'

FLAMMARION   'A rich cluster with a crown of stars scintillating along its border.'

M 69 has an apparent diameter of 3'.8 in the *Handbuch der Physik* catalogue of clusters compiled by Mrs H. S. Hogg; its integrated magnitude is given as 8.94, the spectral class as G5 and a colour index of +0.12 which indicates a fair degree of reddening.

Shapley estimated the ellipticity as 9 and obtained a distance of 18.7 kpc. Later measurements of distance are given in the *Atlas Coeli Katalog* of about 7.2 kpc and T. D. Kinman's revision of the distance of globular clusters gives a mean modulus of 14.3 mag. which is equivalent to a distance of about 7.5 kpc. Kinman's modulus has a rather high uncertainty however.

The linear diameter of the cluster, according to Becvar, is about 21 parsecs or approximately 68 light-years.

Although Shapley found no variable stars in M 69, a few new ones were observed in both M 69 and M 70 by L. Rosino of Asiago, Italy in 1956 when working at the Radcliffe Observatory in South Africa.

A colour-magnitude diagram for a number of the stars in M 69 shows the cluster to be extremely metal-rich in composition. In a search for variables in the cluster during 1969–70, Catchpole, Feast and Menzies, using the Radcliffe reflector at Pretoria, South Africa, found two new Mira-type stars with periods of about 200 days. Work continues on the kinematics and chemistry of this important globular cluster. (For observing details, see next page.)

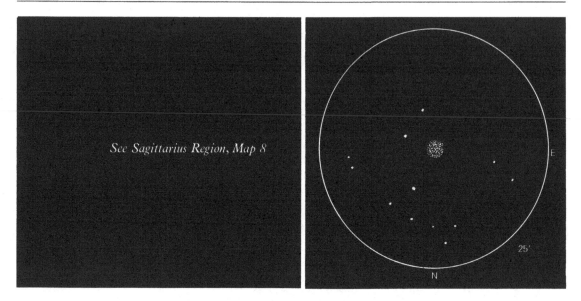

See Sagittarius Region, Map 8

To observe   From δ *Sagittarii* move 2½° S and 2° E, *or*, from ε *Sagittarii* move 2° N and 1½° E.

M 69 is a fairly bright but small globular cluster: it appears quite round although the outer margin is a little irregular. It has a soft, even glow and shows hardly any central condensation. It is not, however, very open and it requires a fairly high power to resolve. With a power of 250× the whole cluster breaks up into a large number of faint, evenly spaced stars.

There is a 9 mag. star about 8' to the NW and most of the nearby stars are to the N.

A very difficult object to see from Great Britain and a latitude lower than about 45° N is necessary to see it bright enough for full resolution.

Objects near by   About one degree to the SE is another globular, NGC 6652. It is a little smaller than M 69, and not quite as dense in structure. At magnitude 8.9 it is not difficult to pick up from a suitable latitude, and is worth examining for a comparison with M 69 itself.

M 70 NGC 6681
A globular cluster, class V in SAGITTARIUS
R.A. 18 h 40.0 m.   Dec. S 32° 21′ (1950).
Size 4′ diam.   Mag. 9.6   Distance 20 kpc, 65 000 lt-yrs.

MESSIER   (Aug. 31st, 1780) 'Nebula without star, near the preceding (M 69) and on the same parallel. Near to it is a 9 mag. star and four small telescopic stars, almost in the same straight line, close to one another and situated below the nebula as seen in a reversing telescope. The (position of) the nebula determined from the star ε Sagittarii. Diam. 2′.'

W. H.   'A miniature of M 3 and pale to the gaze.'

J. H.   (Catalogue of 1864) 'A globular cluster, bright, pretty large, round, gradually brighter in the middle; component stars, 14–17 mag.'

LICK XIII   'Bright condensed cluster, 2′ in diameter; doubtless globular.'

FLAMMARION   'A cluster, decorated with a pretty double star to the NE and a fine 9 mag. star to the NNW.'

In Shapley's 1930 catalogue of globular clusters, M 70 is quoted as having an apparent diameter of 2′.5; an integrated magnitude of 7.5; an ellipticity of 9.5 and a distance of 19.2 kpc.

Helen Sawyer Hogg, in her catalogue of clusters in the *Handbuch der Physik*, gives 4′.1 for the diameter and 8.95 for the integrated magnitude. The spectral class is given as G2 and the colour index −0.04.

The distance of M 70 does not appear to be known with any great accuracy. T. D. Kinman published a distance modulus of 16.5±0.8 mag. which gives a range of distance from about 14 kpc to about 30 kpc. A figure of 20 kpc, which is equivalent to that obtained from the mean value of the modulus, is quoted by Becvar in the *Atlas Coeli Katalog* and this is not far from Shapley's earlier estimate.

The linear diameter of the cluster, according to Becvar, is about 18 parsecs or about 60 light-years. (For observing details see next page.)

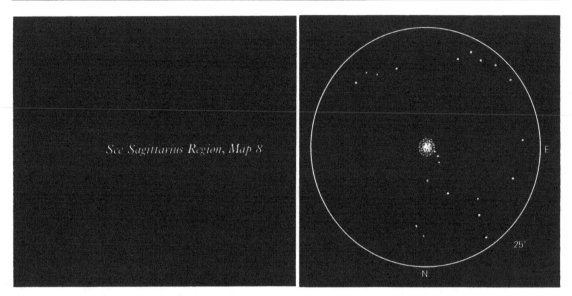

See Sagittarius Region, Map 8

25'

E

N

To observe From M 69 move $2\frac{1}{2}°$ E, *or*, from ζ in the 'dipper' of Sagittarius, move $2\frac{1}{2}°$ S and 4° W.

Although small – it is about the same size as M 69 – it stands out more clearly than the latter because of its sharp central condensation. Around this nucleus is a fainter area which is almost circular except that it is less well defined to the east where it seems slightly flattened.

As if attached to the cluster, there is a little slightly curved 'tail' of small stars, shooting off like sparks to the NNE. These may be the stars mentioned by Messier.

With a power of 250× the outer portion of the cluster can be well resolved but the nucleus is too dense for resolution with this magnification. This gives an interesting comparison with nearby M 69 which, although of the same class V, seems to be wholly resolved with the same power.

As with M 69, it is not easily seen in latitudes above 50° N and is better observed from below 45° N.

M 71   NGC 6838
A globular cluster, Type 'g' in SAGITTA
R.A. 19 h 51.5 m.   Dec. N 18° 39' (1950).
Size 6' diam.   Mag. 9.0   Distance 4.0 kpc, 13 000 lt-yrs.

Discovered by de Chéseaux in 1746?
Discovered independently by Koehler between 1772 and 1779.

KOEHLER   'A very pale nebulous patch in Sagitta, long. 310° 50': lat. 39° N.'
Rediscovered by Méchain June 28th, 1780.

MESSIER   (Oct. 4th, 1780) 'Nebula discovered by M. Méchain 28th June 1780 between the
stars γ and δ Sagittae. On the 4th Oct. following, M. Messier looked for it. The
light is very faint and it contains no star. The least light extinguishes it. It is situated
about 4° below that which M. Messier discovered in Vulpecula, see No. 27. It was
reported on the chart of the comet of 1779. Diam. 3½'.'

J. H.   'Very large and very rich cluster, with stars 11–16 mag.'

SMYTH   'A rich, compressed Milky Way cluster.'

WEBB   'Large and dim; hazy to L.P. with 3.7-inch; yielding to a cloud of faint stars to
higher powers. Interesting specimen of the process of stellar evolution. About 1°
S.p. M 71 is a beautiful L.P. field containing pair and triple group.'

ROBERTS   'Photo shows a cluster in which the curves and arrangements of stars resemble
those of a spiral nebula ... the surrounding region densely crowded with stars
down to about 17 mag. arranged in remarkable curves and lines which are very
suggestive of having been produced by the effects of spiral movements.'

LICK XIII   'Rather sparse globular cluster, 5' in diameter.'

This cluster has been regarded at different times as either a very compact
galactic cluster or a very loose globular cluster.

Both Shapley and Trumpler listed it as a galactic cluster and the former
included it in his type 'g' of the most concentrated clusters.

In 1943, Dr James Cuffey, who has made detailed studies of both kinds of
cluster, found that by comparing M 71 with the globular clusters NGC 5053
(class XI) and M 68 (class X), it resembled these much more than any dense
galactic cluster.

It also has a large radial velocity of −80 km/s. and its brightest stars appear
red: both these factors weigh the evidence in favour of it being a globular.
However, no short period variables have been discovered in the cluster and it
is situated in the Milky Way; points which support its identification with the
galactic clusters.

In 1959, Dr Cuffey, from plates taken with the 82-inch reflector of
Mcdonald Observatory and the 100-inch Mount Wilson telescope, drew up a
colour-magnitude diagram which led him to the revised conclusion that it is
more likely to be an unusual galactic cluster than a globular. However, later
consensus establishes M 71 as a true globular cluster.

M 71 contains at least six M-type giant stars, one of them being the
irregular variable Z Sgr which has a range from 13.5 to 14.9 mag.

M 71 is very rich in faint stars and these have been detected out to a total diameter of 24', but it is not clear whether these are all true members.

In Helen Sawyer Hogg's catalogue of clusters in the *Handbuch der Physik*, M 71 is given an apparent diameter of 6'.1, an integrated magnitude of 8.3 and a spectral class of G2.

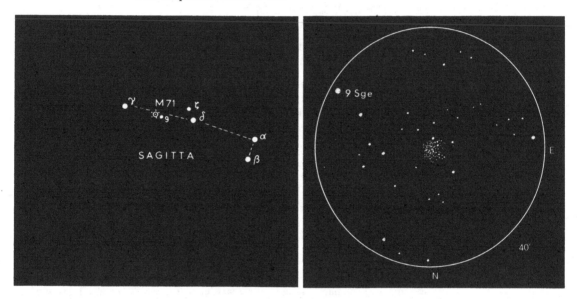

To observe   M 71 is very easy to find; it is a little less than half way from γ to δ *Sagittae. 9 Sagittae* (6 mag.) is 20' away to the SW.

This cluster can be seen as a large compressed mass of stars of almost triangular shape with a distinct angle formed at the SW part. It is fairly sharply terminated on the W side but more open and straggling on the east. For a globular cluster, it is very easily resolved; even the centre being well opened out on M.P.

If it is to be considered as a galactic cluster, then, although it resembles M 11 in shape, it is certainly more compressed even than that dense cluster.

There is a very rich field of stars all around.

Objects near by   30' to the SSW of M 71 is a large, but very faint and sparse cluster, H20. It contains only a few stars and is hardly recognisable as a cluster at all – a marked contrast to M 71.

M 72 NGC 6981
A globular cluster, class IX in AQUARIUS
R.A. 20h 50.7 m.   Dec. S 12° 44′ (1950).
Size 5′ diam.   Mag. 9.8   Distance 19.0 kpc, 62 000 lt-yrs.

Discovered by Méchain in 1780.

MESSIER (Oct. 4th, 1780) 'Nebula seen by M. Méchain on the night of 29–30 Aug. 1780, above the neck of Capricorn. M. Messier looked for it on the 4th and 5th Oct. following: the light as faint as the preceding (M 71). Near to it is a little telescopic star: the position determined from the star ν Aquarii, 5th mag. Diam. 2′.'

W. H. 'Oct. 4th, 1810; 40-foot telescope; Space-penetrating power, 191.68; magnification 280×. Having been a sufficient time at the telescope to prepare the eye properly for seeing minute objects, the 72nd of the *Connaissance des Temps* came into the field. It is a very bright object. It is a cluster of a round figure but the very faint stars on the outside of globular clusters are generally a little dispersed so as to deviate from a perfectly circular form. The telescopes which have the greatest light show this best. It is very gradually extremely condensed in the centre but with much attention, even there the stars may be distinguished.'
'There are many stars in the field with it but they are of several different magnitudes; totally different from the excessively small ones which compose the cluster. It is not possible to form an idea of the number of stars that may be in such a cluster, but I think we cannot estimate them by 100s. The diameter of the cluster is about 1/5 of the field which gives 1′ 53″.6.'

D'ARREST 'Bright, irregularly round; partly resolved with 95×. Very rich heap of the minutest stars: with 123× well resolved. Shape irregular or nearly oval. Diam. 2′.'

SMYTH 'A globular cluster of minute stars. There are many telescopic stars in the field. A small pair closely follows the cluster. This cluster is followed at about 5 min of R.A. and 7′ south by M 73.'

LICK XIII 'Small, bright cluster, 3′ in diameter: globular, comparatively open.'

M 72 is a very open globular cluster; Shapley included it, together with M 4 and M 12 in his class IX of concentration: only M 56, of class X, is in a lower concentration category among the established Messier globulars.

Shapley, in his 1930 catalogue, gives the apparent diameter as 2′.0 and integrated magnitude as 8.6. In Mrs H. S. Hogg's catalogue in the *Handbuch der Physik*, the apparent diameter is given as 5′.1 and integrated magnitude, 10.25. The average magnitude of the 25 brightest stars is put at 15.86, the spectral class as G2 and the colour index is −0.04.

Shapley's early estimate of the distance of M 72 was 23.3 kpc. Becvar, in the *Atlas Coeli Katalog*, gives 18.2 kpc while T. D. Kinman, in his revision of the distance moduli of globular clusters, quotes a modulus of 16.3±0.5 mag. corrected for interstellar absorption. This is equivalent to a mean distance of about 19.0 kpc.

The linear diameter of the cluster, according to Becvar, is about 13 parsecs or about 42 light-years.

Up to 1973 some 42 variables, mostly *RR Lyrae* type, have been found.

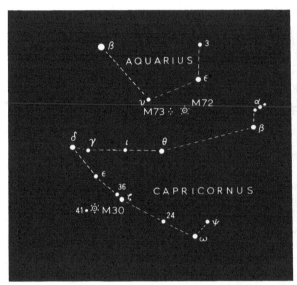

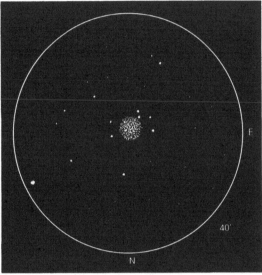

To observe   S from *Delphinus*, first find *3 Aquarii* (4½ mag.) and then move 4½° due S to ε *Aquarii* (4 mag.). M 72 is 3° S and 1½° E of ε *Aquarii*. A 6 mag. star will be seen 40' to the W of the cluster, *or*, from the little line of three or four stars which includes α *Capricorni* (4 mag.) move 9° E.

M 72 is rather pale but fairly large: it is of noticeably even brightness, being only slightly fainter near the edge than at the centre. It seems quite round in its general outline but a few of the stars seem to break away from the circular mass. There is a fairly close pair of stars near the southern edge and a bright, wide pair to the east. Many faint stars can be seen in the background especially to the south and west.

It is surprisingly difficult to resolve for so large and unconcentrated a cluster. In an 8-inch, the extreme edges had merely a suspicion of 'breaking up'. It is difficult to employ enough magnification with moderate apertures as it is not a bright object.

M 73 NGC 6994

An Asterism of four stars in AQUARIUS

R.A. 20 h 56.4 m.   Dec. S 12° 50' (1950).

Diam. 3'   Mag. 9.0 (approx.)   Distance unknown.

MESSIER (Oct. 4th and 5th, 1780) 'A cluster of three or four small stars which looks like a nebula at first sight; it contains a little nebulosity: it is on the same parallel as the preceding nebula (M 72). Position determined from the same star, ν Aquarii.'

J. H. Included in his catalogue as No. 4617; 'A cluster??; extremely poor, very little compressed; no nebula.'

SMYTH 'A trio of 10 mag. stars in a poor field – this is M 73. I give it out of respect to Messier's memory.'

M 73 has always been regarded as an 'asterism', a purely coincidental grouping of four stars along a line of sight. The chances, however, of finding four stars brighter than mag. 12 in an area of that size has been calculated to be no better than one in four. The probability, then, that M 73 is, in fact, a 'real' (if exiguous) grouping, is 0.75, and D. A. Allen has suggested that it would be worth examining the quartet more closely to see if it is composed of closely related stars. As Dr Allen puts it: 'I would gamble that it is a genuine multiple star of some kind.' The issue is perhaps a minor one, but every student of Messier's catalogue would be much interested in the outcome. (For observing details see next page.)

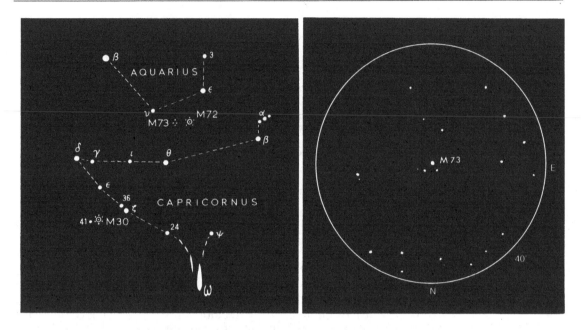

To observe   M 73, although inconspicuous, is not difficult to find as it is only $1\frac{1}{2}°$ E and slightly S from M 72.

There is not very much to see; it first appears as three fairly faint but distinct stars in a small triangle, more or less isolated from the background. The fourth star, which is a good deal fainter than the others, can be made out close to the W of the group.

The whole is easily seen as a small, isolated asterism; it may look a little hazy in very small or poor telescopes.

The magnitudes of the individual stars involved are: A and B, $10\frac{1}{2}$ mag; C, 11 mag. and D, 12 mag., giving a total magnitude of about 9.

Objects near by   2° NE of M 73 and $1\frac{1}{2}°$ W of v *Aquarii* ($4\frac{1}{2}$ mag.) is an object very much more interesting than the asterism. This is the planetary nebula, NGC 7009 – also known as the 'Saturn' Nebula. It is bright ($8\frac{1}{2}$ mag.) and although fairly small ($44''\times36''$) is stands magnification well and can be seen as a greenish oval with faint 'wings' in P.A. about 75°–255° which gives it a very slight resemblance to Saturn.

230

M 74 NGC 628
A spiral galaxy, type Sc in PISCES
R.A. 01 h 34.0 m. Dec. N. 15° 32' (1950).
Size 8' diam. Mag. 10.2 Distance 8 Mpc, 26 million lt-yrs.

Discovered by Méchain in 1780.

MÉCHAIN 'This nebula contains no star; it is fairly large, very obscure and extremely difficult to observe. One can make it out with more certainty in fine, frosty conditions.'

MESSIER (Oct. 18th, 1780) 'Nebula without stars, near the star η Piscium, see by M. Méchain at the end of Sept. 1780. M. Messier looked for it and found it to be as M. Méchain describes. It has been compared directly with the star η Piscium.'

J. H. (Catalogue of 1864) 'A globular cluster; faint, very large; round; very gradually, then pretty suddenly, much brighter in the middle; partially resolved.'

ROSSE (Dec. 14th, 1848) 'Spiral?' (Dec. 15th, 1848) 'Feel confident it is a spiral.' (Oct. 24th, 1851) 'Centre formed of stars: easily seen to be such; several stars seen through the nebula.'

D'ARREST 'Pale and tenuous, very much denser towards the centre: the central part is almost round, diam. 40″. Resolved; uncertain whether it has a true nucleus.'

ROBERTS Photo in Dec. 1893. 'A perfect and beautiful spiral with a central stellar nucleus and numerous star-like condensations in the convolutions of the spiral.'

LICK XIII 'Nearly round, 8' diam. An unusually beautiful and symmetrical spiral, showing numerous, almost stellar condensations. Nucleus bright and small but not stellar.'

M 74 has a mass of only 40 000 million solar masses, according to Holmberg's recent catalogue of external galaxies. This is only one-fifth of that of our own Galaxy. Its diameter, however, is similar to ours, being about 30.2 kpc. Its mass-density is therefore fairly low, being about 0.013 solar masses per cubic parsec or about one solar mass per 75 cubic parsecs.

Its absolute magnitude is −20.6, spectral class F2 and it has a colour index of +0.24.

The distance modulus is given as 29.2 mag. in Becvar's *Atlas Coeli Katalog*: this is equivalent to a distance of about 7.0 Mpc. Holmberg's modulus is 29.7 mag. both by photometric and red shift measurements and this is equivalent to a distance of about 8.7 Mpc.

In the *Hubble Atlas of Galaxies* M 74 is described as having the main spiral arms bordered by thin dust lanes on their outer edges which can be traced deep into the nucleus. (For observing details, see next page.)

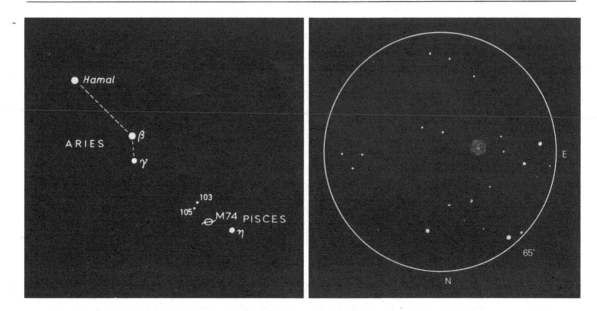

To observe From *Hamal (α Arietis)* follow the line *Hamal-β Arietis* to η *Piscium* (3½ mag.). M 74 is ½° N and 1½° E of η *Piscium*.

This open spiral galaxy often causes the amateur observer considerable trouble and it can be quite a difficult object unless the sky is perfectly dark and the atmosphere clear. The position should be approached carefully from η *Piscium* as described above and cross-checked by reference to the pair of 6 mag. stars, *103* and *105 Piscium* which lie just over 1° to the NE of M 74. If it is not at first visible, look for a pair of 10 mag. stars about 3' apart lying N–S: M 74 is about 6' to the W of this pair. Full dark-adaptation and averted vision may well make the vital difference here.

When seen, it appears round, with a fairly sharp central nucleus which is almost star-like, surrounded by a very diffuse area of about 6' diameter which responds only to averted vision and even then reveals little or no detail. The object is too faint and diffuse to stand anything more than L.P.

The fact that Sir John Herschel considered it to be a globular cluster shows that it was not seen very clearly even by him. However, M 74 was apparently seen in a 3-inch refractor at Bonn with a power of 10× while observations were being made for the *B.D.* catalogue.

Perhaps, like the nebulosity of M 45, it may well be easier to see in small telescopes of wide field than in larger instruments.

Objects near by   η *Piscium* is double: A, 3.7; B, 11.0; P.A. 19°, d. 1".0.

M 75 NGC 6864
A globular cluster, class I in SAGITTARIUS
R.A. 20 h 03.2 m.   Dec. S 22° 04′ (1950).
Size 5′ diam.   Mag. 8.0   Distance 24 kpc, 78 000 lt-yrs.

Discovered by Méchain in 1780. 'A nebula without stars.'

MESSIER (Oct. 18th, 1780) 'Nebula without star between Sagittarius and the head of Capricorn; seen by M. Méchain on 27th and 28th Aug. 1780. M. Messier looked for it on the 5th Oct. following and on the 18th, compared it with the star 4 Capricorni. It seemed to M. Messier to be composed of very small stars and to contain nebulosity. M. Messier saw it on the 5th Oct. but the Moon being above the horizon, it was not until the 18th of the same month that he was able to make out its form and determine its position.'

W. H. 'The 75th of the *Connaissance* is not visible to the eye but may be seen in the finder and telescopic observations of it have ascertained its profundity to be of the 734th order.'

He resolved M 75 in 1784 and described it as 'A miniature of M 3 and pale to the gaze.'

SMYTH 'A lucid white mass among some glimpse stars, with a large one in N.f. verge of the field. One cannot but consider Messier's opinion on an object which, at best is rather faint, bold.'

D'ARREST 'Very bright, globular, much denser in the centre. Accompanied by many small stars, one 12 mag. Diam. 75″–80″.'

WEBB 'Bright nucleus on M.P.'

LICK XIII 'Bright, compact globular cluster, 2′ in diameter; greatly condensed at the centre.'

M 75 is one of the more distant of the globular clusters and Shapley at one time thought it might even be an independent cluster, or a fugitive body, escaping from the control of the Galaxy.

Its distance, once thought to be as great as 42 kpc, has now been revised to about 24 kpc, which still leaves it the most remote of the globular clusters in Messier's list. The most remote of all globular clusters is probably NGC 2419 in Lynx which may be as much as 70 kpc, 225 000 light-years, away.

Harlow Shapley includes M 75 in his class I, which represents the highest concentration. He gives its ellipticity as 9 and detected 11 cluster-type variables among its member stars.

In Helen Sawyer Hogg's catalogue of clusters, the apparent diameter is given as 4′.6; the integrated magnitude as 9.5; the average magnitude of the 25 brightest stars as 17.06; a spectral class of G1 and colour index +0.01.

Antonin Becvar, in the *Atlas Coeli Katalog*, gives the linear diameter of the cluster as 27 parsecs or about 88 light-years. (For observing details see next page.)

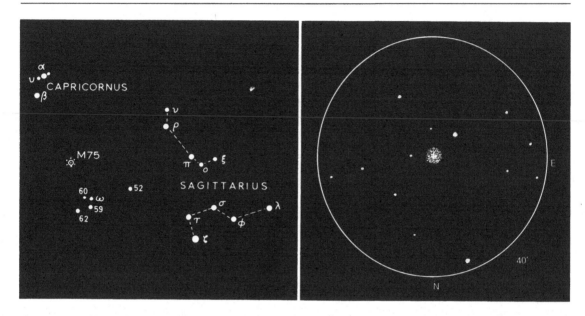

To observe   M 75 is not very easy to find as it is in a fairly barren region. Follow the line γ *Aquilae–Altair*–β S through θ *Aquilae* to α and β *Capricorni*. From β *Capricorni* move 7° S and 4° W, *or*, from π *Sagittarii* move 1° S and a little less than 1 hour of R.A. (about 14°) to the E.

M 75 is a small and very compact globular cluster, only 2′ to 3′ in diameter and it appears to be perfectly round. It is fairly faint but nevertheless quite distinct, having a bright centre which is not quite star-like, being about 1′ across. There is a 10 mag. star about 7′ to the SE but otherwise the field is fairly sparse.

As Admiral Smyth implied, Messier's claim that it was seen to be composed of very small stars is an extraordinary one for an observer whose poor telescope led him to consider that even M 13 contained no star at all.

A slight mottling of the edges was all that could be detected of its globular identity in an 8-inch using a power of 120× which is about the maximum it would tolerate in the latitude of 51½° N. Observers in more southern regions may have more success.

234

M 76 NGC 650 and 651
A planetary nebula, class V in PERSEUS
R.A. 01 h 38.8 m.   Dec. N 51° 19' (1950).
Size 157"×87"   Mag. 12.2   Distance 2.51 kpc, 8200 lt-yrs.

Discovered by Méchain in 1780.

MÉCHAIN — (Sept. 5th, 1780) 'This nebula contains no stars. It is small and faint.'

MESSIER — (Oct. 21st, 1780) 'Nebula on the right foot of Andromeda, seen by M. Méchain 5th Sept. 1780. On the 21st Oct. following, M. Messier looked for it with his achromatic telescope and it seemed to him that it comprised only small stars containing nebulosity and that the least light employed to illuminate the micrometer wires caused it to disappear. The position was determined from the star phi Andromedae, 4th mag. Diam. 2'.'

W. H. — Thought it to be an unresolvable double nebula, hence the two NGC numbers. (651 represents the northern part.)

SMYTH — 'An oval, pearly-white nebula. It trends N–S with 2 stars preceding by 11 s. and 15 s. and following on the same parallel by 19 s. and 36 s.

WEBB — 'Pearly-white nebula; double; curious miniature of M 27 and, like it, gaseous: p. a little brighter.'

ROSSE — 'Spiral: with subordinate nodules and streamers; a system modelled on "reaping-hook" pattern.'

ROBERTS — 'The two nebulae are one and the figure suggests that it is a broad ring seen edge-wise.'

LICK XIII — 'Central star of mag. 16. Quite irregular but evidently to be included as one of the large members of the planetary class. The central and brighter portion of the nebula is an irregular, patchy oblong 87"×42" in P.A. 40° from the ends of which, faint, irregular, ring-like wisps extend: total length, 157" in P.A. 128°. Brightest patch at southern end of central part.'

There are several planetary nebulae which exhibit a 'dumb-bell' appearance; M 27 is the largest and best known of these while M 76 is perhaps the most irregular. It does not appear to have any visible symmetry of revolution as is common with many planetary nebulae.

Strömgren, however, has proposed a model for M 76 and suggests that we can see only part of what is really a symmetrical structure. Due to varying density of material in an inner shell around the central star, some parts are excited by its ultra-violet radiation, while others are blanketed by the denser matter and remain invisible to us.

The central star of M 76 has a magnitude of 16.6 and has a temperature of about 60 000 K. (For observing details, see next page.)

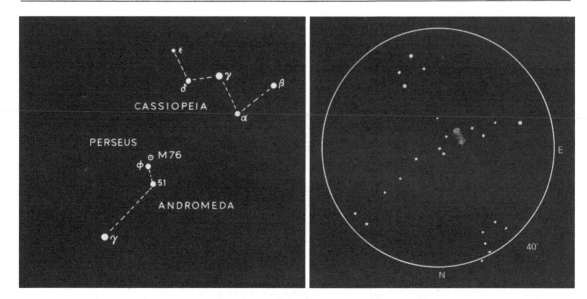

To observe   Although the 'Little Dumb-bell' planetary nebula is faint, it is not difficult to find. Almost half way between γ *Andromedae* and δ *Cassiopeiae* is φ *Persei* (mag. 4). M 76 is just under 1° N and slightly W of this star.

While the best atmospheric conditions are required to observe this nebula, it is much easier to see than its published magnitude of 12.2 would lead one to expect.

It can be seen as a small, hazy, nearly elliptical patch among a straggling group of faint stars, several of which are arranged in pairs. There are many noticeable points of similarity to M 27; the waisted appearance, in particular, though slight, is quite plainly seen.

The P.A. of the major axis is not difficult to make out at about 35° and its dimensions are about 1½′ × 1′ which is a little larger than the 'Ring Nebula' in Lyra (M 57) but it is not as clearly defined as that planetary.

A neat, sickle-shaped group of stars can be seen about 25′ to the NE of the nebula.

Objects near by   φ *Persei* – 50′ to the NNE – is an irregular variable of the R.W. Aurigae type; range 4.3 to 4.4 photographic mag.

Nearly 1° to the NNE of M 76 is Σ 139; A, 8.8 mag., B, 9.0 mag. separation 10″ in P.A. 224° (1866).

M 77 NGC 1068
A spiral galaxy, type Sb in CETUS
R.A. 02 h 40.1 m.   Dec. S. 00° 14′ (1950).
Size 3′ diam.   Mag. 8.9   Distance 16 Mpc, 52 million lt-yrs.

Discovered by Méchain in 1780.

MESSIER   (Dec. 17th, 1780) 'A cluster of small stars which contains some nebulosity in the Whale and on the parallel of the star delta, reported of the 3rd mag. but which M. Messier estimates to be of the 5th. M. Méchain saw this cluster on the 29th Oct. 1780, in the form of a nebula.'

W. H.   (1814) 'Such a cluster of stars as the 77th of the *Connoissance des Temps* will put on the stellar appearance when it is viewed in a very good common telescope.'

J. H.   'Pretty large and irregularly round; brighter in the middle with a nucleus and partly resolved.'

SMYTH   'A round, stellar nebula. Small, bright and exactly in line with 3 small stars, one p. and two f. of which the nearest and largest is 9 mag. S.f.'

ROSSE   (Dec. 22nd, 1848) 'A blue spiral.' (Oct. 29th, 1851) 'The central part is flatter on the f. side.' (Nov. 24th, 1851) 'The central part is, I am nearly sure, spiral.'

D'ARREST   'Splendid, perfectly round, no clustering of stars with 356×; nevertheless the nucleus seems stellar.'

WEBB   '1° f. δ; a little S. Small, faintish, very near 9 mag. star.'

ROBERTS   Photo. 'A stellar nucleus with projecting "ansae" of dense nebulosity ... and surrounding the "ansae" is a zone of faint nebulosity surrounded by a broad nebulous ring which is studded with strong condensations resembling stars with irregular margins.'

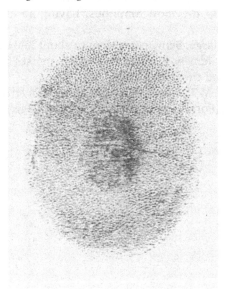

M 77 (Rosse) Drawn by Bindon Stoney (B.A.A.)

LICK XIII   'A very bright and beautiful spiral, 2'.5×1'.7 in P.A. 20°. Several almost stellar condensations on the periphery of the brighter, central portions, near the bright nucleus which is apparently not stellar. The whorls are very compactly arranged.'

M 77 was observed by Fath in 1908 to have a peculiar spectrum: in addition to the usual continuous spectrum, it was seen to contain bright green bands corresponding to the bright green lines observed in planetary nebulae.

Its velocity of recession was measured as +1100 km/s by Slipher in 1914 and as +864 km/s at Lick Observatory in 1918.

In the region of the nucleus of M 77, a large cloud of gas with stars embedded in it was observed in 1959 to be expanding with very high internal velocity. This gas seems to be ejected in directions perpendicular to the galaxy's equatorial plane. An expanding gas cloud has been observed in our own Galaxy but in our case the expansion appears to be mainly in the plane of the disc.

Galaxies of this type, with a small, bright nucleus and having spectra showing broad emission lines are known as *Seyfert galaxies* and these very energetic nuclei have many similarities to the 'quasars'. M 77 is one of the brightest of the Seyfert galaxies and in 1965, Dr Merle F. Walker of Lick Observatory, after a detailed survey of the nucleus, calculated that the energy required to accelerate the gas clouds in the galaxy may be more than that which would be supplied by some millions of supernovae. Dr Walker considers that a new, unknown energy source may be responsible for this phenomenon.

M 77 has been observed to be a radio source, having a radio 'magnitude' of 9.8, compared with radio 'mag.' 1 for M 1.

M 77 is the largest member of a group of galaxies which includes two other spirals, NGC 1073 and 1087, both of photographic magnitude 11.4. M 77 is easily the largest of the Messier galaxies with a linear diameter of 51.3 kpc: it is also intrinsically the most luminous, having an absolute magnitude of −22.5.

Its mass is also large, being equivalent to about 800 000 million solar masses and its mass-density is about 0.05 solar masses per cubic parsec. The spectrum is F0 and the colour index +0.48.

The nucleus of M 77 has been observed to be a strong infra-red emitter, similar to M 82. (For observing details see next page.)

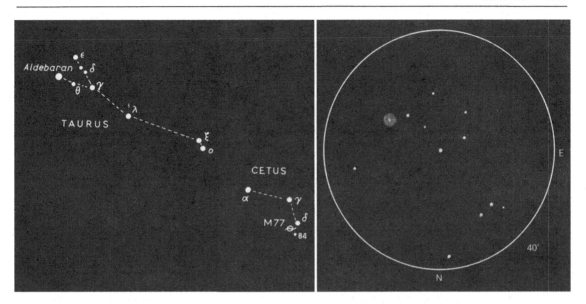

To observe   Bisect the 'V' of the Hyades and continue the line for about another 22° SW to α *Ceti* and then for another 7° in the same direction to δ *Ceti* (4th mag.) M 77 is 15' S and 30' E of δ *Ceti*.

On L.P., M 77 appears as a faint, circular, nebulous patch. It may seem almost star-like at first but its nebulous character can soon be detected by comparing it with the 10 mag. star which is close to the E. It stands higher magnification well, and on M.P. the surrounding area of faint nebulosity appears to grow more distinct, which is not generally the case with diffuse objects. The nucleus is very small and hardly distinguishable from a star.

Objects near by   35' NNW of M 77 is NGC 1055, an Sb galaxy of mag. 12 visual. It is large and elongated (6'.7×1'.5) and needs the best conditions and a good aperture to see well.

1° S of M 77 is a pair of faint spirals: NGC 1087 and 1090, both also mag. 12.

The star, *84 Ceti*, 40' to the SSW is double; Σ 295, A, 5.7 mag.; B, 9.4 mag.; P.A. 316°, d. 3".8 (1930).

M 78   NGC 2068

A gaseous reflection nebula in ORION

R.A. 05 h 44.2 m.   Dec. N 00° 02′ (1950).

Size 8′×6′   Mag. 8.3   Distance 500 pc, 1630 lt-yrs.

Discovered by Méchain in 1780.

MÉCHAIN
'One the right side of Orion; 2′–3′ diam. One can see two fairly bright nuclei, surrounded by nebulosity.'

MESSIER
(Dec. 17th, 1780) 'A cluster of stars with much nebulosity in Orion and on the same parallel as the star delta in the belt which was used to determine its position; the cluster is 3° 41′ f. the star and 27′ 7″ N. M. Méchain had seen this cluster at the beginning of 1780. Diam. 3′.'

SMYTH
'Two stars in a "wispy" nebula. A singular mass of matter, trending from a well defined N disc to the S.f. quadrant, whence it melts away. In its most compressed portion is a double star.'

ROSSE
(Feb. 9th, 1852) 'Spiral arrangement sufficiently seen to confirm former observations.'

WEBB
'Singular "wispy" nebula; best defined N.'

BRODIE
'Very faint but curious.'

KNOTT
'The southern star is brighter.'

D'ARREST
'6′ or 7′ enclosing pair. Shape irregular, approaching oval. N margin clearly terminated; the S one, less so.'

LICK XIII
'A mass of rather irregular, fairly bright, diffuse nebulosity, whose brighter portion is 6′×4′, involving 2, 10 mag. stars. Two fainter patches lie 6′ W apparently separated from the main mass by a wide lane of dark matter; the southerly one of these is NGC 2064 and the northern one, 2067.'

M 78, like M 42 and M 43, is really a brighter portion of the extensive nebulosity involving much of the constellation of Orion. M 78 itself, with neighbouring NGC 2064, 2067 and 2071, are patches of bright reflection nebula containing hot, early B-type stars, all embedded in a large, dark cloud whose dimensions are about 4°×2°. This dark cloud is itself surrounded by a large, broken ring of bright, diffuse nebulosity known as *Barnard's Loop* and which encircles the area containing the stars of Orion's belt and sword.

Barnard's Loop was considered by Öpik, in 1953 to be the remains of an expanding supernova shell, but Menon, in 1958, thought the gas was being accelerated by hot, new stars, near the centre.

In and near M 78, 45 stars of low mass, hydrogen emission-line type were detected similar to the star *T Tauri*. (For observing details see next page.)

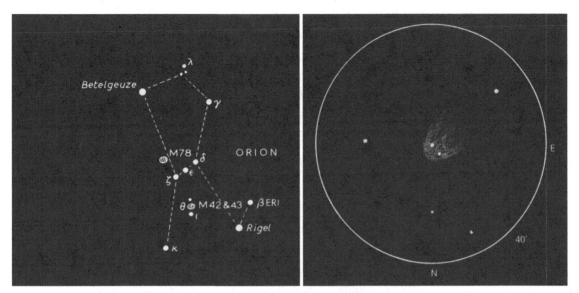

To observe   From δ *Orionis* – The NW star in Orion's belt – move $\frac{1}{2}$° N and sweep $3\frac{3}{4}$° E.
       M 78 is remarkably bright and glowing and the two bright nuclei it contains
       are immediately visible, lying N.f. and S.p. The nebula exhibits a clear-cut,
       semi-circular outline on the NW side, while to the SE it spreads fan-wise and
       becomes gradually more diffuse. This gives it a very distinctive, comet-like
       appearance.
       The background is dark but a few faint stars can be seen, three of them
       forming an almost equilateral triangle around the nebula.

Objects near by   Other portions of bright, diffuse nebulosity are: NGC 2064, 15′ to SW,
       11′×2′; NGC 2067, 15′ to N, 7′×3′; NGC 2071, 20′ to NNE, 4′×3′. All three
       are faint but respond to rich-field telescopes in good conditions.

M 79 NGC 1904
A globular cluster, class V in LEPUS
R.A. 05 h 22.2 m.   Dec. S 24° 34' (1950).
Size 3' diam.   Mag. 7.9   Distance 13.2 kpc, 43 000 lt-yrs.

Discovered by Méchain on Oct. 26th, 1780.*

MESSIER (Dec. 17th, 1780) 'Nebula without star, situated below Lepus and on the same parallel as a star of 6th mag. Seen by M. Méchain, 26th Oct. 1780. M. Messier looked for it on Dec. 17th following. This nebula a fine one, the centre brilliant, the nebulosity a little diffuse; its position determined from the star ε Leporis, 4th mag.'

W. H. 'Beautiful cluster in 20-foot telescope, nearly 3' diam. globular and extremely rich.'

SMYTH 'A bright stellar nebula: a fine object, blazing toward the centre.'

WEBB 'Tolerably bright in 64×, blazing in centre. Higher powers showed it mottled: closely f. 6 mag. star.'

Shapley reported the presence of five variable stars in this globular cluster and gave its ellipticity as 9 in P.A. 45°.

H. S. Hogg gives the apparent diameter as 7'.8 and the integrated magnitude as 8.39. The average magnitude of the 25 brightest stars is quoted as 15.29 and the spectral type, F3.

The distance of the cluster was given by Shapley in 1930 as 20.4 kpc but T. D. Kinman's recent revision of the distance moduli of globular clusters puts the distance at about 14.0 kpc although the range of uncertainty is fairly large. In Becvar's *Atlas Coeli Katalog* the distance is put at 13.2 kpc.

The linear diameter of M 79, according to Becvar, is about 19 parsecs or more than 60 light-years.

M 79 has a radial velocity of recession of +231 km/s. This is the largest value obtained for any globular cluster except for NGC 1851 in Columba which is quoted as +291 km/s. (For observing details, see next page.)

---

*   In Hodierna's catalogue of 1654 he reports an *occulta* (a nebula not resolved into stars) as lying between Lepus and Columba. There is only one bright nebula in this region. M 79, and there is a strong possibility that Hodierna saw this object. He gave no more precise position, however, nor is it marked on his working chart of the area, and this omission denies him any right of discovery.

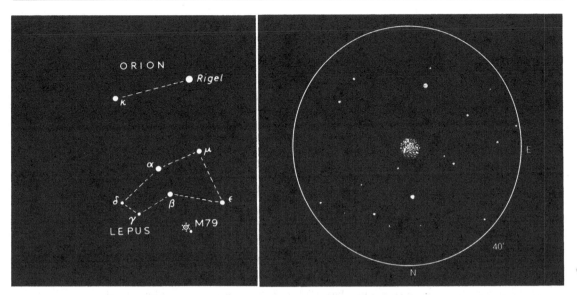

To observe   Extend the line α–β *Leporis* for another 4° to the SSW. M 79 is about ½° NE of a 5th mag. star, *h.3752*.

M 79 is a small globular cluster, 2' to 3' in diameter, with a good bright centre about 1' across, surrounded by a fainter area which appears not quite circular. For observers in latitudes of 40° N or less, the cluster should appear bright enough to stand a moderate magnification and the outer area may then be resolved to some degree. In high latitudes, the cluster does not rise high enough above the horizon even at culmination for more than a slight mottling to be made out, even in the best conditions.

The cluster is equidistant between two 8 mag. stars almost NS and about 20' apart.

Objects near by   *41 Leporis* (= *h.3752*), 35' to SW is double: A, 5.5 mag. yellow; B, 6.7 mag. blue-green; P.A. 98°, d. 3".2 (1947).

M 80 NGC 6093
A globular cluster, class II in SCORPIUS
R.A. 16 h 14.1 m.   Dec. S 22° 52' (1950).
Size 5' diam.   Mag. 7.7   Distance 11 kpc, 36 000 lt-yrs.

MESSIER (Jan. 4th, 1781) 'Nebula without star in the Scorpion, between the stars g. (now rho Ophiuchi) and delta; compared with g. to determine its position. This nebula is round, the centre brilliant and it resembles the nucleus of a little comet, surrounded with nebulosity. M. Méchain saw it on 27th Jan. 1781. Diam. 2'.'

W. H. (*Phil. Trans.* 1785) 'An Opening in the Heavens.' 'This opening is at least 4° broad but its height I have not yet ascertained. It is remarkable that the 80th *Nebuleuse sans Etoiles* of the *Connaissance des Temps*, which is one of the richest and most compressed clusters of small stars I remember to have seen, is on the western border of it and would almost authorise a suspicion that the stars of which it is composed, were collected from that place and had left the vacancy.'

J. H. 'A globular cluster; round, suddenly very much brighter in the middle to a blaze. Stars 14 mag. all resolved. Fine object.'

WEBB 'Like a comet in a beautiful field, ½-way between α and β Scorpii.'

LICK XIII 'Small, bright globular cluster; the diameter of the brightest part is 3'.'

On May 21st, 1860 a very bright nova was discovered near the centre of this cluster by Dr Auwers at Berlin. The nova was of magnitude 7 and outshone the cluster itself and, although by May 28th it had faded a little, it was noticed independently by Pogson.

By June 10th, 1860, the nova, which was designated *T Scorpii*, had almost completely faded away and when last seen some years ago, was at magnitude 12.

Except for one other nova in NGC 6553 in Sagittarius, this is the only case of a nova being *observed* in a globular cluster (but see also M 14).

Shapley observed only four short-period variables in M 80. He considered the cluster to be perfectly circular in outline and to be one of the most concentrated of globular clusters, including it in his class II.

The average magnitude of the 25 brightest stars is quoted by Hogg as 14.85. The spectrum of the cluster is of type F4 and the colour index −0.02.

Shapley's early estimate of the distance of M 80 was 17.5 kpc but later measurements have reduced this considerably. Becvar gives a distance of 11.0 kpc in the *Atlas Coeli Katalog* while Kinman's revised figure for the modulus is 15.2±0.5 mag. The mean value is equivalent to a distance of about 12.0 kpc.

The true linear diameter of the cluster, according to Becvar, is 17 parsecs or about 55 light-years. (For observing details, see next page.)

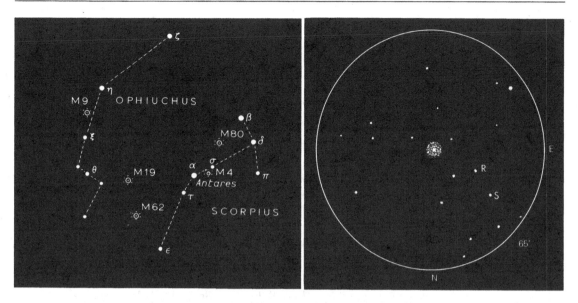

To observe   M 80 is not a very conspicuous cluster but is quite easy to find, being half way between *Antares* and β *Scorpii* and just below the parallel of δ *Scorpii*.

It can be seen as a small, round, concentrated ball with a bright, almost star-like nucleus. It will stand high magnification well but, although it enlarges, it reveals no more than a slight mottling at the edges in an 8-inch. Lower latitude observers will have a better chance of resolving it but good apertures are needed.

It is interesting to compare this very compact globular with M 4 which is only 4° to the SE. The latter is in Shapley's class IX and is one of the most easily resolved.

Objects near by   In the same field with M 80 on L.P. and close to the NE are two variable stars, *R* and *S Scorpii*. These were both discovered to be variable by J. Chacornac in 1854.

   R Scorpii; period 223 days: 10.4 to 15.0 mag.
   S Scorpii; period 117 days: 10.5 to 14.6 mag.

There are extensive areas of dark and bright diffuse nebulosity to the S and SE.

M81 NGC 3031

A spiral galaxy, type Sb in URSA MAJOR
R.A. 09 h 51.5 m.   Dec. N 69° 18' (1950).
Size 16'×10'   Mag. 7.9   Distance 2.6 Mpc, 8.5 million lt-yrs.

Discovered by Bode in 1774.

BODE	(No. 17, Dec. 31st, 1774) 'A nebulous patch, more or less round, with a dense nucleus in the middle.'
MESSIER	(Feb. 9th, 1781) 'A nebula near the ear of the Great Bear, on the parallel of the star d. of 4th or 5th mag. Its position determined from that star. This nebula is a little oval, the centre clear and can be seen well in an ordinary telescope of 3½ feet. It was discovered by M. Bode at Berlin on Dec. 31st, 1774 and by M. Méchain in Aug. 1779.'
J. H.	'A remarkable, extremely bright nebula with a nucleus.'
SMYTH	'A fine, bright, oval nebula of white colour in the Great Bear's ear. Major axis lies N.p. & S.f. and certainly brighter in the middle. A close double in S.p. quadrant, Σ 1386, both 9 mag. P.A. 294°, d. 1".9 (1819). Nearer to M81 is another double, Σ 1387, both 10 mag. P.A. 272°, d. 9" (1864).'
WEBB	'Two nebulae ½° apart. M81 bright with vivid nucleus, finely grouped with small stars, two of which are projected upon the haze to which J. H. gives nearly 15' length. Two little pairs S.p.'
ROBERTS	Photo March 1889. 'A spiral with a nucleus which is not well defined at its boundary and is surrounded by rings of nebulous matter.'
LICK XIII	'This very beautiful spiral is about 16'×10'. Short exposures show the nucleus is almost stellar. Central part very bright.'

M81 was the first spiral to reveal evidence of its rotation. An inclination of the absorption lines in its spectrum was measured by Max Wolf in 1914: he calculated that, at the edge, it had a linear rotation velocity of the order of 300 km/s.

According to Baade, M81 is the chief member of a bright group of galaxies, probably as extensive as our own and containing perhaps 10 or 12 members, of which only three are spirals, the rest being elliptical or irregular. One of the these irregular galaxies is M82 and another, NGC 3077 is a close companion of M81.

Nancy W. Boggis, of the Observatory of the University of Wisconsin has measured the P.A. of the major axis of M81 to be 149°.4 and the inclination of the plane of its disc to our line of sight to be about 35°.

The nucleus of this galaxy appears to be very elliptical, having a major axis which is about twice that of the minor axis. Its mass was determined by G. Münch in 1956 as being between 68 000 and 97 000 million Suns.

It has also been detected as a radio source on 158 Mc/s having a radio 'magnitude' of 6.8.

The mass of the whole galaxy, as given by Holmberg in his catalogue, is about 250 000 million Suns and the linear diameter is thought to be about

29.5 kpc. The mass-density of M 81 is one of the highest determined by Holmberg, being equivalent to about one solar mass per 10 cubic parsecs. This is about twice as dense as M 31 and ten times the density of M 90.

The absolute magnitude is −21.0; the spectral type G3 and the colour index +0.53.

A huge halo of hydrogen, emitting radiation in the 21-cm band has been detected in this galaxy and also in M 51.

In photographs the spiral arms are well defined and separated by narrow dust lanes which lead to within 35″ of the nucleus.

Some 25 novae have been seen in this galaxy together with 7 irregular variables and 3 Cepheids. (For observing details see M 82.)

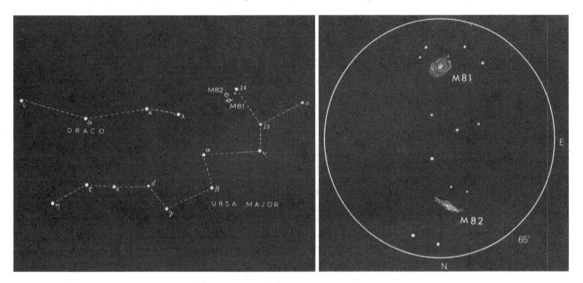

M 82 NGC 3034 (79 H. IV)
An irregular galaxy in URSA MAJOR
R.A. 09 h 51.9 m.   Dec. N 69° 56′ (1950).
Size 7′×1½′   Mag. 8.8   Distance 2.6 Mpc, 8.5 million lt-yrs.

Discovered by Bode in 1774.

BODE (No. 18, Dec. 31st, 1774) 'A nebulous patch, very pale; elongated.'

MESSIER (Feb. 9th, 1781) 'Nebula without star, near the preceding (M 81) both appearing in the same field of the telescope. This one is less distinct than the preceding; the light is faint and elongated with a telescopic star at its extremity. Seen at Berlin by M. Bode on Dec. 31st, 1774 and by M. Méchain in Aug. 1779.'

J. H. 'A very bright ray; very large.'

SMYTH 'Very long, narrow, bright, especially in northern limb, but paler than M 81.'

ROSSE (May 21st, 1871) 'A most extraordinary object, at least 10′ in length and crossed by several dark bands.'

WEBB (M 82) ' "Bode's Nebula" Curious, narrow curved ray.'

D'ARREST 'Fine bright rays, vividly luminous and sparkling; 7′ long and 100″ wide with two nuclei eccentrically disposed in the major axis. It scintillates as if with innumerable brilliant points.'

ROBERTS 'Probably seen edge-wise, with several nuclei of nebulous character involved.'

LICK XIII 'A very pretty and irregular, elongated mass, 7′×1′.5 in P.A. 65° showing numerous rifts: an irregular lane divides it approximately along the shorter axis. It is possibly a very irregular spiral seen edge-wise. Exceedingly bright, the brighter condensations show easily on a 5 min exposure.'

M 82 has excited much lively interest and speculation ever since its strange, irregular form was noticed in Lord Rosse's giant telescope in 1871.

The nebula displays several systems of filaments extending for 3 to 4 kpc above and below the fundamental plane of the galaxy. These filaments have been observed to be expanding at right angles to the major axis at speeds up to 1000 km/s. This appears to be evidence of a gigantic explosion which began about 1½ million years ago and which is still continuing.

Sandage found that the filaments could be resolved into small arcs rather like those which can be seen in the 'Crab' Nebula (M 1) and, like that nebula, the light is strongly polarised and emits 'synchrotron radiation'. Both these facts are evidence of strong magnetic fields being present in the galaxy and M 82 was the first galaxy in which a large-scale magnetic field was detected.

The amount of material which may already have been expelled from M 82 by this process has been estimated at about 5 million times the Sun's mass and it is also thought that it may be a major source of cosmic radiation. An explosion similar to that occurring in M 82 may have begun in our own Galaxy about 10 million years ago for radio observations show that we are surrounded by a spheroidal halo of gas about 30×20 kpc in dimensions. Burbidge and Hoyle suggested in 1963 that this explosion probably originated in the nucleus.

Observations have also shown that the light from M 82 is distinctly reddened, which may be due to large quantities of obscuring matter present in the outer portions of the galaxy.

Holmberg gives the mass of M 82 as equivalent to about 50 000 million solar masses – about one-fifth of the mass of M 81 – and the linear diameter of about 11.2 kpc makes it the smallest of all the galaxies in Messier's list except for M 32, the elliptical companion of the Andromeda galaxy. The spectral class of M 82 is A5 and the colour index +0.54.

The distance modulus obtained by Holmberg for this group of galaxies – which besides M 81 and M 82, includes the Sc galaxy, NGC 2976 and another irregular, type II, NGC 3077 – is given as 27.30 mag. This is a red shift determination only and is equivalent to a distance of about 2.9 Mpc. Becvar gives a modulus of 26.5 mag. or about 2.0 Mpc; Garstang, in the B.A.A. *Handbook* gives 3.0 Mpc.

Dr. Halton C. Arp, in 1964, took photographs of M 81 with the 48-inch Schmidt telescope at Mt Palomar: he found that there was a faint ring around one end of the spiral and that this was brightest in the direction of M 82.

He considers that high-energy electrons emitted from M 82, striking the magnetic field surrounding M 81, are deflected and so radiate energy. Dr Arp also calculates that the explosion postulated in M 82 must have occurred about 400 000 years ago.

In recent years M 82 has been considered to be more 'hyperactive' than 'explosive'. The many radio sources around the nucleus have a high brightness temperature and are thought to be caused by recent Type II supernova events. They are also associated with the strong X-ray emission which has been observed. There is some evidence, too, of strong tidal interaction between M 81 and M 82, and this could be the driving mechanism for the high level of activity in the latter system. Infra-red and radio observations have also revealed more than 40 compact sources which are probably supernovae in an early stage of expansion. This is evidence of M 82 being a classical 'star-burst' galaxy. (For observing details see next page.)

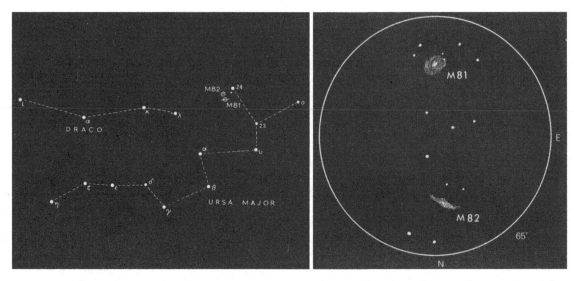

To observe
(M 81 and M 82)

These two galaxies are circumpolar for observers above 30° N. Continue the line γ–α *Ursae Majoris* for 12½° (which is 1½° more than the interval between them) to the NE to the 4½ mag. star *24 Ursae Majoris*. M 82 is on the same parallel as this star and 2° to the E; with M 81 in the same field on L.P., 45' to the S.

The two objects form a spectacular contrast; M 82 being a bright, elongated, slightly curved streak in P.A. about 070° and nearly 10' long while M 81 is a large, diffuse oval with wispy edges and a bright, oval nucleus.

M 81 is manifestly the brighter of the two but M 82 may appear the more distinctive owing to its elongated form and higher surface brightness.

Each object will stand individual magnification when more detail can be made out in each. In this case, the spiral galaxy, M 81, being more 'organised' than the irregular M 82, may respond more readily to close examination in moderate apertures.

Objects near by

45' to the SE of M 81 is NGC 3077, an elliptical galaxy about 2½'×2', mag. 11. It is rather faint and without detail.

1½° to the SW of M 81 is NGC 2976, a spiral galaxy seen nearly edge-on: 3½×1½', mag. 11.5.

M83 NGC 5236
A spiral galaxy, type Sc in HYDRA
R.A. 13 h 34.2 m.   Dec. S 29° 37' (1950).
Size 8'×7'   Mag. 10.1   Distance 2.6 Mpc, 8.5 million lt-yrs.

Discovered by Lacaille at the Cape of Good Hope 1751–1752.

LACAILLE 'A small, shapeless nebula.' (Lac. I. 6.)

MESSIER (Mar. 18th, 1781) 'Nebula without star near the head of Centaurus: it appears as a faint and even glow but it is difficult to see in the telescope as the least illumination of the micrometer wires makes it disappear: only with the greatest concentration is one able to see it at all. It forms a triangle with two stars estimated at 6th and 7th mag. Position determined from the stars i, k and h in the head of Centaurus. M de la Caille has already determined this nebula. See the end of this list.'

J. H. 'A very remarkable object, both W. H. and J. H. agree – very bright, very large, extended in direction 55°.1: extremely suddenly brighter in the middle where there is a nucleus.'

LASSELL 'A 3-branched spiral.'

LICK XIII 'A bright and unusually beautiful spiral, 10'×8'. The nucleus is 20" in diam. and very bright. In a 2 min exposure it shows as a faint disc with a bright peripheral streak or whorl. A large number of almost stellar condensations in the rather open whorls of this fine object.'

M83 is intrinsically one of the brighter galaxies: with an absolute magnitude of −19.5, it is of similar luminosity to M31 and our own Galaxy.

A supernova was observed in M83 by C. O. Lampland of Lowell Observatory in 1923 at magnitude 14: a second supernova was observed by G. Haro in 1950 which had a maximum brightness of about 14.5 mag. A third faint supernova or very bright nova was discovered by H. S. Gates on photographic plates taken at Mt Wilson: this was in one of the spiral arms, about 3' NNE of the nucleus. On July 6th, 1968, J. C. Bennett of Pretoria, South Africa, while sweeping for comets, discovered a fourth supernova in M83. Spectra taken at Radcliffe Observatory showed it to be of Type I, approximate magnitude 11 and 5" preceding the nucleus. A fifth supernova, of type undetermined, appeared on July 20th 1986 at mag. 11.2. This tally is the highest for any galaxy to date except NGC 6946, a 9th mag. spiral in Cepheus, which has produced no less than six supernovae between 1917 and 1980.

M83 was classified as Sc by Hubble but more recent observations show indications of an incipient bar in the inner portion of the system. G. de Vaucouleurs describes the galaxy as a typical transition type between ordinary and barred spirals of intermediate or late type. He classes it under a new, detailed scheme, as SAB (s)c.

The nucleus of M83 is about 20" in diameter, showing strong emission lines, and the spiral arms are seen to be separated by narrow dust lanes. (For observing details, see next page.)

251

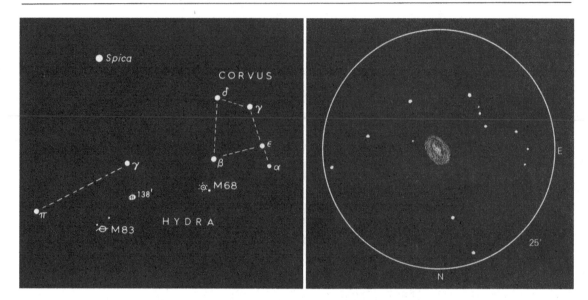

To observe   M 83 is a little difficult to find as there are no bright stars very near. The nearer ones in Centaurus are further south than the nebula and these may be out of sight for observers in latitudes above 45° N.

From β *Corvi* move 10° E to 3½ mag. γ *Hydrae*. From γ *Hydrae* move 6½° S (the elliptical galaxy 138 H. I may be seen at 3½° S) and then 3¼° E. There is a 5½ mag. star about 40′ to the NE of the nebula.

M 83 appears as a hazy star at first and seems much brighter than its 'advertised' magnitude of 10.1. The nucleus can then be seen to be surrounded with a faint, nebulous area which is elongated in P.A. about 50° and extends to about 8′×6′ and almost overlapping a distinct 10–11 mag. star to the SW.

This outer area is difficult to define but in a 12-inch a faint shading can be made out which suggests its branching shape.

To the SE is a bright, close double, about 5″ apart in P.A. 200°.

M 83 is a difficult object to see from Great Britain, and in the U.S.A. a latitude lower than about 45° N is preferable if the faint outer nebulosity is to be seen.

Objects near by   138 H. I (NGC 5061) mentioned above is a small, pearly, oval patch, slightly elongated in P.A. about 130°. It is hardly distinguished from a star on L.P.

M 84 NGC 4374
A galaxy, type SO in VIRGO
R.A. 12 h 22.5 m.   Dec. N 13° 10' (1950).
Size 2' diam.   Mag. 9.3   Distance 12.5 Mpc, 41 million lt-yrs.

MESSIER (Mar. 18th, 1781) 'Nebula without star in Virgo. In the centre it is pretty bright and surrounded with a slight nebulosity. The brightness and general appearance are similar to M 59 and M 60.'

J. H. (Catalogue of 1864) 'Very bright, pretty large, round, pretty suddenly brighter in the middle, scarcely resolvable.'

D'ARREST 'Shining with a brilliant nucleus. Round; increasing gradually towards the centre. At least 3' diam. The nucleus is equivalent to a 10 or 11 mag. star. Another nebula, smaller and paler, closely following to south.'

FLAMMARION 'A nebulous spot with a brilliant nucleus, gradually decreasing in brightness until vanishing at a distance of 35".'

ASTROPHYSICAL JOURNAL 1917 'Practically the same type and size as M 86; nucleus possibly a little larger.'

LICK XIII 'Very bright, round, 1' in diameter. Centre large. No spiral structure discernible.'

M 84 is an SO galaxy, which is an intermediate type between elliptical and spiral galaxies – a kind of spiral galaxy without any spiral arms. On long exposure photographs M 84 looks a typical SO galaxy but on shorter exposures it can be seen to have a ring of obscuring matter surrounding the nucleus, which is an unusual feature.

A supernova was observed 1½' N of the centre of M 84 by G. Romano in Italy on May 18th, 1957: its magnitude was then 13. On examining plates taken about a month earlier at Mt Palomar, it was seen to have been present then at a photographic magnitude of 13.4.

M 84 was identified as a radio source by H. C. Ko in 1957 and at one time the existence of a 'bridge' of radio emission was thought to connect this galaxy with M 87. This, however, has since been shown to be undetectable even by the 260-foot radio telescope of Ohio State Observatory.

Examples of such 'bridges' have been observed elsewhere, an example being that between IC 3481 and IC 3483 which are about 72 000 light-years apart, and there is a possibility that such a link of radio-emissive connecting material may exist between our Galaxy and the Large Magellanic Cloud.

Holmberg gives the mass of M 84 as 500 000 million solar masses and estimates its linear diameter to be about 38.9 kpc. The mass-density of the system is nearly as high as M 81, being approximately 0.08 solar masses per cubic parsec. The absolute magnitude of M 84 is −20.5, the spectrum of type G5 and colour index +0.77. Its distance modulus of 30.5 mag. makes it a true member of the Virgo Group of galaxies.

Observations with the Explorer satellite showed M 84 to be an X-ray emitter with about half the intensity of M 87. (For observing details, see next page.)

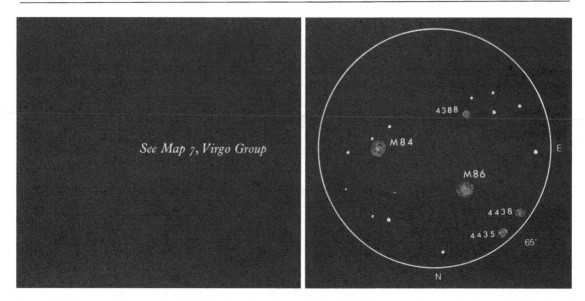

See Map 7, Virgo Group

To find   (M 84 and M 86) See *Guide to Virgo Group* (p. 387).

To observe   M 84 and M 86 appear together in the field of an L.P. eyepiece, M 86 being about 20′ to the E and slightly N of M 84. They are both about the same size and of similar brightness with M 84 slightly the clearer. M 84 appears quite round while M 86 can be seen to be slightly elongated in P.A. about 080°. Both galaxies brighten towards the centre but the nucleus is not very well defined in either.

Objects near by   About 20′ S of M 86 can be seen the distinct but faint nucleus of another galaxy; this is NGC 4388 which is actually a large edge-on spiral, but its outer portions are too faint to be seen in small instruments.

About 20′ to the NE of M 86 is another pair of galaxies; NGC 4435 and 4438, about 7′ apart. They are both quite bright and easily seen, 4435 being an E4 galaxy of mag. 10.3 and 4438 a 10.8 mag. spiral.

A 1° field here discloses no less than five galaxies, all of which can be seen in a 6-inch when conditions are favourable.

M 85 NGC 4382
A galaxy, type SO in COMA BERENICES
R.A. 12 h 22.9 m.   Dec. N 18° 28' (1950).
Size 3' diam.   Mag. 9.3   Distance 12.5 Mpc, 41 million lt-yrs.

Discovered by Méchain in 1781.

MESSIER   (March 18th, 1781) 'Nebula without star, above and near to the ear of Virgo between the two stars in Coma Berenices, Nos. 11 and 14 of Flamsteed's catalogue: this nebula is very faint. M. Méchain had determined its position on 4th Mar. 1781.'

SMYTH   'A bright and rather large nebula, brightening in the middle. 11 Comae Berenicis, mag. 5 precedes a little to S.'

WEBB   'Fair specimen of the many nebulae in this region: midway from 24 to 11: the nearest bright star p. a little S.'

D'ARREST   'Large, bright, strongly dense near the centre. Another nebula immediately following (55 H. II).'

LICK XIII   'Very bright oval, 4'×2'. Very slight traces of spiral structure.'

M 85 is very nearly a 'twin' with M 84 in almost all its essential statistics. It is an SO galaxy with the same visual magnitude of 9.3 and similar apparent dimensions: its mass, equivalent to about 400 000 million solar masses, is only a little less than M 84 while the linear diameter, of 39.8 kpc, is slightly larger.

The spectral classes of the two galaxies are the same, both being of type G5 while the absolute magnitude of M 85 is very slightly the brighter at −20.7. The colour index of M 85 is +0.07 which is a little less reddened than M 84.

The mass-density of M 85 is about 0.06 solar masses per cubic parsec which is a little less than that of M 84, as would be expected from its smaller mass and greater dimensions.

M 85 is also an accepted member of the Virgo Group.

A supernova Type I reached 11.7 mag. on Dec. 20th, 1960. (For observing details see next page.)

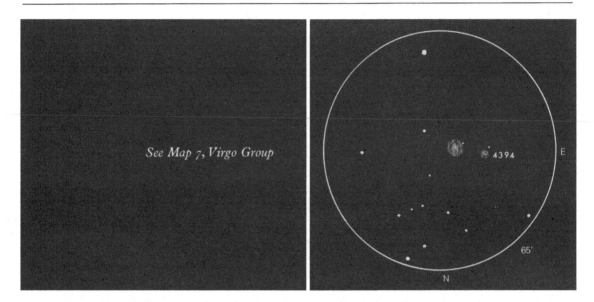

See Map 7, Virgo Group

To find   See *Guide to Virgo Group* (p. 387).

To observe   M 85 is very small, about 3′ diam., but it is bright and concentrated. On some occasions it can be seen to be extending to about 5′ in P.A. approximately 30°. It has a small, fairly bright nucleus which is also elongated. This SO galaxy will stand M.P. quite well but no further detail can be made out with the higher magnification.

Objects near by   About 8′ to the E of M 85 can be seen NGC 4394, looking like a hazy star. It is an SB galaxy of mag. 11.2. It is round and distinct and a higher magnification will show it to be distinctly nebulous. This is 55 H. II, mentioned by D'Arrest.

M 86 NGC 4406
An elliptical galaxy, type E3 in VIRGO
R.A. 12 h 23.7 m.   Dec. N 13° 13' (1950).
Size 3'×2'   Mag. 9.7   Distance 6 Mpc, 20 million lt-yrs.

MESSIER (Mar. 18th, 1781) 'Nebula without star in Virgo on the same parallel and very near the nebula No. 84 above: they both have the same appearance and are seen together in the same field in the telescope.'

J. H. (Catalogue of 1864) 'Very bright, large, round, gradually brighter in the middle where it exhibits a nucleus; scarcely resolvable.'

D'ARREST 'Large, shining nebula, circular; nucleus equivalent to an 11 or 12 mag. star. The whole region is abundant with true and false nebulae.'

WEBB 'In this neighbourhood, Lowe sees 7 nebulae in large field with fine 15½-inch Calver.'

LICK XIII 'Slightly oval 2' long; bright centre which is not stellar in short exposures. No spiral structure discernible.'

Although M 86 is often included in the Virgo Group of galaxies and appears well in the middle of the group as we observe it, Holmberg found the distance modulus of this galaxy to be only 28.9 mag. compared with a modulus of 30.5 mag. for the rest.

This puts M 86, together with the Sap galaxy, NGC 4438, well in the foreground of the Virgo Group at a distance of about 6.0 Mpc – approximately half way between us and the main body of the Virgo galaxies.

The mass of M 86, in Holmberg's catalogue, is equivalent to about 130 000 million solar masses and its linear diameter about 20.9 kpc. Its absolute magnitude is −19.1 and the spectrum is of type G7.

Its colour index, of +0.78, shows a considerable amount of reddening, being only exceeded by M 60 among the Messier galaxies.

Markarian has pointed out that the Virgo Group contains a conspicuous 'chain' of bright galaxies of which M 84 and M 86 form prominent 'links'. This arrangement seems similar to the chain-like patterns of stars in some galactic associations and may have a cosmological significance. (For observing details see M 84.)

M 87 NGC 4486
Giant elliptical galaxy, type E0 in VIRGO
R.A. 12 h 28.3 m.   Dec. N 12° 40' (1950).
Size 3' diam.   Mag. 9.2   Distance 19.3 Mpc, 63 million lt-yrs.

MESSIER (March 18th, 1781) 'Nebula without star in Virgo, below and very near an 8 mag. star; the star having the same R.A. as the nebula, and its Dec. was 13° 42' 21" north. This nebula appears to have the same light as the two nebulae, Nos. 84 and 86.'

J. H. (Catalogue of 1864) 'Very bright, pretty large, round, very suddenly brighter in the middle where there is a nucleus.'

ROSSE 'Seen on Apr. 3rd, 1861. Very bright, round; fades off gradually.'

D'ARREST 'Very large and bright, perfectly circular, diam. 85". A little brighter towards the middle and, in the absolute centre, almost resolved to a star of 9 or 10 mag.'

LICK XIII 'Exceedingly bright; the sharp nucleus shows well in a 5 min. exposure. The brighter central portion is about 0'.5 in diameter and the total diameter about 2', nearly round. No spiral structure is discernible. A curious, straight ray lies in a gap in the nebulosity in P.A. 20° apparently connected with the nucleus by a thin line of matter. The ray is brightest at its inner end which is 11" from the nucleus.'

M 87 is almost certainly the best known of the elliptical galaxies: it is one of the largest and brightest and it also displays a curious and spectacular 'jet'.

Photographs taken with the 200-inch Hale reflector at Mt Palomar show that it is surrounded by a halo of more than 1000 globular clusters of absolute magnitude −6 to −8 which are too distant to be resolved into stars.

The famous jet is straight and has a spectrum which is much bluer than that of the galaxy itself. It extends for a distance of at least 1500 parsecs in a direction directly towards the SO galaxy M 84 and has been thought to be part of a 'bridge' of matter joining the two galaxies (see M 84).

The light from the jet is strongly polarised and several strong condensations of material have been observed in its outer regions.

The galaxy, as an E0 type, is considered to be in a later stage of development than the spirals such as our own but several novae have been observed in it and the jet itself is evidence of some vigorous activity taking place.

In addition, a supernova was discovered by I. Balanowski in 1922. From photographic plates it was found that its maximum brightness of 11.5 mag. had occurred in Feb. 1919.

M 87 is a strong radio source on 18.3 Mc/s and, in 1956, J. E. Baldwin and F. G. Smith of Cambridge detected a weak halo of radiation on 81.5 Mc/s similar to that observed in our Galaxy and M 31.

As a result of co-operation between American and Russian astronomers, radio observations using a baseline linking Goldstone Observatory and a similar facility in the Crimea in 1971 were made in an attempt to measure the angular dimensions of the nucleus. This gave a value of 0.001 arcsec which, at the distance of the galaxy, amounts to a quarter of a light-year.

From an analysis, made in 1966 of an Aerobee rocket fired in April 1965,

M 87 has also been found to be an X-ray emitter with an intensity between 10 and 100 times its combined radio and optical emission.

In Holmberg's *Catalogue of External Galaxies*, M 87 is quoted as having a mass equivalent to about 790 000 million solar masses and a diameter of 38.9 kpc. Holmberg gives the absolute magnitude as −21.2; the spectral class as G5 and colour index +0.73.

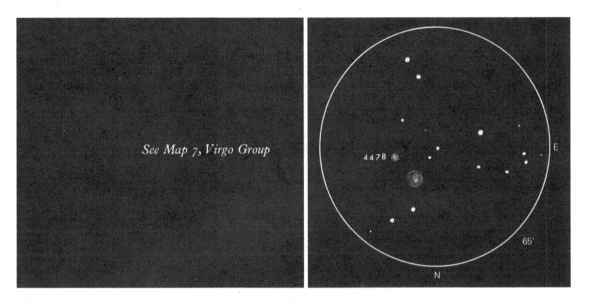

To find   See *Guide to Virgo Group* (p. 387).

To observe   This well-known giant elliptical galaxy is an easy object to find, being very bright and distinct. It has a brilliant nucleus which is not quite star-like. From the centre of the nebula the brightness falls off gradually towards the edge and although it stands higher magnification well, no greater detail can be made out. There appears to be no easily visible indication of the famous 'jet'.

Objects near by   About 10' to the SW of M 87 is another hazy object of very similar appearance but very much smaller, looking, at first, merely like a nebulous star. This is the elliptical galaxy NGC 4478, of magnitude 10.9 and a diameter of only about 1'. A fair magnification may be needed to distinguish it from a star.

M 88 NGC 4501

A spiral galaxy, type Sc in COMA BERENICES
R.A. 12 h 29.5 m.   Dec. N 14° 42' (1950).
Size 5'×2'   Mag. 10.2   Distance 12.5 Mpc, 41 million lt-yrs.

MESSIER   (March 18th, 1781) 'Nebula without star in Virgo, between two small stars and one of 6 mag. which appear at the same time as the nebula in the field of the telescope. It is one of the fainter nebulae and resembles the one reported in Virgo as No. 58.'

SMYTH   'A long elliptical nebula. Pale white and trends in a line bearing N.p. and S.f. and with its attendant stars forms a pretty pageant. The N part is brighter than the S – a circumstance which, with its spindle figure, opens a large field for conjecture.'

ROSSE   (Mar. 9th, 1850) 'Spiral; dark spaces, especially one S.f. nucleus.'

D'ARREST   '7'×1½' with a nucleus. In wonderful nebulous region.'

WEBB   'Long, pale: marvellous region as swept with 64×. Identification difficult.'

LICK XIII   'A bright, beautiful spiral, 5'×2'.5 in P.A. 140°. Bright, elongated nuclear region, including a bright, almost stellar nucleus. The whorls are rather close and show numerous condensations.'

In Erik Holmberg's *Catalogue of External Galaxies*, M 88 is quoted as having a mass equivalent to about 100 000 million solar masses: this is about one half the mass of our own Galaxy. Its linear diameter, however, is considerably larger, being about 34.7 kpc, compared with about 30.0 kpc for ours.

Its mass-density is about 0.025 solar masses per cubic parsec or one solar mass per 40 cubic parsecs. Its absolute magnitude is given as −21.3, a spectral class of G5 and colour index +0.31.

Its distance modulus is given by Holmberg as 30.5 mag., obtained by both photometric and red shift methods which confirms it as a true member of the Virgo Group.

Becvar, in the *Atlas Coeli Katalog*, gives the apparent dimensions as 5'.5×2'.4 and apparent brightness as 10.1 mag. photographic and 10.2 mag. visual. (For observing details see next page.)

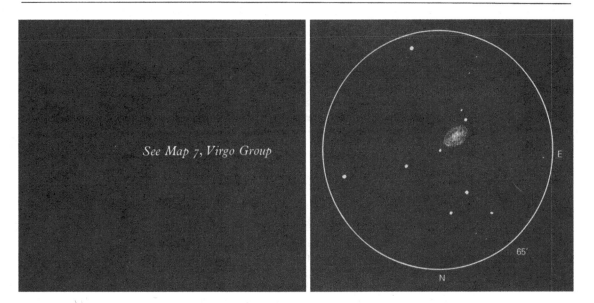

See Map 7, Virgo Group

To find   See *Guide to Virgo Group*.

To observe   M 88 appears a good deal brighter than its reported magnitude of 10.2. It stands out well as a bright oval extended in P.A. about 135° and has a distinct, elongated nucleus.

It is one of the best of the Virgo group galaxies for viewing in small telescopes and it can be immediately recognised by the close pair of stars at its SE edge and the single star close to the NW. It will stand M.P. well and, with averted vision, a considerable amount of faint detail can be made out from time to time.

Objects near by   There are no other bright galaxies in the same field but a little over 1° to the SW – towards M 86 – there are more than half-a-dozen brighter than mag. 13, some of which can be seen in a 6-inch or 8-inch.

Wm Herschel reported another nebula, 'just following M 88' and this he numbered 118 H. II. This was omitted from the NGC on the grounds that 'nobody after W. H. has seen any neb. f. M 88' (Dreyer's note).

M 89 NGC 4552
An elliptical galaxy, type E0 in VIRGO
R.A. 12 h 33.1 m.   Dec. N 12° 50′ (1950).
Size 3′ diam.   Mag. 9.5   Distance 12.5 Mpc, 41 million lt-yrs.

MESSIER

(March 18th, 1781) 'Nebula without star in Virgo, a little distance from and on the same parallel as the nebula reported above, No. 87. Its light was extremely faint and pale and it can be seen only with difficulty.'

J. H.

(Catalogue of 1864) 'Pretty bright, pretty small; round, gradually much brighter in the middle.'

D'ARREST

'Very fine: round; 45″ diam. A good deal brighter in the centre. Nucleus equivalent to a 10 mag. star. 14 mag. star follows at 7 s, 30 and 16″ to S, 9 mag. star 21 s 40 and 461″ to N (Bonn 2552).'

FLAMMARION

'A slight nebulosity, less brilliant and less extended than M 87. Increasing in brightness towards the centre.'

M 89 is an elliptical galaxy with no spiral structure: it is of type E0 which indicates that its outline is effectively circular, and in Becvar's *Atlas Coeli Katalog* its apparent dimensions are given as 1′.3×1′.3. The galaxy, however, may well be an oblate spheroid whose axis of rotation is directed towards us, appearing circular to us but elliptical in outline if seen from a different angle.

Its mass has been estimated as equivalent to about 250 000 million solar masses which is about 25 per cent greater than the probable mass of our own galaxy.

Becvar gives its photographic magnitude as 11.0 and visual magnitude 9.5. Its spectral class is G7. (For observing details see M 90.)

M 90 NGC 4569
A spiral galaxy, type Sb in VIRGO
R.A. 12 h 34.3 m.   Dec. N 13° 26' (1950).
Size 7'×3'   Mag. 10.0   Distance 12.5 Mpc, 41 million lt-yrs.

MESSIER   (March 18th, 1781) 'Nebula without star in Virgo. Its light as faint as the preceding, No. 89.'

J. H.   (Catalogue of 1864) 'Pretty large, brighter in the middle, where it exhibits a nucleus.'

D'ARREST   (April 21st, 1862 – air calm and clear) 'Splendid, immense: in the centre of an elliptical nebula, 7'×90" is a shining and sparkling 11 mag. star. The nucleus is a true star and no more than a point of light.'
(May 1st, 1864 – sky exceedingly bright) 'Very large and elongated nebula surrounding a 12 mag. star; 7'×2'. Nucleus rather as if it were a star gleaming through the nebula, nearer the preceding limb.'

FLAMMARION   'Fairly large nebulosity nearly 10' in extent, elongated SSW–NNE. Of a faint light but with a brilliant nucleus at the centre.'

M 90 is one of the larger spiral galaxies: its linear diameter, as given in Holmberg's catalogue, is 42.7 kpc, which is the same as that given for the giant elliptical galaxy M 49. On the other hand, its mass is equivalent to about 79 000 million solar masses. Its overall mass-density, therefore, is rather low, being equivalent to about one solar mass per 100 cubic parsecs.

The absolute magnitude of M 90 is −21.6, its colour index is +0.20 and spectral class G0.

In Antonin Becvar's *Atlas Coeli Katalog*, the apparent dimensions are 7'.5×2'.2 and apparent magnitude 10.1 photographic and 10.0 visual. It is a true member of the Virgo Group of galaxies. (For observing details see next page.)

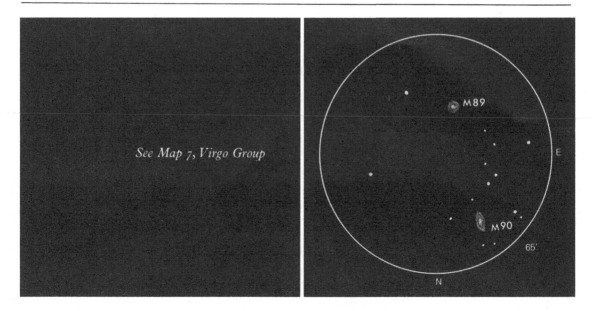

To find   (M 89 and M 90) See *Guide to Virgo Group* (p. 387).

To observe   M 89 and M 90 can both be seen together in the field of a L.P. eyepiece, M 90 being about 40′ N of M 89. The two galaxies are connected by a long, zig-zag chain of faint stars.

M 89 is very small, being only about 3′ in diameter, perfectly round and having a well-marked central condensation which is almost, but not quite, star-like. It is a bright object and stands out clearly, exhibiting the typical 'pearly' appearance of an E0 galaxy.

M 90 is a good deal larger than M 89 and extends to about 7′×3′ with the major axis in P.A. about 20°. It has a bright central nucleus also elongated and although the spiral M 90 is slightly fainter than M 89, it is very distinct and easily seen.

Objects near by   There are no other bright nebulae in the same field but about 20′ S of M 89 is a small, faint elliptical galaxy, NGC 4550, of magnitude 11.7.

M 91   NGC 4548 (120 H. II)
A spiral galaxy, type SBb in COMA BERENICES
R.A. 12 h 32.9 m.   Dec. N 14° 46′ (1950).
Size 3′.7×3′.2   Mag. 10.8   Distance 11.48 Mpc, 37 million lt-yrs.

One of the 'missing' Messier Objects, here identified as NGC 4548. Discovered by Messier in 1781.

MESSIER   (March 18th, 1781) 'Nebula without star in Virgo, above the preceding No. 90: its light again fainter than that above.'

(After the entry for M 91 in the *Connaissance des Temps* for 1784, Messier added the following note:

'The constellation of Virgo and especially the northern Wing is one of the constellations which encloses the most nebulae. This catalogue contains thirteen which have been determined, viz. Nos. 49, 58, 59, 60, 61, 84, 85, 86, 88, 87, 89, 90 and 91. All these nebulae appear to be without stars and can be seen only in a good sky and near meridian passage. Most of these nebulae have been pointed out to me by M. Méchain.')

W.H.   Identified in *NGC* as '602 H. III (NGC 4571) M 91??'

J. H.   Notes in *NGC*, 'NGC 4571, M 91 must have been a comet.' (Dreyer's note.)

NGC   (4548) 'Bright, Large, little extended, little brighter in middle.'

D'ARREST   (1863) 'M 91?? JH 1367, the place is that of a star of 8–9 mag. A spot described by Herschel as "a small, well-defined body ... The nebulae M 91 in this position, no longer exists in the heavens.'

FLAMMARION   Referred to M 91 as an enigma, and 'In all probability M 91 was a passing comet.'

SHAPLEY AND
DAVIS   *P.A.S. Pacific*, Vol. 29, 1917, 'M 91 probably a comet.'

O. GINGERICH   *Sky and Telescope*, Oct. 1960, 'Although the evidence is circumstantial, I believe we can reasonably assume that the listing of M 91 resulted from a duplicate observation of M 58.'

W. C. WILLIAMS   *Sky and Telescope*, Dec. 1969, 'It can be simply demonstrated that M 91 is very probably the galaxy NGC 4548.'

G. L. FORTIER   *Skyward*, (R.A.S. of Canada Newsletter, Jan. 1972) 'I propose that NGC 4548 be placed on the Messier list as M 91.'

NGC 4548 is seen almost face on with the distinctive bar lying in P.A. 65°/245°. Two broad spiral arms emerge as tangents to the bar, and in photographs can be traced through more than 200° of arc.

In Holmberg's catalogue the abs. mag. is given as −20.5, the colour index as +0.32 and spectral class G5. De Vaucouleurs, in his *Second Reference Catalogue of Bright Galaxies* (1976), classes the NGC 4548 nucleus as D3 (moderately bright but diffuse).

The mass density of the system is almost the same as M 63. The radial velocity is 403 km/s, which confirms its membership of the Virgo Group. (For observing details, see next page.)

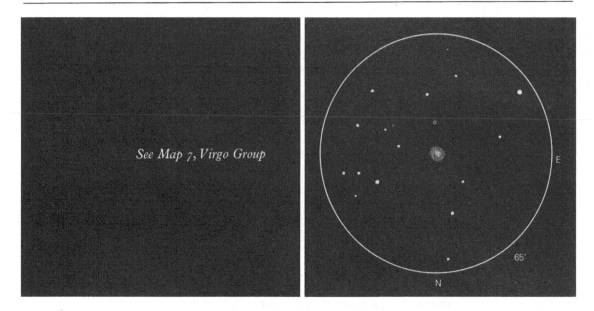

See Map 7, Virgo Group

To find   See *Guide to Virgo Group* (p. 387).

To observe   M 91 is 1° E and slightly N of M 88. It can be seen easily in apertures of 6 to 8 inches in good conditions, when it appears as quite large (about 3' to 4' diam.) although rather formless at first. With close attention it can be seen to glow brighter towards the centre, but no clear-cut nucleus is apparent. It stands M.P. quite well and this should reveal a slight elongation in P.A. about 60° which is the axis of the bar. To make out the bar itself, however, would require an aperture of at least 12 inches.

A comparison with the nearby objects, M 88 (about 1° to the W) and M 90 (about 1½° to the SSE) shows M 91 to be almost one magnitude fainter than M 88 and almost ½ mag. fainter than M 90. (This fits reasonably well with Messier's descriptions of these objects.)

There are a few brightish stars in a L.P. field around M 91, and NGC 4571 (once thought to be a candidate for the 'missing' object) may be glimpsed just to the south of a distinct star about 30' away in P.A. 120°. In an 8-inch it is rather pale and faint, with a small, star-like nucleus, and the outline seems irregular but slightly elongated in P.A. 100°. Seeing both 4548 and 4571 in the same field leaves one in no doubt as to which is the more likely to have been seen by Messier, as the former appears distinctly the brighter.

M 92 NGC 6341
A globular cluster, Class IV in HERCULES
R.A. 17 h 15.6 m.   Dec. N 43° 12' (1950).
Size 10' diam.   Mag. 6.1   Distance 8.6 kpc, 28 000 lt-yrs.

Discovered by J. E. Bode in 1777.

BODE (Dec. 27th, 1777) 'More or less round, with a pale glow.'

MESSIER (Mar. 18th, 1781) 'A fine, distinct and very bright nebula in Hercules: it can be easily seen in a telescope of one foot. It contains no star, the centre is clear and bright, surrounded by nebulosity and it resembles the nucleus of a large comet. In size and brightness it closely resembles the nebula which is in the belt of Hercules – see No. 13 of this catalogue. Its position has been determined by direct comparison with the star σ Herculis, 4th mag.: the nebula and the star on the same parallel. Diam. 5'.'

W. H. (1783) 'A brilliant cluster, 7'–8' in diam.'

J. H. 'A globular cluster, very bright and large: well resolved into small stars.'

SMYTH 'A globular cluster of minute stars; large, bright and resolvable with a very luminous centre and, under the best vision, has irregular, streamy edges.'

ROSSE 'Nucleus possibly spiral; darker spaces: nucleus barely, if at all resolved.'

WEBB 'Very fine cluster, though not equal to M 13, less resolvable, intensely bright in centre.'

LICK XIII 'Bright, unusually condensed globular cluster. Cluster about 10' diam. the bright central portion 2' in diam.'

Shapley detected only 13 or 14 short-period variables in this globular cluster whose ellipticity he found to be 8 in P.A. 25°.

M 92 (Smyth) From A Cycle of Celestial Objects *(B.A.A.)*

M 92 (Trouvelot) Drawn at Harvard College Observatory, 1874 (B.A.A.)

Its absolute magnitude has been estimated as −7.8. Swartzchild and Bernstein at Princeton University Observatory obtained a value for its mass of 140 000 Suns while another estimate gives 330 000 Suns.

Studies of M 92 were made by Baade when investigating stellar evolution and he put the age of the cluster at about 2000–3000 million years.

It is not a very rewarding cluster for spectroscopic observations as the metallic lines in its spectrum are very weak and difficult to detect. Many observations of its distance have been made and most show a reasonable agreement. Its distance modulus, obtained in 1964, was 14.62 mag. which is equivalent to a distance of about 9 kpc.

The weakness of the metallic lines in the spectrum of M 92 reveals a metal deficiency in its component stars and for one K-type giant in this cluster the total metal content was found to be only about 1 per cent of that of the Sun.

Helen Sawyer Hogg, in her catalogue of clusters, gives the spectrum of M 92 as A5, the colour index as −0.09 and the average magnitude of the 25 brightest stars as 13.86.

In the *Atlas Coeli Katalog*, the linear diameter of the cluster is given as 27 parsecs or about 88 light-years. Kinman's figure for the distance of M 92 is 11.5 kpc.

From more recent measurements, using the few known *RR Lyrae* variables, the distance of M 92 has been found to be 8.6 kpc or about 28 000 light-years. (For observing details see next page.)

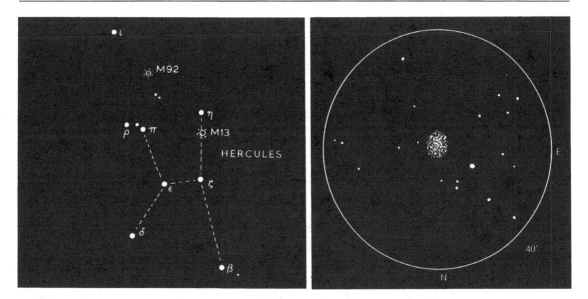

To observe   Although M 92 is large and bright, it is a little distance apart from any of the brighter stars and it may prove more elusive than expected if one's approach is 'hit or miss'.

It lies on the line between ι and η *Herculis* and a little east of north from π *Herculis, or,* move $2\frac{1}{2}°$ S and $3\frac{1}{2}°$ W from ι *Herculis.* It can be seen quite well with the naked eye on a clear night.

M 92 is very similar to the other great globular cluster in Hercules, M 13, but it is slightly smaller and a little more concentrated (class IV as compared with class V for M 13).

It does not appear to be either circular or symmetrical, being slightly flattened on the eastern side and a little elongated N–S. It is easily resolved in the outer portions but the central nucleus requires a pretty high magnification to see the individual stars. However, the cluster is very bright and with a good aperture and fine seeing, a full resolution can be accomplished in an 8-inch. There is no doubt that it is a fine cluster and is only a little inferior to M 13 in splendour.

M 93 NGC 2447
A galactic cluster, type 'g' in ARGO PUPPIS
R.A. 07 h 42.4 m.   Dec. S 23° 45' (1950).
Size 18' diam.   Mag. 6.0   Distance 1.1 kpc, 3600 lt-yrs.

MESSIER (March 20th, 1781) 'A cluster of small stars without nebulosity between Canis Major and the prow of Navis. Diam. 8'.'

W. H. Resolved into stars of 8–13 mag.

SMYTH 'A neat group of star-fish shape. S.p. portion being brightest, with individuals of 7–12 mag. Mistaken for a comet by Chevalier d'Angos of the Grand Master's Observatory in Malta. Baron de Zach called such egregious astronomical blunders, "Angosiades".'

WEBB 'Bright cluster in a rich neighbourhood.'

ROBERTS 'A cluster of widely scattered stars.'

Trumpler's classification for this galactic cluster is I,3,r. According to Helen Sawyer Hogg, the apparent diameter of the cluster is 18', the integrated magnitude 6.5 and the magnitude of the brightest star, which has a spectrum of type B9, is given as 9.7.

Åke Wallenquist, in his catalogue of galactic clusters, gives a count of 63 probable member stars within a diameter of 18' and estimates the density of stars to be 26 stars per cubic parsec in the central portion and an overall density of about 0.75 stars per cubic parsec.

Wallenquist gives a value for the linear diameter of M 93 as 5.5 parsecs or about 18 light-years but in the *Atlas Coeli Katalog* the diameter is given as 8.0 parsecs or 26 light-years.

Hogg gives a distance of 700 parsecs but Becvar's figure is 1100 parsecs and Wallenquist is in close agreement with the latter, giving 1040 parsecs. (For observing details see next page.)

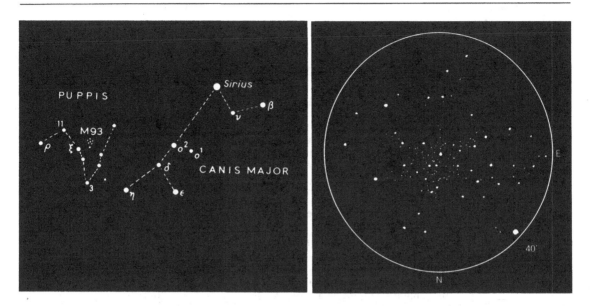

To observe   This bright cluster is easy to find but for observers in the higher northern latitudes it needs to be taken at or near culmination.

From *Sirius* move 8° SE to *o² Canis Majoris* which is on the same parallel as M 93, and then sweep 10° E.

M 93 is a bright, fairly condensed cluster, containing 20 or more brightish stars and many fainter still, concentrated toward the centre. The brighter stars are spread out in straggling arms which could be held to resemble something like a star-fish as Smyth suggested. To me, however, the overall pattern strongly suggested a butterfly with open wings.

The two brightest stars in the cluster are to the SW and there is a 6 mag. star, separate from the cluster, 20' to the NE of the centre.

Several observers in favourable latitudes have reported M 93 as being visible to the naked eye in good conditions.

M 94 NGC 4736
A spiral galaxy, type Sb in CANES VENATICI
R.A. 12 h 48.5 m.   Dec. N 41° 24' (1950).
Size 3'×2'   Mag. 7.9   Distance 10 Mpc, 32.6 million lt-yrs.

Discovered by Méchain in 1781.

MESSIER   (March 24th, 1781) 'Nebula without star above Charles' Heart. On the same parallel as the star beta. It is brilliant in the centre and the nebulosity is a little diffuse. It resembles the nebula which is below Lepus, No. 79, but is finer and brighter. M. Méchain made its discovery on Mar. 22nd, 1781.'

J. H.   'Large, very bright, irregularly round, with a bright nucleus; barely resolvable.'

SMYTH   'A comet-like nebula; a fine, pale-white object with evident symptoms of being a compressed cluster of small stars. It brightens toward the middle. Several small stars in field, of which one in S.f. quadrant is double.'

ROSSE   (Apr. 13th, 1855) 'A dark ring round the nucleus; then bright ring exterior to this. The annulus, however is not perfect, but broken up and patchy, and the object will probably turn out to be a spiral. Much faint outlying nebulosity.'

D'ARREST   'Very fine; an almost circular disc with a very bright nucleus, shaped like a planetary; sky-blue, like an 8 or 9 mag. star surrounded by a limpid, elongated nebulosity, $1\frac{1}{2}'$ to 2' diam.

WEBB   'Small bright nebula like a comet.'

ROBERTS   Photo May, 1892. 'A stellar nucleus and rings. I am unable to trace any spiral structure on the photograph.'

LICK XIII   'A beautiful object; from the very bright, large nucleus spring many bright, closely-packed whorls, forming a bright inner oval $2'\times1\frac{1}{2}'$ in P.A. about 110°. These inner whorls show many stellar condensations whose sharpness and proximity to the nucleus would seem to make this one of the most favourable examples known for the investigation of motion in spirals. Fainter, closely packed outer whorls bring the nebula to a size of $5'\times3\frac{1}{2}'$.'

M 94 is a face-on galaxy in which the spiral arms are tightly wound: the very compact nucleus is seen to contain dark bands and bright matter intermingled. It is sometimes classed as Sa but is more probably a peculiar type of Sb.

E. M. and G. R. Burbidge attempted to obtain a rotation curve for this galaxy in 1961. The evidence showed that there may be considerable departure from circular motion in the disc or that the inner and outer sections may have different planes of rotation. On plates taken with the 48-inch Schmidt camera, a fainter, outer ring can be detected which has a major axis about 30° different in P.A. from that of the inner, brighter portion.

M 94 has a true velocity of recession, corrected for our movement in the Galaxy, between 359 and 398 km/s.

Holmberg's figure for the mass of M 94 is equivalent to 250 000 million solar masses and he gives it the large linear diameter of 41.7 kpc. The mass-density is approximately one solar mass per 30 cubic parsecs. Its absolute magnitude is given as −21.7, the spectral class is G0 and colour index +0.43.

Becvar quotes the apparent dimensions as 5'.0×3'.5; the photographic magnitude as 8.9 and visual magnitude 7.9. His distance modulus is given as 28.2 mag., equivalent to about 4.4 Mpc, but Holmberg's figure for the modulus is 29.9 mag. from photometric measurements, which works out at close to 10 Mpc.

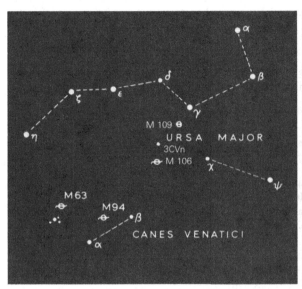

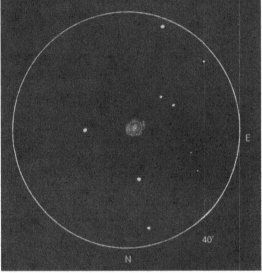

To observe   Although small, M 94 is not difficult to pick up as its nucleus, at least, is bright and distinct. From α *Canem Venaticorum* (3rd mag.) move 2¾° N and 1° W.

M 94 exhibits a very bright, intense nucleus which is surrounded by a hazy, slightly elliptical halo. This outer portion is very diffuse and its extent is very difficult to determine but it can be seen to cover an area of at least 3'×2'.

The major axis of the nebula is also rather difficult to assess but it can be made out, with averted vision, to lie between 90° and 150°. There are very few background stars, the brightest being about 20' to the south.

Objects near by   1° 35' E of M 94 is the Sc galaxy, NGC 4618. It is rather faint (mag. 11.7 visual), large (3½'×3') and very diffuse.

M 95 NGC 3351
A barred spiral galaxy, type SBb in LEO
R.A. 10 h 41.3 m.   Dec. N 11° 58' (1950).
Size 3' diam.   Mag. 10.4   Distance 9 Mpc, 29.3 million lt-yrs.

Discovered by Méchain in 1781.

MESSIER (March 24th, 1781) 'Nebula without star in Leo above the star "l" (53 Leonis): its light is very faint.'

J. H. 'Bright, large, round; pretty gradually much brighter in the middle to a nucleus.'

SMYTH 'A lucid white nebula with only 2 stars N.p. & N.f. in field. This nebula is round and bright and perhaps better defined on S than N limb.'

ROSSE 'Two ellipses: centre perhaps resolvable.'

D'ARREST 'Beautiful and very bright, slightly oval, much denser in the middle.'

LICK XIII 'A beautiful object, nearly round, 3' in diam. The whorls are rather faint and form an almost complete ring: a wide band of matter extends across the nebula from one side to the other: an example of the φ-type spiral. The centre is exceedingly bright and of unusual structure. It is about 12" in diameter and appears tri-nuclear; the centre of the disc is not marked by any condensation but on the periphery are two stellar nuclei and a short, very bright, slightly curved mass.'

M 95 is a member of the Leo Group of galaxies which includes M 65, M 66, M 96 and four other bright spirals, NGC 3338, 3423, 3628 and 3810. Holmberg has obtained a distance modulus for all these galaxies of 29.8 mag. by both photometric and red shift methods with a mean error of ±0.13 mag. This is equivalent to a distance of about 9.0 Mpc.

The mass of M 95, given in Holmberg's catalogue, is equivalent to 100 000 million solar masses and its linear diameter, 24.6 kpc. The mass-density of the system works out at about one solar mass per 16 cubic parsecs.

Its absolute magnitude is −20.3, its spectral class F5 and colour index 0.48.

Becvar gives the apparent magnitude as 10.5 photographic and 10.4 visual. The apparent dimensions are 6'.1×3'.9. (For observing details see M 96.)

M 96   NGC 3368
A spiral galaxy, type Sa in LEO
R.A. 10 h 44.1 m.   Dec. N 12° 05′ (1950).
Size 4′×3′   Mag. 9.1   Distance 9 Mpc, 29.3 million lt-yrs.

Discovered by Méchain in 1781.

MESSIER   (March 24th, 1781) 'Nebula without star near the preceding (No. 95): this one is less distinct; both are on the same parallel as Regulus. They resemble the two nebulae in Virgo, Nos. 84 and 86. M. Méchain saw them both on March 20th, 1781.'

J. H.   'Very bright, very large, a little elongated, very suddenly very much brighter in the middle; barely resolvable.'

SMYTH   'Round; not as well defined as M 95; large and pale white.'

D'ARREST   'A twin nebula with M 95 although perhaps exceeding it in splendour. A heap, poorly terminated at the margin. Resolved with magnification 226×.'

LICK XIII   'A fine, strong spiral, brightest part 2′.6 long. Bright stellar nucleus; the outer whorl of the brighter structure forms a nearly complete oval ring: a much fainter whorl outside this brings the total length to about 7′. A clear-cut rift goes in to the nucleus on N.p. side.'

M 96 is a near neighbour of M 95 in the Leo Group of galaxies. It is both larger and more massive than the latter but has about the same overall density. In Holmberg's catalogue, its linear diameter is given as 28.8 kpc and its mass as equivalent to about 160 000 million solar masses. The mass density is quoted as about 0.063 solar masses per cubic parsec or, in reciprocal terms, one solar mass to every 16 cubic parsecs.

It is a little more intrinsically luminous than M 95, having an absolute magnitude of −20.7. Its spectral class is G0 and colour index +0.51.

In Becvar's *Atlas Coeli Katalog*, the apparent magnitude is 10.1 photographic and 9.1 visual and the apparent dimensions 5′.0×4′.0.

According to de Vaucouleurs, the plane of M 96 is inclined at about 35° to our line of sight and rotates with the spiral arms trailing. (For observing details see next page.)

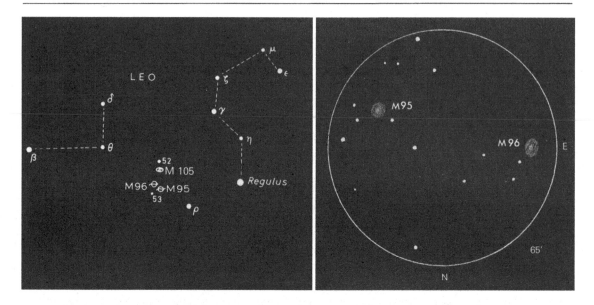

To observe   (M 95 and M 96)

From *Regulus* move 9° E, *or*, from ϱ *Leonis* (4th mag.) move 2½° N and 3° E. Both galaxies can be seen in the same field of view on L.P. but, of the pair, M 96 may be seen first as it is slightly larger and a good deal brighter than its neighbour. M 95 is about 50′ W of M 96.

M 95 is a faint, round, glowing patch with a nucleus which is also faint and almost star-like. The outer, faint portion becomes a little more obvious with averted vision but it is difficult to make out any detail. There are three fairly bright stars in a N–S line to the W of the galaxy.

M 96 is slightly oval and glows with a soft light: it is distinctly brighter than M 95 and the nucleus, too, is more distinct and shows a definite area which is not star-like. The background is much sparser than that around M 95 although there is an 8 mag. star about 25′ to the ENE.

Smyth seemed to think that M 96 was less distinct than M 95 but there can be little doubt that M 96 is easily the brighter and clearer.

Objects near by   About 55′ NNE of M 96 lies a pair of galaxies about 8′ apart in P.A. about 50°, both quite bright and slightly oval. The brighter and more westerly of the two is NGC 3379 which has been entitled M 105 (see Chapter 1). The other is NGC 3384.

A third galaxy, NGC 3389, makes a right-angled triangle with the pair to the SE but it is a good deal fainter than the others.

M 97 NGC 3587

The 'Owl' Nebula.   A planetary nebula, type IIIa in URSA MAJOR

R.A. 1 h 12.0 m.   Dec. N 55° 18' (1950).

Size 3' diam.   Mag. 12.0   Distance 3 kpc, 10 000 lt-yrs.

Discovered by Méchain in 1781.

MESSIER
(March 24th, 1781) 'Nebula in Ursa Major, near beta. "It is difficult to see," reports M. Méchain, "especially when one illuminates the micrometer wires: its light is faint, without a star." M. Méchain saw it for the first time on Feb. 16th, 1781, and the position is that given by him. Near this nebula he saw another which has not yet been determined, also a third which is near gamma Ursae Majoris. Diam. 2'.'
    (These two have been entitled M 108 and M 109 by O. Gingerich; see Chapter 1, *The Additional Objects*.)

W. H.
'A globular body of equal light throughout.' (Being unable to resolve it into stars, he thought it belonged outside the galaxy.')

J. H.
'Diam. 2' 40", light equable, with a softened edge and faintly bi-central.'

ROSSE
(Mar. 11th, 1848) 'Two stars considerably apart in the central region; dark penumbra around each spiral arrangement.' (On many occasions only one star seen and spiral form doubtful.)

ROBINSON
(In Lord Rosse's 6-foot telescope, 1848) 'A most intricate group of spiral arcs disposed round two starry centres, looking like the visage of a monkey.'

SMYTH
'This very singular object is circular and uniform and after long inspection, looks like a condensed mass of attenuated light seemingly the size of Jupiter.'

WEBB
'Large pale planetary nebula. Very remarkable object: two stars, one in each opening – only one seen since 1850.'

HUGGINS
'Spectrum gaseous.'

ROBERTS
Photo. Apr. 1895. 'Elliptical, 203" diam. 15 mag. star in centre but no others.'

LICK XIII
'Central star about mag. 14 vis. and about mag. 12 photo. The brighter central oval lies in P.A. 12° and the diameter along this line is 199". At right angles to this

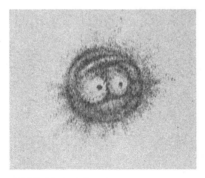

M 97 (Rosse) From
Collected Scientific Papers
(B.A.A.)

277

direction it is 203″ in diameter to the outside of the whorls. Aside from the outer whorls, all structural details are very vague and indistinct.'

M 97 is one of the largest and also one of the faintest of the planetary nebulae. It has received a lot of attention from amateur and professional astronomers but there is little that can be made out visually. Photographic examination shows that it does not consist of a simple shell, and micro-densitometer tracings of emission density reveal two bright, smooth condensations which are symmetrically placed on opposite sides of the faint, central star. These condensations lie inside a faint, almost circular torus with a still fainter, hollow shell enclosing the whole.

Minkowski and Aller at Mt Palomar found that the structure of M 97 is very uniform and is free from the filamentary wisps which are often found in large planetary nebulae. There is also a marked absence of stratification of the material of the nebula and its average density is as low as that of the gaseous nebulae. Its mass has been estimated as about 15 per cent that of the Sun, which is large for a planetary nebula.

The structure of M 97 appears to be very similar to another faint planetary discovered by Jones and Emberson in 1939 in Lynx which, with dimensions of 348″×435″ is more than twice as large as M 97. Another similar planetary nebula is NGC 6058 in Hercules.

The central star has a temperature of 85 000 K and a density which approaches that of a white dwarf. (For observing details see next page.)

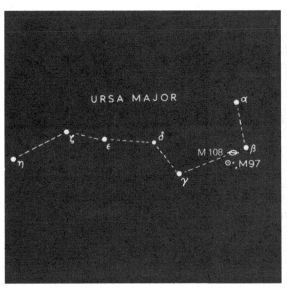

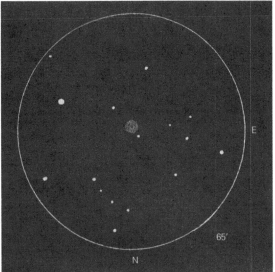

To observe   M 97 is a fairly difficult object to find as it is small and faint and will probably elude instruments smaller than 6-inches aperture. The very best atmospheric conditions are essential and the Plough needs to be high in the sky.

The correct location is not difficult to arrive at: from *Merak (β Ursae Majoris)* move $1\frac{1}{2}°$ S and 2° f. There is a distinct, $6\frac{1}{2}$ mag. star about 25′ WSW from the nebula.

Even when M 97 is found, there is little that can be made out except in large apertures. All that can be seen is a faint, round patch emitting a soft, even glow which does, however, seem a little brighter than expected from its quoted magnitude of 12. There are several faint stars in the field and one of about mag. 11 is very close to the nebula to the NE.

The central star of mag. 14 needs a 12-inch aperture to see and the 'owl' perforations might be seen in good conditions in an instrument of this size.

Objects near by   45′ to the NW of M 97 is NGC 3556. This is an edge-on spiral 7′.7 × 1′.3 mag. 10.7 and is brighter and more easily seen than M 97 itself. It is nearer to *Merak* than M 97 and may be picked up first when looking for the planetary, but its distinctly elongated shape distinguishes it at once from the virtually circular M 97. This object was mentioned by Messier as being near M 97 and Gingerich has designated it M 108 (see Chapter 1 and M 108.)

M 98 NGC 4192
A spiral galaxy, type Sb in COMA BERENICES
R.A. 12 h 11.3 m.   Dec. N 15° 11' (1950).
Size 8"×2'   Mag. 10.7   Distance 11 Mpc, 36 million lt-yrs.

Discovered by Méchain in 1781.

MESSIER   (April 13th, 1781) 'Nebula without star, of an extremely faint light, above the northern wing of Virgo; on the same parallel as, and near to the star No. 6, 5th mag. of Coma Berenices, after Flamsteed. M. Méchain saw it Mar. 15th, 1781.'

W. H.   (Dec. 30th, 1783) 'A large, extended, fine nebula. Its situation shows it to be M. Messier's 98th; but from the description it appears that that gentleman has not seen the whole of it, for its feeble branches extend above a quarter of a degree, of which no notice is taken. Near the middle of it are a few stars visible and more suspected. My field of view will not quite take in the whole nebula.'

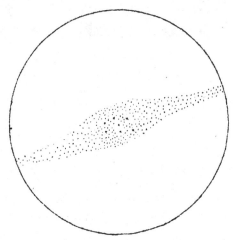

M 98 (William Herschel)
From Phil. Trans. *of the Royal Society (R.A.S.)*

ROSSE   (Mar. 9th, 1850) 'A ray; diminution of light in neighbourhood of nucleus; edges parallel; night bad; remarkable object.'

SMYTH   'A fine, large, but rather pale nebula with bright star, 6 Comae Berenicis following in next field exactly on parallel. On keeping a fixed gaze, it brightens up toward the centre. It is elongated in the direction of two stars, one N.p. and the other S.f. with another star in N.f. quadrant, pretty close.'

D'ARREST   'Very large and bright, elongated, 9'×1'. Shining nucleus.'

WEBB   'Long ray (J.H. 10').'

LICK XIII   'An open, elongated spiral, 8'×2' in P.A. 150°. Bright stellar nucleus, numerous almost stellar condensations. Absorption effects on east side.'

M 98 is another galaxy which Holmberg has found to be in the foreground of the Virgo Group and possibly not a true member of that association. Compared with M 86, the other non-member, the difference in distance is, however, much less. Holmberg gives a distance modulus for M 98 of 30.2 mag. by photometric means, compared with a modulus of 30.5 for the main Virgo Group. This is equivalent to a distance of about 11 Mpc for M 98, compared with 12.5 Mpc for the other galaxies in the Virgo association.

The mass of M 98, given in Holmberg's catalogue, is 130 000 million solar masses and the linear diameter is 37.2 kpc. The mass-density works out to be equivalent to about one solar mass per 40 cubic parsecs. The spectral class is G0, the colour index +0.38 and absolute magnitude −21.2.

In the *Atlas Coeli Katalog* the apparent dimensions are 8′.4×1′.9 and apparent magnitude 10.9 photographic and 10.7 visual.

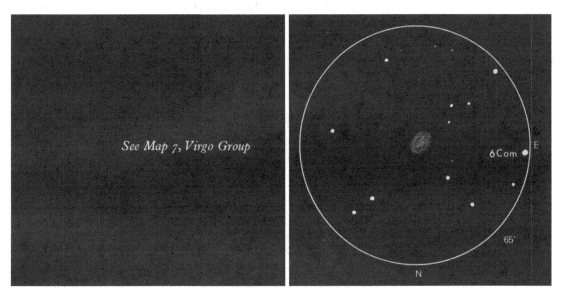

See Map 7, Virgo Group

To find   See *Guide to Virgo Group* (p. 387).

To observe   M 98 is a fairly difficult object for a small telescope and the best conditions are necessary to make much of it.

Although quite large, it is fairly diffuse and so thin-looking that it often seems to disappear into the background. It is clearly elongated in P.A. about 150° and a slightly brighter central area can be detected.

This galaxy may be more easily seen if the 5th mag. star to the E (*6 Comae Berenices*) is kept out of the L.P. field. Averted vision is essential if any detail is to be made out.

There are no other bright galaxies in the same field and M 98 lies, in fact, on the NW edge of the Virgo Group.

M 99 NGC 4254
A spiral galaxy, type Sc in COMA BERENICES
R.A. 12 h 16.3 m.   Dec. N 14° 42' (1950).
Size 5'×4'   Mag. 10.1   Distance 12.5 Mpc, 41 million lt-yrs.

Discovered by Méchain in 1781.

MESSIER   (April 13th, 1781) 'Nebula without star, of a very pale light, nevertheless a little clearer than the preceding, M 98. On the northern wing of Virgo and near the same star, 6 Comae Berenicis. The nebula is between two stars of 7 and 8 mag. M. Méchain saw it on Mar. 15th, 1781.'

J. H.   'A very remarkable object, observed by W. H. and J. H. Bright, large, round, gradually brighter in the middle, resolvable: three-branched spiral.'

ROSSE   (Mar. 11th, 1848) 'Spiral with a bright star above; a thin portion of the nebula reaches across this star and some distance past it. Principal spiral at the bottom and turning towards the right.'

D'ARREST   'Large, round with vividly sparkling light: nucleus more or less resolvable. Obviously diffuse around the edges.'

WEBB   'Though pale, it is well defined in my instrument.'

ROBERTS   'Spiral: shows many star-like condensations in convolutions.'

A. M. CLERKE   'A right-handed spiral: branches turning opposite to M 51. Their tendency to form nodes and angles strikingly shown in photograph by von Gothard, Apr. 12th, 1888.'

LICK XIII   'A very bright, approximately round spiral, 4'.5 in diameter. Nucleus almost stellar. There are two main whorls, rather open, which show many almost stellar condensations.'

M 99 is one of the smaller and less massive members of the Virgo Group of galaxies. Holmberg, in his 1964 *Catalogue of External Galaxies*, gives the mass of M 99 as equivalent to about 50 000 million solar masses which is only a quarter of that of our own Galaxy. Its linear diameter is approximately 26.6 kpc which is less than the estimated size of our system.

The average density of material in M 99 has been estimated to be equivalent to about one solar mass per 40 cubic parsecs.

M 99 has an absolute magnitude of -20.7, a spectrum of type G2 and a colour index of $+0.27$.

In the *Atlas Coeli Katalog*, the apparent dimensions are given as 4'.6×3'.9 and the apparent magnitude as 10.4 photographic and 10.1 visual.

M 99 produced a supernova Type II of 14 mag. in June 1967, another Type II of 15.6 mag. on Dec. 16th, 1972, and a third, of Type I in 1986. (For observing details, see page 284.)

*M 99 (Rosse) The second nebula to
be recognised as spiral (B.A.A.)*

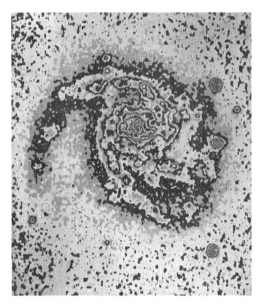

*M 99 Not a drawing but an isophote
tracing of M 99 obtained by electronic
scanning of a plate taken with the
37-inch Cassegrain Schmidt telescope
of University Observatory, St Andrews
(Photo by C. W. Fraser)*

283

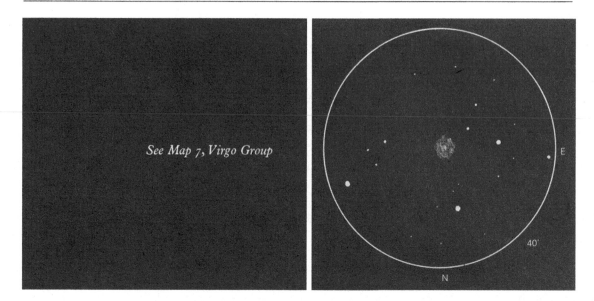

To find   See *Guide to Virgo Group* (p. 387).

To observe   M 99 is almost as faint as its neighbouring galaxy M 98 but, although diffuse, it is not quite so pale. It is quite large, being about 5′ in diameter, but it is very indistinctly defined towards the outer edges. There is a fairly distinct but broad central condensation and with averted vision, several small, brighter patches can sometimes be glimpsed in the south and east borders of the nebula. This appearance has led to its description as the 'Pin-wheel' Nebula by some observers.

There are three bright stars in the field but otherwise the background is sparse. There are no other bright galaxies in the field but there are several about 1° to the NE which can be seen in an 8-inch.

M 100   NGC 4321
A spiral galaxy, type Sc in COMA BERENICES
R.A. 12 h 20.4 m.   Dec. N 16° 06' (1950).
Size 5' diam.   Mag. 10.6   Distance 12.5 Mpc, 41 million lt-yrs.

Discovered by Méchain in 1781.

MESSIER   (April 13th, 1781) 'Nebula without star; it has the same light as the preceding: in the ear of Andromeda. Seen by M. Méchain Mar. 15th, 1781. The three nebulae, Nos. 98, 99 and 100, are very difficult to recognise because of their feeble light: one can observe them only in good weather and near meridian passage.'

W. H.   (*Phil. Trans.* 1814) 'Connoiss. 100 is a nebula of about 10' in diameter, but there is in the middle of it, a small, bright cluster of supposed stars.'

SMYTH   'A round nebula, pearly white: this is a large but pale object of little character though it brightens from its attenuated edges toward the centre and is therefore thought to be globular. Accompanied by four small stars at a little distance around it besides minute points of light in the field, seen by occasional gleams.'

ROSSE   'Spiral, with the centre a planetary nebula.'

ROBERTS   Photo, May 1896 'Strikingly perfect spiral with a sharp nucleus in the midst of faint nebulosity.'

LICK XIII   'A bright, regular, nearly round spiral, 5' in diameter. Very faint stellar nucleus surrounded by bright, short whorls, forming a central oval. The outer whorls are rather open, quite regular and show many stellar condensations. Two novae have appeared in this spiral.'

In the faint spiral arms of this galaxy there can be seen several of the usual condensations which at first appear to be clusters of stars. These, however, have been recently shown to be really patches of emission nebulosity containing hydrogen and early type stars.

Earlier estimates of the distance of M 100 (which is one of the Virgo Group) gave a figure of about 6 Mpc but this was increased to 11 Mpc in about 1950. Sandage, in 1958, made the distance 13 Mpc and in 1964, Erik Holmberg found a figure in close agreement at 12.5 Mpc for M 100 and a mean distance for the whole of the Virgo Group between 12 and 13 Mpc.

Baade discovered that M 100 is surrounded by many dwarf elliptical galaxies.

Among the first novae to be discovered in spiral nebulae were two in M 100. Both were found by H. D. Curtis in 1917: one nova had a maximum of about 13.5 mag. in March 1901 and the other, about 14 mag. some time before Mar. 2nd, 1914.

Holmberg estimates the mass of M 100 to be equivalent to about 160 000 million solar masses and its linear diameter to be about 36.3 kpc. The resulting density of material in the galaxy works out to be about one solar mass to 30 cubic parsecs.

M 100 has an abs. mag. of −21.0, a spectral type of F5 and a colour index of +0.44.

In the *Atlas Coeli Katalog*, the apparent dimensions are given as 5'.3×4'.5 and the apparent magnitude as 10.1 photographic and 10.6 visual.

To date, four supernovae have appeared in M 100; a Type I, mag. 15.6 in March 1901, another of type not determined, mag. 15.7 in Feb./Mar. 1914, a Type I, mag. 17.5 in Aug./Sept. 1959, and a Type II, mag. 11.6 (which faded quickly) on Apr. 15th, 1979.

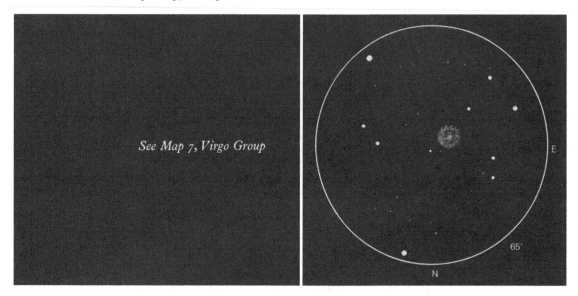

To find   See *Guide to Virgo Group* (p. 387).

To observe   M 100 is not difficult to see: it appears as a circular, distinct patch with a much brighter, almost stellar nucleus and an unusual, mottled or brush-like perimeter. This appearance led Admiral Smyth to the conjecture that it might be 'globular' but Lord Rosse's telescope plainly showed it to be spiral, introducing at the same time the diversion of describing the nucleus as 'a planetary nebula'.

In small or moderate telescopes it does resemble, in its outline at least, a globular cluster, but the area between the mottled circumference and the nucleus is plainly seen to be diffuse as in most Sc galaxies.

The object seen in a 6-inch or 8-inch telescope is, in fact, only the nucleus with its immediate surrounding area, for even in photographs, the spiral arms are very faint and sparse.

M 101 NGC 5457
A spiral galaxy, type Sc in URSA MAJOR
R.A. 14 h 01.4 m.   Dec. N 54° 35' (1950).
Size 10'×8'   Mag. 9.6   Distance 7.1 Mpc, 23 million lt-yrs.

Discovered by Méchain in 1781.

MESSIER   (March 27th, 1781) 'Nebula without star, very obscure and pretty large, 6' or 7'
diam. between the left hand of Bootes and the tail of the Great Bear. Difficult to
distinguish when graticule lit.'

W. H.   'A mottled nebulosity.'

J. H.   'Pretty bright, very large and irregularly round.'

SMYTH   (1844) 'It is one of those globular nebulae that seem to be caused by a vast
agglomeration of stars rather than by a mass of diffused, luminous matter ... the
paleness tells of its inconceivable distance.'
(1855) 'A pale white nebula: under a favourable view it is large and well spread,
though somewhat faint except towards the centre where it brightens (may be the
same as 214 H. I).'

ROSSE   'Large, spiral, faintish; several arms and knots. 14' diam. at least.'

D'ARREST   'Outer margin dilute and poorly terminated. Not perfectly round, nucleus not a true
star. Complex, with two intertwining nuclei, though not well distinguished.'

ROBERTS   Photo, May 1892 'A well defined stellar nucleus and star-like condensations.'

BRODIE   'Very large but very faint.'

LICK XIII   'This unusually beautiful spiral is about 16' diam. There is an almost stellar nucleus
with two bright condensations very close, which give it a tri-nuclear appearance.
The open whorls show a multitude of stellar condensations. 5449, 5450, 5451,
5453, 5455, 5458, 5461, 5462 are simply knots in the great nebula.'

M 101 (Rosse) Drawn by S. Hunter (B.A.A.)

287

M 101 is a very open spiral which is face-on to our view. The nucleus is very small and bright while the spiral arms are seen to contain dark filaments together with resolved stellar condensations. In addition, bright nodules of emission nebulosity, surrounded by early-type stars have been identified. Its visual appearance is very similar to M 74 in Pisces and it is of almost the same spectral type also.

M 101 is a member of a group of galaxies containing perhaps 10 or more members, of which the more conspicuous are the spirals NGC 5474 and 5485. According to Baade, other members include two dwarf elliptical systems, like those in Sculptor which belong to our Local Group, and also an irregular galaxy similar to the Magellanic Clouds.

In Holmberg's catalogue, M 101 is quoted as having a mass of only 16 000 million Suns, which is only one-tenth of that of M 100 and is the least of any of the Messier galaxies other than M 32 and M 33. Its linear diameter, however, is quoted as about 28.2 kpc which is not much less than that of our own Galaxy. The mass-density of M 101 is therefore very low, being equivalent to about one solar mass for every 160 cubic parsecs.

The spectrum of M 101 is of type F8 and its colour index of +0.16 is the lowest of all the galaxies in Messier's list. The absolute magnitude is −20.1.

Becvar gives the apparent diameter as 22'.0, the apparent magnitude as 8.2 photographic and 9.6 visual. Becvar's distance for M 101 is about 4.2 Mpc but Holmberg separates M 101 from the group of galaxies mentioned above, giving a distance of only 3.5 Mpc.

In 1986 two very faint (mag. 23) Cepheid variables of periods 37 and 47 days were found in M 101. Preliminary results give a distance modulus for the galaxy of 29.0 to 29.5, which works out between 6.3 Mpc (20.6 M lt-yrs.) and 7.9 Mpc (25.9 M lt-yrs.).

M 101 has produced three supernovae to date; the first, a peculiar type, mag. 13.5, discovered by Max Wolf in Feb. 1909, the second, Type II, of mag. 17.5 in Sept./Oct. 1950, and a third, also of Type II, mag. 11.7, on Aug. 2nd, 1970. (For observing details see next page.)

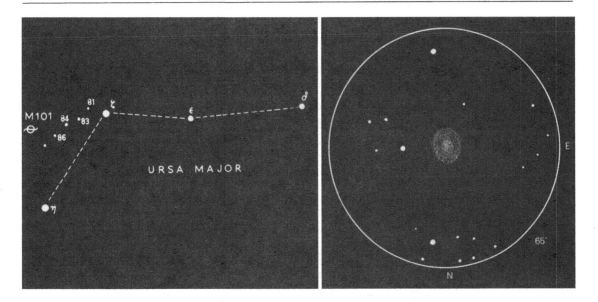

To observe   This galaxy is rather faint and diffuse and may be a little difficult to find at first. It forms an almost equilateral triangle to the NE with *Benetnash* and *Mizar*, being 5° N and 2° E of *Benetnash*. A more careful approach can be made by following, from *Mizar* to the SE, the little chain of 5 mag. stars, *81, 83, 84* and *86 Ursae Majoris*. M 101 is *45′* N and *80′* E of *86 Ursae Majoris*.

M 101 is intermediate in difficulty of perception between M 33 and M 74, both of which it resembles. It is smaller and much fainter than M 33 but larger and brighter than M 74.

The very best conditions are necessary to make the most of this object which appears as a faint, fairly large misty patch which is not quite circular. It looks about 7′×5′ at first seeing but, as with many objects of this kind, averted vision reveals a larger, diffuse area outside this. A very small but definite bright nucleus can be made out and with apertures of 10 to 12 inches one or two of the bright condensations in the spiral arms might be made out to the E and SW.

There is an 8 mag. star about 12′ to the W, and a curve of fainter stars about 25′ to the N.

Objects near by   45′ to the SSE of M 101 is an Sc galaxy, NGC 5474, mag. 11.4, 4′×3′. It is this object, and not M 101, which Wm Herschel designated as 214 H. I (see Smyth's description).

M 102   NGC 5866? NGC 5879?

One of the 'missing' Messier Objects (see Chapter 1).

'Discovered' by Méchain in 1781 and 'disowned' by him in a letter to Bernouilli at Berlin in 1783 as a duplicate observation of M 101.

MESSIER   (Méchain) 'Nebula between the stars omicron Bootis and iota Draconis: it is very faint, near it is a star of 6th magnitude.'

SMYTH   'A small but brightish nebula on the belly of Draco with four small stars spreading across field N of it. Doubt as to whether this is the nebula discovered by Méchain in 1781 since Messier describes it as "very faint", situated between omicron Bootis and iota Draconis. If omicron Bootis should be theta Bootis, this is probably the object seen by Méchain and J. H.'s 1910, being the brightest nebula of five in that vicinity.'

If, despite Méchain's admission of a mistake about the identity of M 102, the observer still wishes to do some 'detective' work on this object he will, no doubt, investigate the little group of galaxies centred about 3° SW from ι *Draconis*.

The 'possibles' in this area are:

NGC 5866 R.A. 15h 05.1 m. Dec. N 55° 57' (1950).
    Size 2'.8×1'.0 Mag. 10.8 Type E6p.
    This is 215 H. I. J.H. 1909 'vB, cL, pmE 146°, gbM'.
    Identified as M 102 in *Atlas Coeli*. The brightest in the area.

NGC 5879 R.A. 15 h 08.4 m. Dec. N 57° 12' (1950).
    Size 4'.5×1'.1 Mag. 12.2 Type Sb.
    This is 757 H. II. J.H. 1910 'cB, S, E, mbM, RNr'.
    The object named by Smyth but *not* the brightest object.

NGC 5907 R.A. 15 h 14.6 m. Dec. N 56° 31' (1950).
    Size 11'.1×0'.7 Mag. 11.3 Type Sb.
    This is 759 H. II. J.H. 1917 'cB, vL, vmE 155°, vg, psbMN'.
    This largest of the group and very little fainter than 5866.

NGC 5908 R.A. 15 h 15.4 m. Dec. N 55° 36' (1950).
    Size 2'.4×0'.4 Mag. 13.0 photo. Type Sb.
    This is 760 H. II. 'pF, pS, R'.
    The faintest of the four and hardly a possible candidate.

J. L. E. Dreyer, in *Notes and Corrections* to the NGC (p. 283), suggested NGC 5928, R.A. 15 h 23.9 m. Dec. N 18° 15' (1950) as possibly the missing M 102. He assumed that ι *Draconis* was an error for ι *Serpentis*.

M 103 NGC 581
A galactic cluster, type 'd' in CASSIOPEIA
R.A. 01 h 29.9 m.   Dec. N 60° 27' (1950).
Size 12'×5' Mag. 7.4 Distance 2.6 kpc, 8500 lt-yrs.

Discovered by Méchain in 1781.

MESSIER (Méchain) 'A cluster of stars between ε and δ of the leg of Cassiopeia.'

J. H. 'Bright cluster, pretty large; round and rich, with stars 10 and 11 mag.'

SMYTH 'A neat double star in a cluster. Double, Σ 131, straw and dusky blue, P.A. 141° d. 14".4 (1832).
Cluster a fan-shaped group diverging from a sharp star in N.f. quadrant.
Brilliant from the flash of a score of its larger members, four principal ones of which are from 7 to 9 mag. Under the largest in the S.f. is a red star of 8 mag.'

WEBB 'Beautiful field 1° f. a little N of delta, containing Σ 131 and a red star.'

D'ARREST 'A double star on the northern limb of a cluster in Messier's catalogue. A beautiful 10 mag. reddish star prominent, its colour is rose-tinted. An irregular cluster of 9–10–11 mag. stars. Size 9' approx.'

Although included in most catalogues of galactic clusters, M 103 raises some doubts about its true nature. Shapley, who classified it at first among his type 'd' of loose and poor clusters, later considered that it may, in fact, be nothing more than an accidental grouping of stars which may have no real association as a cluster.

R. J. Trumpler gave it the classification of II,3,m and made the spectral class of the brightest star B3.

In Mrs H. S. Hogg's catalogue of clusters, M 103 is quoted as having an apparent diameter of 6'.5, an integrated magnitude of 6.9 and the magnitude of the brightest star, 9.8.

Dr Wallenquist, in his catalogue, gives a total of 40 probable member stars in a diameter of 6'.5 and the star density as 5.76 stars per cubic parsec in the central region and 0.69 stars per cubic parsec for the cluster as a whole.

The linear diameter of the cluster, according to Wallenquist, is 4.8 parsecs or about 15½ light-years and the distance, 2.54 kpc.

Becvar, in the *Atlas Coeli Katalog* gives a linear diameter of 1.7 parsecs and a distance of 1150 parsecs with the number of cluster stars as 60. (For observing details see next page.)

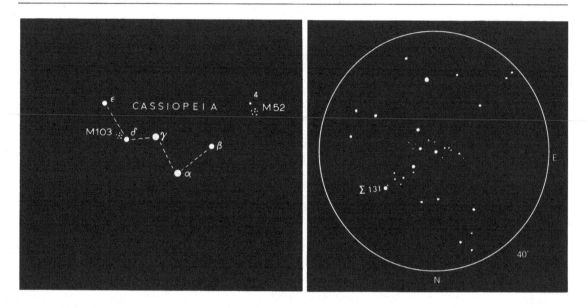

To observe   From δ *Cassiopeiae* move ½° N and 1° E.

M 103 is not all that easy to pick up as it is very loose and poor and there are many other groups in the vicinity which could be confused with it.

It has the outline of an isosceles triangle with the apex to the NW and a clear, red star near the mid-point of the NE side. It contains about half a dozen stars of mag. 10 and about 25 stars of mag. 11 and 12.

It is not a very impressive cluster and there are several others in the vicinity of at least equal claim to attention.

The double star noted by Smyth, Σ 131, is the brighter star at the NW apex of the cluster. (A, 7.3 mag. B, 10.5 mag. Separation 13″.8 in·P.A. 142°, 1956).

A wide field and low power, by reason of the smaller scale, may make the cluster easier to identify. Observers with small telescopes may well be at an advantage when looking for clusters like this.

Objects near by   1½° E of M 103 is NGC 659, a faint, small and rather poor cluster: it is circular and looks rather like a faint open globular. On its SW edge is a double star, *44 Cassiopeiae*, 5½ mag. and 12½ mag. P.A. 5° d. 1″.5.

1¾° E and ½° N of M 103 is NGC 663, a bright open cluster, more or less of a diamond shape and containing several bright pairs and many fainter stars.

About 45′ N and slightly W of 663 is another small cluster; it is quite bright and triangular in shape: it has an 8 mag. star on the south edge. This is NGC 654.

M 104 NGC 4594 (43 H. I)
The 'Sombrero' Nebula. A spiral galaxy, type Sb in VIRGO
R.A. 12 h 37.4 m. Dec. S 11° 21' (1950).
Size 6'×2' Mag. 8.7 Distance 12.5 Mpc, 41 million lt-yrs.

Discovered by Méchain in 1781.

MÉCHAIN (In a letter to Bernoulli, May 6th 1783) 'On 11th May 1781 I discovered a nebula above Corvus which did not appear to contain a single star. It emits a weak light and is difficult to find if the micrometer wires are illuminated. This does not appear in the *Connaissance des Temps*.'

MESSIER 'In a hand-written note to his copy of the *Connaissance des Temps* for 1784) '11th May, 1781. Very faint nebula.' This was followed by a position which was later found by Flammarion to coincide with Wm Herschel's 43 H. I.

W. H. 'A faint, diffused oval light all about it: almost positive that there is a dark interval or stratum separating the nucleus and the general mass of the nebula from the light above it.'

SMYTH (Under 43 H. I) 'A lucid white elliptical nebula in an elegant field of small stars. Nearly parallel to equatorial plane and on intense attention may be seen to blaze in the middle. The half-dozen principal stars form a great "Y" with the nebula as the centre. A mere wisp of subdued light.'

D'ARREST 'Luminous ray, 4'×50"; smallish nucleus like a 10 mag. star.'

WEBB (Under 4594) 'Beautiful low power field: fine and singular 7 mag. group N.p. Long. (J.H. 5'×30", nucleus and dark cleft.)'

LICK XIII '7'×1'.5 in P.A. 92°, very bright. A remarkable, slightly curved, clear-cut dark lane runs along the entire length to the south of the nucleus; probably the finest known example of this phenomenon. There are very slight traces of spiral whorls.'

FLAMMARION 'It has the position of the nebula 43 H. I found by Wm Herschel and is No. 4594 of the NGC of Dreyer. We can add it to Messier's catalogue and give it the number 104. The result is that Messier's catalogue from now on is reckoned as numbering 104 instead of 103.'

M 104 was one of the first nebulae in which large radial velocities were measured by means of the spectroscope. In 1913, V. M. Slipher, using a short-focus camera in conjunction with the 24-inch refractor at Lowell Observatory, obtained the 'enormous' velocity of recession (red shift) of about 1000 km/s. Slipher was also able to detect that the nebula was in rotation.

This famous galaxy can be seen to be surrounded by a halo of globular clusters much as in our own Galaxy and in M 31. In addition, photographs taken with red-sensitive plates show an almost spherical distribution of bright, red Population II stars. The dark streak across the nucleus, which is an outstanding feature of the many photographs taken of this object, is an absorption band of dust and gas in the galactic disc.

In M 104, the plane of the disc, according to de Vaucouleurs, is inclined at only 6° to our line of sight.

Lindblad made a careful photographic examination of bright condensations in the spiral arms of M 104 and was able to determine its orientation with respect to our line of sight. With the additional information of radial velocities in different parts of the galaxy made by Pease, the direction of rotation was ascertained. This has turned out to be with the spiral arms concave to the spin, as if the galaxy were 'unwinding' which is contrary to most other known instances.

M 104, until recently, was one of the brightest nebulae in the heavens in which radio emission had not been detected: it has now been found to be a weak radio source with a radio 'magnitude' of about 6.3 which is 2.4 magnitudes greater than its visual brightness.

Among its other distinctions, M 104 can claim to be the most massive of the Messier galaxies. In his catalogue, Holmberg puts the mass at the equivalent of 1.3 million million solar masses. The linear diameter he gives as 43.7 kpc and absolute magnitude as -22.1. The spectrum is of type G3 and colour index $+0.67$.

Becvar gives the apparent dimensions as $6'.0 \times 2'.5$ and apparent magnitude as 9.2 photographic and 8.7 visual.

Holmberg confirms M 104 as a true member of the Virgo Group. (For observing details see next page.)

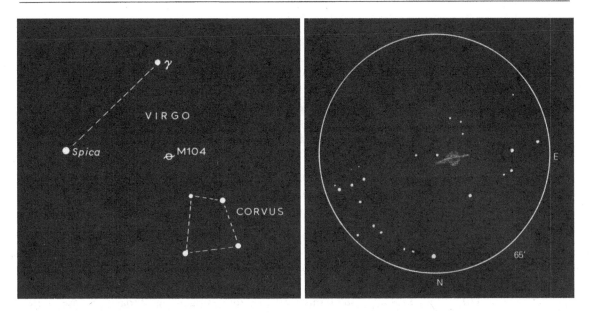

To observe On low power set the telescope so that *Spica* is in the N of the field. Then sweep to the W for $11\frac{1}{2}°$ until just past S of 3rd mag. γ *Virginis*. M 104 is a very conspicuous object and easy to find.

With a wide field of view one obtains a truly beautiful sight: a bright, glowing lenticular nebula with a striking, pretty curve of stars to the NW providing (as Smyth says) 'an elegant field'.

M 104 stands higher magnification well and with 8 inches aperture a faint *bright* line can be seen running E–W just S of the nucleus. No doubt this is produced by contrast with the actual dark absorption band which probably needs a 10-inch or 12-inch to see well.

The 'glow' of the spherical halo of Population II stars can also be made out in an 8-inch and possibly in a 6-inch. Undoubtedly one of the finest sights to be seen in small telescopes.

Sir William Herschel did not consider it to be one of Messier's Objects and gave it his own designation of 43 H. I – the 43rd object found in his class I (*bright nebulae*).

Objects near by 80' to the SSE of M 104 is the double star Σ 1669, 6.1 mag. and 6.0 mag., P.A. 130°, d. 5″.2.

M 105 NGC 3379 (17 H. I)
An elliptical galaxy, type E1 in LEO
R.A. 10 h 45.2 m.   Dec. N 12° 51′ (1950).
Size 4′.5×4′.0   Mag. 9.2   Distance 6.6 Mpc, 21.5 million lt-yrs.

Discovered by Méchain in 1781.

MÉCHAIN   'Near M 95 and M 96 there is a third, somewhat more northerly and even brighter than the others. I discovered this on March 24th 1781, four or five days after the other two.'

J. H.   'Very bright, considerably large, round, pretty suddenly brighter in the middle; resolvable.'

WEBB   (With NGC 3384) 'Two faint nebulae, p. much brighter and larger, with stellar nucleus.'

D'ARREST   'Nucleus exceptionally clear, like a 10th mag. star, easy to see and most brilliant at the centre. Two others compared in the same field. First of these much brighter in the middle.'

LICK XIII   'With 3384 and 3389 forms a striking group, a right-angled triangle whose shorter sides are 7′ long. 3379 is nearly round, very bright, no spiral structure. (3384) A replica of 3379.'

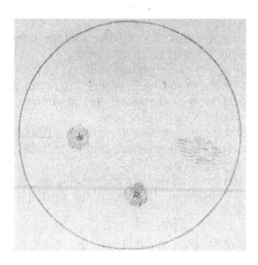

M 105 (left) with NGC 3384 (below) and NGC 3389 (right) drawn by Wm Herschel for Philosophical Transactions. *(R.A.S.)*

M 105, with its companions in the same field, is a member of the M 95 group of galaxies in Leo. Its nucleus is category 4 in de Vaucouleurs' classification, indicating brightness in a scale increasing from 1 to 5. (For observing details, see next page.)

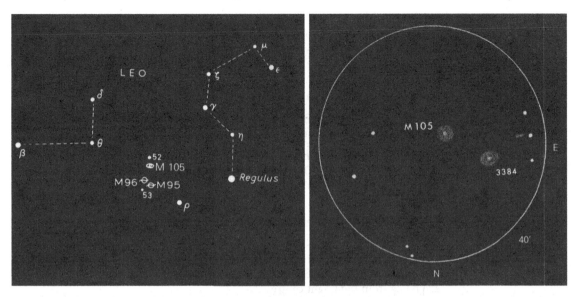

To observe   From M 96 move 1° N and a little E. At first sight, two objects looking like typical 'nebulous stars' will be seen, with NGC 3379 slightly the larger and distinctly the brighter. Both have bright centres with small glowing areas around them. In an 8-inch NGC 3379 appears quite circular, with a diameter about 4', while 3384 seems a little smaller and slightly elongated in P.A. approximately 110°/290°. In a 12-inch M 105 itself will be seen to be more oval, with the major axis in P.A. 15°/195°. With the larger aperture the third member of the group (NGC 3389, mag. 12.5) may be glimpsed about 7' NE of 3384 as a tiny streak elongated almost E–W.

M 106 NGC 4258 (43 H. V)
A spiral galaxy, type Sbp in CANES VENATICI
R.A. 12 h 16.5 m.   Dec. N 47° 34' (1950).
Size 18'×8'   Mag. 8.6   Distance 6.3 Mpc, 20 million lt-yrs.

Discovered by Méchain in 1781.

MÉCHAIN   'In July 1781 I found another nebula close to Ursa Major, near star No. 3 of Canes Venatici and 1° further south.'

SMYTH   'A large white nebula. A noble-sized oval trending rather from the vertical in direction N.p. and S.f. with a brightish nucleus in the S portion. The lateral edges are better defined than the ends. P. by 2 stars, 10 mag., and followed by two others.'

WEBB   'Large, bright oval best defined at sides, nucleus small like M 31. Size 10'×2½'. Continuous spectrum.'

D'ARREST   'Extremely large and bright, extending to 9 minutes of arc. The nucleus is like a 10 mag. star with extensive branching N–S, the northern arm longer and brighter extending from the bright centre.'

LICK XIII   'With the very faint extensions this spiral is nearly 20'×6' in P.A. 165°. Very bright elongated nuclear portion on W of which is a short dark lane. Numerous almost stellar condensations in 2 principal spiral branches.'

M 106 has many visual similarities to M 31, particularly because it is tilted by some 25° to our line of sight. (M 31 is tilted to about 16°.) It has only about half the mass of the Andromeda Nebula, however, being equivalent to about 190 billion solar masses, compared with 300 billion for M 31.

The spiral arms in this galaxy, as traced in H. I emission, seem to cut almost radially through the disc in opposite directions. This may be due to matter being ejected from the nucleus, which is itself classified as 'peculiar'.

In 1950 M 106 was found to be a radio source in which the emission covered an area about twice as large as its visual extent.

A supernova, type undetermined, reached 16th mag. in Aug. 1981. (For observing details, see next page.)

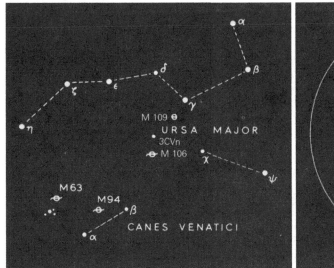

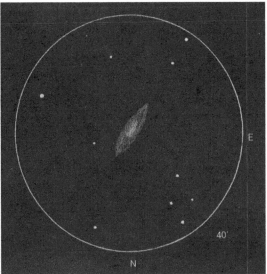

To observe   First locate the 5½ mag. star *3 CVn* which forms to the E a right-angled triangle with γ and χ *UMa*. M 106 is 2° S of *3 CVn*.

This object, being both bright and large, is much more distinct than many other Messier galaxies. In an 8-inch it is seen to extend to at least 10′×4′ with major axis in P.A. 165°/345°. The bright nucleus seems to be either curved or bi-central. The outer areas are fairly diffuse, but with averted vision can be seen to extend a good deal further than the dimensions above. All the near-by stars are very faint, except for one 10 mag. star to the SW.

This object reveals considerable detail in larger apertures and is well worth a meticulous survey and drawing at the eyepiece of a 16-inch.

Objects near by   One degree to the NW of M 106 is NGC 4420, an 11.7 mag. Sa galaxy of dimensions 4′×1½′.

M 107 NGC 6171 (40 H. VI)
A globular cluster, class X in OPHIUCHUS
R.A. 16 h 29.7 m.   Dec. S 12° 57′ (1950).
Size 8′ diam.   Mag. 8.5   Distance 5.9 kpc, 20 000 lt-yrs.

Discovered by Méchain in 1782

MÉCHAIN   'In April 1782 I discovered a small nebula on the left flank of Ophiuchus between the stars ζ and φ Oph.'

SMYTH   'A large but pale granulated cluster of small stars. 5 tel. stars around it in the form of a crucifix. Region beyond a comparative desert. After long gazing becomes more compressed in centre.'

WEBB   'Best with low power.'

LICK XIII   'Bright globular cluster. Brighter part 3′ diam. with fainter extensions about 8′.'

Although globular clusters do not usually contain much in the way of dust, photographs of M 107 have revealed several possible dark obscured regions. The very open distribution of stars in this globular (Class X) enables the interstellar regions to be examined more easily, and globular clusters are important 'laboratories' in which to study the processes by which galaxies evolve.

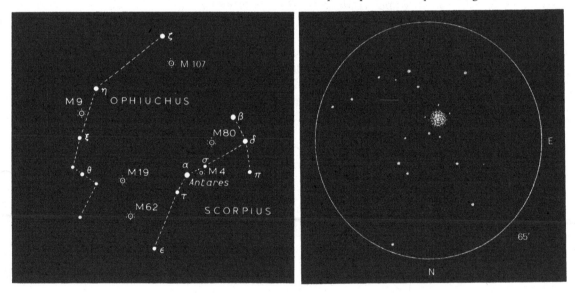

To observe   From ζ *Oph.* (2½ mag.) move 3° SSW. An almost equilateral triangle of 9 mag. stars to the N of the cluster is a useful guide.

M 107 is rather small and faint at first sight but rapidly improves with closer attention. The outline seems a little irregular but can be seen to extend to 5′ or 6′ from a distinct but not quite sharp centre. In an 8-inch it was partly resolved on M.P., but was rather too faint at my latitude (52° N) to stand higher magnification.

Smyth's 'crucifix' of stars was recognisable, and a long chain of fainter stars could be seen extending N–S on the W side of the cluster.

M 108 NGC 3556 (46 H. V)
A spiral galaxy, type Sc in URSA MAJOR
R.A. 11 h 08.7 m.   Dec. N 55° 57' (1950).
Size 8'×2'   Mag. 10.1   Distance 7200 kpc. 23.5 M lt-yrs.

Discovered by Méchain in 1781 or 1782.

MÉCHAIN   'A nebula close to β in Ursa Major.'

SMYTH   'A large milky-white nebula with small star at Sp. apex and 8 mag. p. at double the distance, also a brightish group in N.p. quadrant. Faint but well defined, being much elongated with axis-major trending N.p. and S.f. with a small star like a nucleus in centre.'

WEBB   'Large, faint, well defined, elongated, rather curved. Small star in centre.'

LICK XIII   'An irregular patchy spiral 8'×1'.5 in P.A. 84°, quite bright. Has a faint nucleus and shows a number of condensations, 3 of which are almost stellar. A star of mag. 11 near nucleus.'

This edge-on spiral shows almost no sign of any central 'bulge'. The spiral arms display heavy obscuration along the major axis, with a few H. II regions emerging as bright condensations against a mottled background. The mass has been determined from Doppler-shift velocities to be about 14 billion solar masses, which is only about one-twentieth of that of M 31.

M 108 produced a supernova Type II, mag. 13.9 on Jan. 23rd, 1969. (For observing details see next page.)

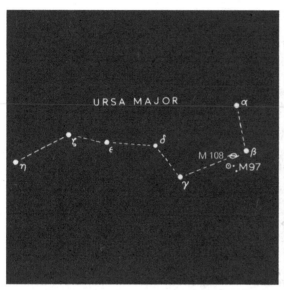

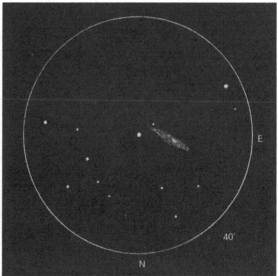

To observe   M 108 lies about 1½° SE of β *UMa* and a little less than 1° NW of M 97 which can be seen in the same L.P. field. Although catalogued as mag. 10, this object is more easily seen than one might expect. This is probably due to the contrast of the bright patches as seen against the dark obscuration along the nebula's major axis which extends to 8′×2′ in P.A. 75°/255°. A distinct star-like spot is positioned close to the N edge of the galaxy, but is evidently not the true nucleus. The E portion of the nebulosity seems slightly brighter than the W part.

This is certainly a most interesting object and should repay closer inspection in telescopes larger than 8 inches aperture.

M 109 NGC 3992 (61 H. IV)
A barred spiral galaxy, type SBc in URSA MAJOR
R.A. 11 h 55.0 m.   Dec. N 53° 39' (1950).
Size 7½'×5'   Mag. 9.8   Distance 11.5 Mpc, 27 million lt-yrs.

Discovered by Méchain in 1781 or 1782.

MÉCHAIN 'A nebula close to γ in Ursa Major.'

NGC 'Considerably bright, very large, pretty much extended, suddenly brighter in the middle to a bright mottled nucleus.'

LICK XIII 'A beautiful, slightly oval spiral 7' in length. Bright almost stellar nucleus: whorls are rather open and show a number of faint condensations; the central portions show traces of φ type formation.'

In M 109, the bar, which is rather short, terminates in club-like features to which the narrow spiral arms are attached as tangent arcs. The arms themselves are clearly delineated as if separated by dark dust lanes, and can be traced on photographs to nearly 300° of arc. The elliptical outline evidently indicates a degree of foreshortening, and gives a strong impression of a system slightly oblique to our view.

The nucleus, in de Vaucouleur's classification, is D4 (diffuse and fairly bright).

A Type I supernova was found in this galaxy on March 17th, 1956 and reached a maximum of 12.8 mag. (For observing details, see next page.)

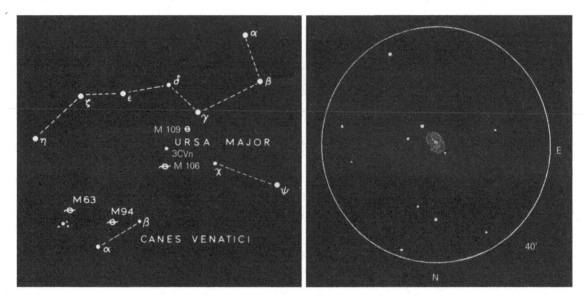

To observe M 109 is easily found, being only 1° SE of γ *UMa*.

It first shows up as a large nebula with a soft, amorphous glow. Glimpses of the nucleus can be made out with averted vision, but appearing neither sharp nor clear. With longer perception the oval shape becomes quite evident, covering an area of some 6′×4′ with the major axis in P.A. about 40°/220°. A 12 mag. star lies very close to the NE edge. No trace of the bar could be made out in an 8-inch, but some degree of mottling was apparent in the region close to the centre. This galaxy would repay examination with larger apertures.

M 110 NGC 205 (18 H. V)
An elliptical galaxy, type E5p in ANDROMEDA
R.A. 0 h 37.6 m.   Dec. N 41° 25′ (1950).
Size 10′×5′   Mag. 8.5   Distance 680 kpc, 2.2 M. lt-yrs.

Discovered by Messier on Aug. 10th, 1773.

MESSIER   (On a drawing of M 31 and companion nebulae published in 1807) '*Petite Nebuleuse, plus faible*'. (See p. 128.)
    Independently re-discovered by Caroline Herschel, Aug. 27th, 1783.

W. H.   (1785) 'The nebula in the girdle of Andromeda ... [has] a very considerable, broad, pretty faint small nebula near it; my sister discovered it August 27th, 1783 with a Newtonian 2-feet sweeper. It shows the same faint colour with the great one, and is, no doubt, in the neighbourhood of it. It is not ... [M 32] but this is about two-thirds of a degree north preceding it in a line parallel to β and ν Andromedae.'

SMYTH   'A large faintish nebula of an oval form with major axis N–S. Registered by W. H. as 30′×12′ but only half that size by his son; and there was a faint suspicion of a nucleus.'

WEBB   'Large faint oval nebula, best on L.P. Seems to sparkle. A very large field includes it with 221 and 224.'

D'ARREST   'First discovered by Messier in the year 1772 (*sic*); and anew by Caroline Herschel in 1784 (*sic*).'

LICK XIII   'The bright central portion is about 2′ diam. showing traces of rather irregular spiral structure. Nucleus almost stellar. Two small dark patches near brighter central portion. Very much fainter matter forms the outer portion of the nebula in an oval about 8′×3′. No whorls can be made out in this outer portion. Doubtless a spiral of the Andromeda type.'

M 110 is classified as E5p, the epithet 'peculiar' being justified by the fact that compared with most ellipticals it exhibits a good deal of structure. H. D. Curtis was, of course, mistaken in assuming it to be a spiral, but the two dark patches noted by him are most unusual. Baade considered them to be extensive dust clouds, and as he also found about a dozen very luminous B-type stars to be present, it shows that the galaxy is far from being purely 'elliptical' in structure and composition. A large number of globular clusters also appear in the outer halo.

Estimates of the mass of M 110 range from 3600 million to 15 000 million solar masses, the larger figure being similar to that of M 32. The radial velocity, measured by both optical and radio-frequency Doppler shifts, appears to be very much the same as that of M 31 itself. (For observing details see next page.)

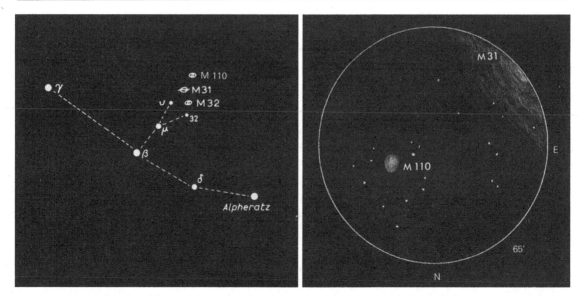

To observe   M 110 is about 40′ NW of the centre of M 31. This puts it well outside the usual telescopic field when centred on the Great Nebula, and perhaps explains why it was not discovered before 1773.

If an L.P. ocular is used and the telescope positioned so that the centre of M 31 is moved to the SE border of the field, M 110 will appear at the SW edge.

In an 8-inch it shows up as a fairly faint but distinctly glowing ellipse (but not 'sparkling', as Webb remarked). It is obviously oval, extending to about 10′×5′ in P.A. 340°/160°. No detail could be made out, and the brighter portion of the nebula (which seemed slightly off-set to the south) gave no indication of an evident nucleus.

5 Biographical and historical

NOTES ON SOURCES

Early observers   The descriptions by their discoverers, of those objects which were known before Messier observed them, are not all available from the original. Some, however, are quoted by other observers and Messier himself includes, for example, Maraldi's description of M 15 and Darquier's description of M 57 as looking 'like a fading planet'.

The long lost observations of G. B. Hodierna are derived from an important article by Serio, Indorato and Nastasi in *Journal for the History of Astronomy*, No. 45, (Feb. 1985), augmented by further identifications by myself in the same publication, No. 50, (Aug. 1986).

Halley's original descriptions of M 11, M 13 and M 22 are taken from his *Memoir, Of Nebulae or Lucid Spots among the Fixt Stars* in the *Phil. Trans.* of the Royal Society, No. 347, of 1715.

On reference to William Derham's *Memoir* in the *Phil. Trans.* No. 428 of 1730, it turned out that he did not actually observe any of the 16 objects he listed, except M 31. The objects were those of Hevelius, and Derham in fact states, 'The catalogue of Hevelius (which I transcribed from Hevelius' *Prodromus Astronomiae*) may be of good use to such as are minded to enquire into them.' Derham did, in fact, observe five of the six objects listed by Halley and briefly described M 11 and M 42.

Lacaille's descriptions of the objects which he had observed at the Cape of Good Hope in 1751–52, are taken from his *Catalogue of Nebulae of the Southern Sky* which appeared in the *Memoirs* of the Royal Academy of Sciences for 1755. The complete list of Lacaille's 42 southern nebulae was reprinted by Messier with his own catalogue of 103 objects in the *Connaissance des Temps* for 1784.

The numbers and descriptions of those objects observed by de Chéseaux are taken from a list which he included in a letter to his grandfather, M. de Crouzas in about 1746. De Crouzas, while visiting Paris gave the letter to the eminent French scientist, de Réaumur who read it before the Academy of Sciences on Aug. 6th, 1746. The list was not otherwise published and was rediscovered by Bigourdan and reprinted in the *Paris Annales* (Obs.) 1884, published in 1891.

The 'Connaissance des Temps' for 1784   The *Connaissance des Temps ou des Mouvements Célestes* – to give it its full title – is the senior of the great national ephemerides, being founded in 1679 by the Abbé Jean Picard. The British *Nautical Almanac* was introduced by Nevil Maskelyne, then Astronomer Royal, in 1766; the German *Astronomisches Jahrbuch* first appeared in 1774 and the *American Ephemeris and Nautical Almanac* in 1852.

The *Connaissance des Temps*, like other almanacs, was published three years in advance of its epoch date and contained positions of stars, tables of the Sun and Moon, details of predicted eclipses, positions for Jupiter's major satellites, together with other astronomical data and, on occasions, various memoirs on astronomical subjects.

The *Connaissance des Temps* for 1784 – published in 1781 – contained, as an additional item, Messier's complete catalogue of 103 nebulae and star clusters and also a reprint of Lacaille's *Catalogue of Nebulae of the Southern Sky* which had first appeared in the *Memoirs* of the Royal Academy of Sciences in 1755.

The title page of this volume of the ephemeris reads:*

'CONNOISSANCE DES TEMPS'
ou
CONNOISSANCE DES MOUVEMENTS CÉLESTES,
Pour l'Année bissextile 1784
Publiée pour la première fois en 1679,
Par l'ordre de l'Academie Royale des Sciences
Et calculée par M. Jeaurat, de la même Academie

A PARIS,
De l'Imprimerie Royale.
M. DCCLXXXI.

A translation of Messier's introduction to his catalogue and his account of the nebulae of other observers which he had been unable to find, is given below. The translation of his descriptions of the individual objects is included in the pages for each individual object.

'Catalogue of Nebulae and Star Clusters.'
Observed at Paris by M. Messier.
At the Observatory of the Marine.'
Hotel de Clugni. Rue des Mathurins.

'M. Messier has observed with the utmost care, the nebulae and star clusters which are found above the horizon of Paris: he has determined their Right Ascension and Declination and given their diameters, together with the circumstantial details of each one: a work which was lacking in astronomy.

'He also gives details of the searches he has made for those other nebulae which other astronomers have discovered but which he has been unable to find.

'The catalogue of nebulae and star clusters of M. Messier was included in the volume of the Academy of Sciences for the year 1771 on page 435. He reported at the end of his *Memoir*, an outline drawing, carefully drawn, of the

---

* After the Revolution the title included the phrase *'A l'Usage des Astronomes et des Navigateurs'* and publication was by the Bureau des Longitudes.

fine nebula in the sword of Orion with the stars it contains. This drawing will be of assistance to show whether it is subject to any change in the future.

'Comparing the present drawing with those of Ms. Huyghens, Picard, de Mairan and le Gentil, it is surprising to find such a change in it that, considering its shape alone, one would have difficulty in recognising it as the same nebula. These drawings given by M. le Gentil can be seen on page 470, plate XXI of the volume of the Academy for 1759.

'To the catalogue, printed by M. Messier which we give here, we have added a large number of nebulae and star clusters which he has discovered since the printing of his *Memoir* and which he has communicated to us. With the positions of the nebulae, M. Messier gives their numbers which are the same on the following pages where the details of each of the observed nebulae are given.'

At the end of his catalogue, Messier gives an account of the nebulae reported by earlier observers, but which he had failed to find.

'*Nebulae discovered by different Astronomers which M. Messier has looked for without success.'*

'Hevelius, in his *Prodrome d'Astronomie,* reported the position of a nebula, situated on the top of the head of Hercules, having R.A. 252° 24' 3" and Dec. N. 13° 18' 37". On 20th June 1764, in a good sky, M. Messier looked for this nebula without being able to discover it.

'Hevelius, in the same work, reported the positions of four nebulae; one on the forehead of Capricorn, the second preceding the eye, the third following the second and the fourth above that and attending the eye of Capricorn. M. de Maupertuis reported the position of the four nebulae in his *Discours sur la Figure des Astres,* 2nd edition, p. 109. M. Derham mentions them also in his memoir printed in the *Philosophical Transactions,* No. 428, p. 70. One also finds these nebulae on many planispheres and celestial globes.

'M. Messier looked for these four nebulae on 27th July, 3rd Aug., 17th and 18th Oct. 1764, without being able to find them and he doubts that they exist.

'In the same work, Hevelius reported the position of two other nebulae; one on the near side of the star which is above the tail of Cygnus and the other on the far side of the same star.

'On the 24th and 28th Oct. 1764, M. Messier looked carefully for these two nebulae without being able to find them. M. Messier saw clearly at the end of the tail of Cygnus, near the star, pi, a cluster of small stars but the position he determined was different from that which Hevelius reported in his work.

'Hevelius also reported in the same work, the position of a nebula situated in the ear of Pegasus. M. Messier looked for it in fine weather on the night of 24th and 25th Oct. 1764 without being able to find it: unless it is the same one as the nebula which M. Messier has observed between the head of Pegasus and that of Equuleus. See No. 15 of this Catalogue.

'M. l'Abbé de la Caille, in his *Memoir* on the nebulae of the southern pole, printed in the *Memoirs* of the Academy of 1755, p. 194, reported the position of a nebula which resembled, he said, the small nucleus of a comet: its R.A. Jan. 1st 1752, 18° 13' 41" and Dec. S 33° 37' 5". On July 27th 1764, in an entirely serene sky, M. Messier looked for this nebula in vain: perhaps the instrument which M. Messier employed for the search was not adequate to find it. It was later seen by M. Messier. (See No. 69.)

'M. de Cassini reported in his *Elements of Astronomy* p. 79, that his father discovered a nebula in the area between Canis Major and Canis Minor and which was one of the finest to be seen in the telescope. M. Messier looked several times for this nebula in a serene sky without finding it and he presumes it to have been a comet just becoming visible or disappearing, for nothing resembles a nebula more than a comet which is only just becoming visible in an instrument.

'In the great *Catalogue of Stars* of Flamsteed, there is reported the position of a nebula situated in the right leg of Andromeda, having R.A. 23° 44' and Polar Distance 50° 49' 15". M. Messier looked for it on Oct. 21st 1780 with his achromatic telescope without finding it.'

Bode   Bode's numbers and descriptions are taken from his *Memoir, Upon some newly-discovered Nebulous Stars and a complete Catalogue of those hitherto known*, printed in the *Astronomisches Jahrbuch* for 1779, published at Berlin in 1777. He observed with a 7-foot heliometer (split-objective) telescope and also used a 14-foot refractor.

W. H. (William Herschel) and J. H. (John Herschel)   Wm Herschel's first *Catalogue of 1000 new Nebulae and Star Clusters* appeared in the *Philosophical Transactions* of the Royal Society in 1786. Additions to these were contributed to the same publication in 1789 and 1802.

John Herschel's Slough Observations appeared in the *Phil. Trans.* (Nos. 1–2306) in 1833 and his *Cape Observations* (Nos. 2307–4021) in 1847. The same observer's *General Catalogue of Nebulae* was published in the *Phil. Trans.* for 1864.

The notes and descriptions of the two Herschels have been widely quoted by many writers and Smyth uses them extensively in his *Cycle of Celestial Objects*.

Wm Herschel's classification of:

    I. Bright Nebulae
    II. Faint Nebulae
    III. Very faint Nebulae
    IV. Planetary Nebulae
    V. Very large Nebulae
    VI. Very compressed and rich star clusters
    VII. Compressed clusters of small and large stars
    VIII. Coarsely scattered clusters of stars

were not generally given to Messier's previously discovered objects but there are a few exceptions. M 33 was classed as 17 H. V, M 43 as possibly 17 H. III, the two 'missing objects' M 47 and M 48 were included as 38 H. VIII and 22 H. VI respectively. M 61 was listed under 139 H. I and M 82 as 79 H. IV. All the 'additional' objects, M 104 to M 110 were given Herschel designations and these will be found on the relevant pages in Chapter 4.

All the Messier Objects were given numbers in John Herschel's *General Catalogue* of 1864 except M 25, M 40, M 45 and M 102. The asterism, M 73, was included and M 91 was included in the form 'M 91??=G.C. 3113=602 H. III'.

The NGC has the full title of: – *A New General Catalogue of Nebulae and Clusters of Stars, being the Catalogue of the late Sir John F. W. Herschel, Bart., revised, corrected, and enlarged,* by J. L. E. Dreyer, Ph.D.

This was first published in 1888 and contained all the Messier Objects with the exception of those omitted from John Herschel's catalogue as noted above. M 25 was later included in the second *Index Catalogue*, published in 1908, the observer being Bailey.

Smyth and Webb The origins of the descriptions by Admiral Smyth and the Rev. T. W. Webb are accounted for in the introduction to Chapter 4. Smyth describes all the Messier Objects except M 40, M 47 and M 91, but for 16 others, merely quotes John Herschel's 1864 catalogue entries.

Webb's descriptions are generally brief and 22 of Messier's Objects are not included.

Rosse Many of the notes of the third Earl of Rosse were published in the *Phil. Trans.* for 1844, 1850 and 1861. These were brought together and added to by his son, the fourth Earl, and published as a catalogue in the *Transactions* of the Royal Dublin Society in 1880.

D'Arrest The descriptions given by D'Arrest are short extracts from his great catalogue, *Siderum Nebulosorum, Observationes Havniensis*, published by the Danish Royal Society of Science at Copenhagen in 1867. This work is written entirely in Latin and gives detailed descriptions and accurate positions (epoch 1861) for some 500 nebulae and clusters down to declination 28° S observed with a 16-foot telescope of 11-inch aperture.

Lick XIII The descriptions under this heading are taken from the *Lick Observatory Publication No. XIII, Studies of the Nebulae* published by the University of California Press, Berkeley, 1918. The author of this important work was the eminent American astronomer, Heber Doust Curtis (1872–1942), who made great contributions to the study of variable stars and excelled in the photography and analysis of nebulae and star clusters.

The observations were made at the Lick Observatory, Mount Hamilton and at Santiago, Chile from 1898 to Feb. 1918 and cover, altogether, 762 objects. These include 513 spiral nebulae, 56 diffuse nebulae, 36 globular clusters, 24 open clusters, 78 planetary nebulae, 8 dark nebulae and 47 unclassified objects. The descriptions given are not those of the visual objects but of photographs taken mostly with the 36-inch Crossley Reflector.

The mirror for this famous telescope, made by Calver, and at one time owned by Dr A. A. Common, was later re-figured by Sir Howard Grubb and then presented to Lick Observatory by Edward Crossley of Halifax. In a modern mounting, the mirror is still much in use at Lick Observatory.

Another *Lick Publication*, Vol. VIII, published in 1908, contains a fine series of photographs of nebulae, including many of the Messier Objects.

Flammarion A few of Flammarion's descriptions are included where they provide additional information of interest or where other material is scarce. These items are taken from the series of articles which Flammarion contributed to the *Bulletin* of the Astronomical Society of France, Vols. 31–35, 1917 to 1921. Flammarion quotes Messier's original descriptions and also some of that observer's additional manuscript notes. These notes appeared on Messier's own copy of his catalogue of 1781 which Flammarion found, apparently by chance, on a Paris bookstall.

Drawings and photographs of nearly all the objects are included together with descriptions by other observers such as Smyth and D'Arrest.

Further information concerning other observers will be found in the biographical notes which follow and additional references are provided in the *Bibliography*.

BIOGRAPHICAL NOTES

The subjects of the following biographical notes have been selected for their connection with Messier himself or as observers, at one time or another, of the objects in his catalogue. Those who were more closely associated with Messier, such as Méchain, Delisle and others, have been given a little more space but some early observers, like Ihle and Palitzch, are obscure figures and little appears to be known about them.

No 'modern' astronomers (in this case, those who lived after 1925) have been included. This restriction is quite arbitrary and purely due to the obvious difficulty of deciding where to stop.

In no way do the notes pretend to be anything more than a brief summary of the careers of those concerned. More complete biographies can be found, in most cases, in works exclusively devoted to the history of astronomy; although many of these may be available only in French. A short list of those works consulted is given in the *Bibliography*.

Barnard Edward Emerson Barnard (1857–1923)
An American astronomer who graduated in 1887 at Vanderbilt University where he had directed the observatory for four years. He was appointed astronomer at Lick Observatory in 1887 and in 1895 was professor of practical astronomy at Yerkes Observatory: he also took part in the U.S. Naval Eclipse expedition to Sumatra in 1901. He made great contributions to celestial photography and discovered 16 comets. He was also the discoverer of the fifth satellite of Jupiter and the star with the large proper motion of 10".3 per annum, a 10th mag. red dwarf, known as Barnard's star.

He detected a ninth star, itself double, in the 'trapezium' in the Orion

Nebula, M 42, in 1888. He made visual observations of the nebulosity around *Merope* in the Pleiades, M 45, in 1890. In 1893 he succeeded in obtaining a photograph of the nebula. He also photographed M 11, M 35 and other nebulae and clusters in his various sky-surveys from 1892 to 1895.

He is perhaps best known for his photographic atlas of selected regions of the Milky Way which was published in 1917 by the Carnegie Institution of Washington. These photographs, taken on blue-sensitive plates, were accompanied by Barnard's descriptions of 349 dark nebulae in this region of the Galaxy and he put forward the conclusion that these dark areas were due to masses of obscuring gas and dust, and not merely regions where stars were absent.

Bessel Friedrich Wilhelm Bessel (1784–1846)

He was interested in astronomy from an early age and in 1806, when only 21, recalculated the orbit of Halley's comet from observations going back to 1607. In 1810 he became the director of Konigsberg Observatory and began a revision of Bradley's star catalogue which was extended to a total of 63 000 stars by 1818.

The accurate positions obtained in this survey led him to detect irregularities in the proper motion of *Sirius* and *Procyon* which suggested that both these stars might have invisible companions. After Bessel's death, the German astronomer Peters (1806–82) calculated a 50-year orbit for the companion of *Sirius* and this was discovered by the famous American optical worker and astronomer, Alvan G. Clark (1832–97), when testing a newly-figured 18-inch objective in 1862. *Procyon's* companion was investigated by Auwers in 1862 who calculated its elements and this was found by Schaeberle, using the 36-inch telescope of Lick Observatory in 1892.

Bessel started observations of stellar parallax with a Fraunhofer heliometer in 1837 and in Dec. 1838 obtained a parallax for *61 Cygni* of 0″.31. This was a remarkable achievement – the modern value is about 0″.30 – and although both F. G. W. Sturve at Dorpat and Henderson at the Cape of Good Hope also obtained parallaxes (Struve of *Vega*, and Henderson of α *Centauri*) Bessel is generally credited with the first successful measurement of the distance of a star.

Bevis John Bevis (1695–1771)

An English physician and amateur astronomer, he was born in Old Sarum in Wiltshire. He is credited with the discovery of the 'Crab' Nebula in Taurus, M 1, in 1731. He observed an occultation of Mercury by Venus on May 28th, 1737, and was also the inventor of an ingenious slide rule with which he was able to predict the eclipses of Jupiter's satellites.

In 1738 he set up a private observatory at Stoke Newington in North London where he began observations for his star atlas, *Uranographia Britannica*. This major work was completed in 1750 but unfortunately the publisher went bankrupt soon after the 52 elaborate plates were engraved and only one or two printings were made. The catalogue to complement the atlas was never published. One copy of Bevis' Atlas is in University Library, Cambridge,

England and, according to Jerome Lalande, Messier also possessed a copy in 1792. The Cambridge copy was printed from the original plates in 1818. There is an earlier copy of the same atlas in the British Museum: this was published in 1786 and is entitled *Atlas Celeste* or *Celestial Atlas*.

Messier, in his catalogue of 1784, makes several references to *l'atlas Celeste Anglais* or *le grand atlas Anglais*. As he appears to have found there the objects M 1, M 11, M 13, M 22, M 31 and M 35 already printed as having been discovered before, this may well mean that the atlas he refers to was that of Bevis, because Flamsteed's atlas, which Messier also used, has none of these objects, except M 31, included in it. Bevis' atlas, on the other hand, contains only those nebulae mentioned above (with the addition of the southerly objects, ω *Centauri* and the Magellanic Clouds) and Messier must have had at least access to a copy of this rare and beautiful volume.

In recent years several other copies of Bevis' atlas have come to light and about a dozen more or less complete versions are now known to exist, the copy in the Library of the American Philosophical Society having the only known version of Bevis' star catalogue included.

Bevis was recognised as a most assiduous observer and was one of only two persons in Britain known to have seen Halley's comet on its first predicted return in 1758–59. Bevis observed it on May 1st and 2nd, 1759 (after perihelion): the other observer was another 'medical astronomer', Nicholas Munckley, MD, FRS on the same dates.

John Bevis died from injuries received during a fall while working in his observatory on 6 Nov. 1771, aged 76.

Bode   Johan Elert Bode (1747–1826)
Born in Hamburg, he became a member of the Berlin Academy of Science and later director of the Berlin Observatory. He founded the famous German astronomical ephemeris *Berliner Astronomisches Jahrbuch* and continued to publish it until 1829.

He discovered M 81 and M 82 on Dec. 31st, 1774 and the latter galaxy was sometimes known as 'Bode's' Nebula. He also discovered M 53 on Feb. 3rd, 1775, and in 1777 published in the *Astronomisches Jahrbuch* (for 1779), 'A Complete Catalogue of hitherto observed Nebulous Stars and Star-Clusters'. This contained 75 entries but is a somewhat inflated list which contains many asterisms and 'non-objects' taken from Hevelius and others: no more than 50 of the objects are true nebulae or clusters and there are several mistakes in the positions of these. Two additions to this catalogue in 1780 shows that Bode had discovered M 92 on Dec. 27th, 1777, and M 64 on Apr. 4th, 1779.

The new planet discovered by Wm Herschel in 1781 was named by its discoverer 'Georgium Sidus' in honour of George III; Lalande suggested that it should be called 'Herschel' but it was Bode's proposal of the name Uranus which was finally adopted.

Bode was greatly interested in the discovery of Uranus: he collected together practically all the observations which had been made by different astronomers and published them in the *Astronomisches Jahrbuch*. By working through older star catalogues he found that Uranus had been observed, as a star, by Tobias Mayer in 1756 and this observation was used by calculators to obtain a better orbit for the planet. In 1785 he was able to show, with the help of Méchain and Zach, that Flamsteed's observation of the 'star' *34 Tauri* in Dec. 1690 was actually the first recorded sighting of Uranus.

Bode's book, *The Knowledge of the Starry Heavens* attained great popularity and was reprinted many times. In this book he put forward the well-known *Bode's Law* for obtaining the relative distances of the planets; however, the real discoverer of this fortuitous rule was J. D. Titius (1729–96), a professor of Mathematics and Physics at Wittemberg University.

In 1801, Bode published his famous star atlas, *Uranographia*: this large and popular work contained many new and strange constellations such as Officina Typographica, Apparatus Chemica, Globus Aerostaticus, Honores Frederici, Felis and Custos Messium. These have all completely vanished from modern charts although one of them, Quadrans Muralis (formed from stars in the northern part of Bootes), still lingers on as the name of the Quadrantid meteors which have their radiant in this area.

After having directed the Berlin observatory for nearly 40 years, Bode retired in 1825, to be succeeded by J. F. Encke.

Bond William Cranch Bond (1789–1859)

A successful businessman and talented astronomer, he was invited to become the first director of the new Harvard College Observatory in 1837. After J. W. Draper had taken a first successful photograph of the Moon on Mar. 23rd, 1840, Bond and his son, with the Boston photographers Whipple and Black, obtained a daguerreotype of the Moon on Dec. 18th, 1849, which was widely exhibited.

Bond George P. Bond (1826–65)

The son of W. C. Bond, he succeeded his father as director of Harvard College Observatory. He made many detailed drawings of the Andromeda Nebula, M 31, in 1847 which showed several of the 'dark rifts' later to be revealed in Isaac Roberts' photographs in 1888. He also made many observations and sketches of the Orion Nebula, M 42. The nucleus of the adjoining nebula, M 43, was often known as 'Bond's Star'.

In 1848 he discovered an eighth satellite of Saturn, Hyperion, at about the same time as Lassell observed it in England. The inner, *crape* ring of Saturn, first noticed by Galle at Berlin in 1838 was independently rediscovered by G. P. Bond in 1850, again at about the same time as an English observer, W. R. Dawes.

Much of Geo. P. Bond's work was with celestial photography and one of his remarkable pioneering achievements was in the use of wet-plate photographs to obtain the accurate registration of double stars.

He also published, in 1863, a *List of New Nebulae and Star-Clusters found at Harvard College Observatory*, compiled from observations made by himself, S. Coolidge, and J. H. Safford.

Cassini I   Giovanni Domenico Cassini (1625–1712)

The founder of a dynasty of astronomers, he studied mathematics and astronomy under the Jesuits and by the age of 25 was professor of astronomy at the University of Bologna. He determined the obliquity of the ecliptic at 23° 29' and made studies of refraction and solar parallax. He observed comets in 1652 and 1664 and solved a long standing problem in planetary orbits in 1652.

His fame was known far outside Italy and he was invited to come to France by Louis XIV who offered attractive remuneration to foreign scientists when he created the *Academie Royale des Sciences*. Cassini arrived in Paris on April 4th, 1669, and became the first director of the Paris Observatory where he set up, out of doors, his immensely long air telescopes. Cassini worked with another imported scientist, Christiaan Huyghens, and together they made many important advances in astronomy. Cassini was the first to observe the division in the rings of Saturn which now bears his name; he obtained an accurate value for the rotation of Jupiter and discovered the 2nd, 3rd, 4th and 5th moons of Saturn.

Unlike Huyghens, who later returned to Holland, Cassini became a naturalised Frenchman in 1673 and married a well-connected Frenchwoman who brought him a useful dowry which included the Chateau de Thury. In 1683, together with La Hire, he surveyed the meridian of Paris and devised a projection for a large-scale map of France.

Cassini published new tables of the satellites of Jupiter in 1693 and was an active member of the Academy of Sciences until his blindness in 1711.

Cassini II   Jacques Cassini (1677–1756) – Son of Cassini I

He was, at one time, tutor to the Duc de Burgoyne, grandson of Louis XIV. He was one of the first to detect stellar proper motion by showing that the latitude of *Arcturus* had altered by 5' in 159 years while η *Bootis*, nearby, had remained unchanged. He published *Elements of Astronomy* in 1740 and spent much of his career in geodesic surveys. He was killed in a carriage accident at the age of 79.

Cassini III César François Cassini (1714–84) – Son of Cassini II

Also known as Cassini de Thury, he assisted his father in much of his work in geodesic surveys. He also made surveys in Germany and went to Vienna to determine the longitude difference of that city from Paris. He also produced a new map of France to a scale of 1/86 000, some portions of which were drawn by Messier.

Cassini IV Jean Dominique Cassini (1748–1845) – Son of Cassini III

Later created Comte de Cassini. He made a voyage at the age of 20 in the frigate *l'Enjouée* to study the behaviour of Le Roy's chronometers in the Atlantic and published an account of this in 1770. In 1771 he published an account of 100 years of observations at the Paris Observatory and he succeeded his father as director there in 1784. During the Revolution he was imprisoned as a Royalist for $7\frac{1}{2}$ months and was fortunate to escape the guillotine. After his release he retired to his estate at Thury and in 1810 published his *Memoirs pour servir a l'histoire des Sciences* which included eulogies of Le Gentil, De Saron and others.

D'Angos The Chevalier Jean Auguste D'Angos (1744–1833)
A captain in the Navarre regiment of the French Army, he became one of the Knights of Malta and dabbled in chemistry and astronomy, achieving disastrous results in both sciences. He was invited by the Grand Master of the Knights of Malta to set up an observatory in the Palace of Valetta, and in 1789 an experiment by D'Angos with phosphorus ended in a fire which wrecked the observatory and destroyed much of the building.

In astronomy, D'Angos was an undoubted charlatan: in 1784 he claimed to have discovered a comet in Vulpecula on April 11th and wrote a letter to Messier in Paris giving the elements of its orbit. Messier looked for it on several nights but was unable to find it. D'Angos also claimed to have discovered a comet in Taurus in 1793 but this, too, was seen by no one else. In 1798 he wrote to several astronomers in Paris that he had seen a comet cross the face of the Sun on Jan. 18th.

All these observations were later found to have been falsified and the Chevalier was roundly denounced by Encke who discovered the deception in 1820.

Admiral Smyth also relates that D'Angos mistook the cluster, M 93, in Puppis, for a comet on one occasion.

Darquier Antoine Darquier de Pellepoix (1718–1802)
A French astronomer who worked at Toulouse. He made an extensive catalogue of star positions between 1791 and 1798 which Lalande later incorporated in his important catalogue of nearly 50 000 stars, published in 1801. Darquier was the author of a translation of Lambert's *Lettres cosmologiques sur l'ordre du monde*, published in 1761. He was also the first to discover, in 1779, the 'Ring' Nebula in Lyra, M 57.

He was one of the early observers of the new planet Uranus in 1781, using an achromatic refractor of about $2\frac{1}{2}$ inches aperture and 42 inches F.L. He calculated its orbit and found, like de Saron, Lexell and others, that it had only a slight eccentricity and lay outside that of Saturn.

D'Arrest Heinrich Ludwig D'Arrest (1822–75)
When a student under J. F. Encke at the Berlin Observatory in 1846, he suggested that the new planet, whose probable position had been computed by Le Verrier, might be found by comparing the designated part of the sky with some star charts of the region which had recently been prepared. He was able to find the chart and, by comparison, the planet Neptune was discovered.

On June 27th, 1851, at Leipzig, he discovered a faint telescopic comet in Pisces. It was visible for more than three months and its orbit was found to be elliptical. It was predicted that the comet would return in the winter of 1857–58 and this apparition was seen by only one observer, Maclear, at the Cape of Good Hope.

D'Arrest went to Copenhagen in 1857 and built an observatory in a bastion of the old wall of the city where he made many observations of nebulae and star clusters with a 4½-inch refractor. Later, with an 11-inch refractor, he compiled his great *Siderium Nebulosorum Observationes Havnienses* which was published in 1867. This catalogue provided accurate positions for a large number of nebulae and clusters and was the basis for Dreyer's revision of John Herschel's *General Catalogue* which was later published as the *New General Catalogue*.

De Chéseaux Philippe Loys de Chéseaux (1718–51)
A Swiss astronomer and mathematician, he was the son of a wealthy land-owner living near Lausanne and became interested in astronomy at an early age. He had a small private observatory and observed with a 14-foot simple refractor and a Gregorian reflector of 2 feet F.L. He also possessed an 18-inch quadrant made by Bion that could measure to 15 seconds of arc.

He is perhaps best known for the comet which he observed in December 1743 and which was visible until March 1744. The comet became a brilliant object, outshining even Jupiter, and in a treatise he wrote on its apparition, de Chéseaux claimed that at one time it exhibited no less than six tails and this was later confirmed by Delisle. Although de Chéseaux found the comet independently, the first discoverer was Klinkenberg, after whom the comet is named.

De Chéseaux was a correspondent with many astronomers including Jacques Cassini; he was a member of the Paris Academy of Sciences and also of the Royal Society of London. He was offered the directorship of the St Petersburg Observatory but was unable to take it up owing to ill-health. He also seems to have been the first to propose the problem of why the night sky is dark in a near-infinite universe of bright, radiating bodies. This puzzle is generally known as *Olbers' paradox* after the German astronomer, Heinrich Olbers (1758–1840), who reformulated the problem in 1823.

De Chéseaux also made many observations of nebulae and clusters and in a letter to his grandfather, written in about 1746, he gave a list of 20 such objects. These, he claimed, were 'truly nebulous stars' and not 'simple star-groups' like the objects in Derham's list of 1733. The list was read at a meeting of the French Academy of Sciences by the celebrated scientist Réaumur on Aug. 6th, 1746, but was not otherwise published. This important but unpublished catalogue was analysed by Bigourdan in the *Paris Annales (Obs.)* for 1884 who identified most of de Chéseaux' objects with clusters and nebulae later included in the lists of Messier and Herschel.

Bigourdan could find no objects in the NGC to correspond with Nos. 2, 5 and 8 of de Chéseaux' list and he did not attempt to trace Nos. 12 and 13 as, having been given no R.A. or Dec., he considered 'their indications too vague'.

The position of de Chéseaux' No. 5, however, is very close to that of M 25 which was omitted from the NGC proper and added in the *Index Catalogue* as IC 4725. Had Bigourdan succeeded in tracing this he would certainly have credited the discovery of M 25 to de Chéseaux.

No. 12 was quoted by de Chéseaux as 'above the northern feet of Gemini': this location is certainly not precise but it is in the area of the position of the conspicuous cluster, M 35. As this object is included in Bevis' Atlas *Uranographia Britannica* in 1750 and no other observer has claimed its discovery, it may well be credited as having been found by de Chéseaux in about 1745.

Altogether, de Chéseaux' list shows him to have been the probable first observer of M 4, M 16, M 17, M 25, M 35 and M 71.

Delisle

Joseph Nicholas Delisle (1688–1768)

Delisle was one of 12 children; his father, Claude Delisle, was a teacher of history and geography in Paris. The oldest son, Guillaume Delisle (1675–1726) was a professor of geography and an eminent member of the Academy of Sciences.

J. N. Delisle was one of the last pupils of the celebrated G. D. Cassini and, after graduating, taught mathematics and astronomy at the College de France. He was allowed to set up an observatory in the dome of the Palais de Luxembourg from 1710 to 1715 but later had to move his equipment to the Hotel Taranne.

In 1721 he was invited by Tzar Peter I to go to St Petersburg to establish a school of astronomy but he did not take this up until 1725 when, accompanied by his younger brother Simon and an instrument maker named Vignon, he set out on the long journey to Russia. He stayed in Russia for 21 years; he was a member of the Academy of Sciences and director of the St Petersburg Observatory and also travelled widely throughout Russia and Siberia on survey work. Simon Delisle ventured further afield into eastern Siberia: he died at sea off the coast of Kamchatka in 1741 and was buried near Avachka.

J. N. Delisle returned to Paris in Sept. 1747 and resumed his chair of astronomy at the College de France where, among his pupils he taught Jerome Lalande. During the whole of his career he had been a passionate collector of books and old registers of astronomical observations and these he made over to the State in return for a pension of 3000 livres and the post of Astronomer to the Navy.

He set up a small observatory on top of the tower of the old Hôtel de Cluny where he also had permanent quarters. He had a secretary named Libour and in 1752 he engaged the young Charles Messier as clerk and observing assistant. The débâcle over the search by Messier for Halley's comet in 1758–59 was undoubtedly due to Delisle's mismanagement but by this time he was over 70 and in reality, already in semi-retirement. He finally retired in about 1761 and died of apoplexy on Sept. 12th, 1768.

He had visited England in 1724 and was there received into the Royal Society. He proposed, in a letter to Jacques Cassini in 1743, a more accurate method of timing the transits of Mercury and Venus from the instants of contact. While at St Petersburg he devised a method for determining the heliocentric co-ordinates of sunspots and he was also an innovator in the use of a sidereal time clock in the observatory.

Derham   William Derham, DD, FRS (1657–1735)

An English clergyman and amateur astronomer, he went up to Trinity College Oxford in 1675 and obtained his BA in 1679. He was appointed vicar of Wargrave in Berkshire in 1682 and later rector of Upminster in Essex in 1689. He was made a canon of Windsor in 1716 and became chaplain to the Prince of Wales, afterwards George II.

Derham had many scientific interests including that of paper making but his main interest was astronomy: he was one of the first to observe the 'ashen light' of Venus on May 2nd, 1715.

The list of 16 nebulous stars which he published in the *Philosophical Transactions* in 1733 was not of his own discovery but a transcription of the objects in Hevelius' *Prodromus Astronomiae* and the only one of these he had seen was '. . . that in Andromeda's Girdle which is as considerable as any I have seen'. He describes the appearance of the others only as they were shown on Hevelius' *Maps of the Constellations*.

Derham, however, did observe with an 8-foot reflector, five of the objects reported by Halley in 1716: the sixth, ω *Centauri*, being invisible from England. He was aged 75 when he made these observations and concluded: 'I fear I shall not be able to pursue my observations much further by reason my Reflector loseth its Excellence and Power by beginning to be tarnished.'

Flammarion   Nicholas Camille Flammarion (1842–1925)

He first studied theology but became interested in astronomy at an early age and when only 16 wrote a 500-page manuscript entitled *Cosmologie Universelle*. He became an assistant to Le Verrier at the Paris Observatory in 1858 and worked as a calculator there until 1862 when he transferred to the Bureau des Longitudes. In 1867 he returned to the Observatory to take charge of a programme of double star observations. The results of this work were published in a catalogue of 10 000 double and multiple stars in 1878.

He also made many observations of the Moon and Mars but he is best known as an author and populariser of astronomy. His *L'astronomie Populaire*, published in 1879, eventually sold over 100 000 copies and was also widely read in an English translation made in 1894 by J. E. Gore.

In 1882, an admirer of his work, a M. Méret, gave Flammarion a chateau and estate at Juvisy near Paris where he erected a private observatory and meteorological station. He encouraged amateur observers and founded the Astronomical Society of France in 1877.

His book on the planet Mars, published in 1892, supported the existence of 'canals': he also thought them artificial and therefore proved that Mars was inhabited by people in an advanced state of technology.

Flammarion possessed a copy of the *Connaissance des Temps* for 1784 which had belonged to Messier himself and which contained several manuscript additions by the author of the famous catalogue. Flammarion published by instalments in the *Bulletin* of the Astronomical Society of France between 1917 and 1921, a detailed resurvey of the Messier Objects and, on evidence from his copy of the catalogue, added M 104 to the list.

320

In 1922 he was made a Commander of the Legion of Honour for his services to astronomy.

Flamsteed   John Flamsteed (1646–1719)

The first Astronomer Royal and director of the Royal Observatory, established at Greenwich in 1675. He was one of the first, after the Abbé Picard, to combine a magnifying telescope with a graduated arc for measuring angles, and spent 30 years making precise observations of star positions. The first part of his famous star catalogue, *Historia Coelestis Britannica*, was published in 1712 but this was done without his full consent and he later suppressed most of this edition. The second, revised and extended edition was published in three volumes in 1725, six years after Flamsteed's death. An equally famous star atlas, *Atlas Coelestis* was published in 1729.

In his catalogue he noted some 16 nebulous objects although many of these were later found to be 'asterisms'. Flamsteed was the first to discover the fine galactic cluster surrounding the star *12 Monocerotis* (NGC 2244) – not in Messier's list – and also made the first recorded observations of the cluster NGC 6530 in M 8, and M 41.

Among his observations, Flamsteed obtained positions for Uranus in 1690, 1712 and 1715 but logged them as stars; the discovery of the planet falling to Wm Herschel in 1781.

Messier makes many references to Flamsteed's star numbers when obtaining the positions for his nebulae and almost certainly used Flamsteed's catalogue and atlas.

Flamsteed suffered from poor health; his salary as Astronomer Royal was meagre and yet he had to provide his own instruments and publish his work at his own expense. He quarrelled with both Hooke and Halley, who put pressure on him to publish his work before he considered it ready; Hooke, in fact, during one of their disputes, became so incensed that he intemperately referred to Flamsteed as, 'A cockscombe and conceited of nothing: an ignorant, impudent Asse!'

Galileo   Galileo Galilei (1564–1642)

He was well known as a teacher of mathematics in the Universities of Pisa and Padua. On hearing of the discovery of the telescope by Dutch opticians, he constructed an improved version for himself and first turned the instrument on the heavens in 1609. Using a magnification of about 30×, he saw the mountains and craters of the Moon and discovered that the Milky Way was composed of innumerable faint stars. He observed the four major satellites of Jupiter in Jan. 1610 and in the same year published *The Sidereal Messenger* to reveal his discoveries and to support the Copernican system of an heliocentric universe. Galileo also observed that the planet Venus exhibited phases like the Moon, which also supported the Copernican theory.

He published *The Assayer* in 1624 which, although it criticised the Ptolemaic, Earth-centred system accepted by the Church, escaped any strong censure and was warmly praised by many savants of the time.

His later work, *Dialogue of the two Worlds,* published in 1632, although submitted first to Pope Urban VIII and cleared for printing by the Holy Office examiners, was later used by his adversaries to bring about his official recantation in 1633 and subsequent 'restriction of movement' at Rome and Florence.

Galileo was the first to observe that Praesepe was not a 'nebula' but 'a mass of more than 40 small stars'. He also examined and sketched the Pleiades, the false cluster in Orion's head and in his drawing of the belt and sword of Orion, he plotted no less than 83 stars. The nebulosity around θ *Orionis,* M 42, however, seems to have escaped his notice.

The last eight years of his life at Florence were devoted to the study of his newly created science of dynamics.

Gore John Ellard Gore (1845–1910)
He was an Irishman who became a civil engineer and who was a keen amateur astronomer all his life. He wrote many books, including *Studies in Astronomy,* 1904, *Astronomical Essays,* 1907 and also translated Flammarion's *Popular Astronomy* into English. He published a large number of articles in various journals, in which he made several successful excursions into astronomical theory.

While working as a civil engineer in the Punjab, he was able to observe most of the Messier Objects, including those too far south to be seen from England. He wrote brief descriptions of each of the objects except M 40, M 45, M 48 and M 102 which he was unable to identify: M 47 and M 91 he mentioned but did not describe. This list was published in the journal, *The Observatory* in 1902 and also included in his *Studies in Astronomy.*

He was a great admirer of Sir Wm Herschel and of the little known Belgian astronomer, Jean Charles Houzeau (1820–88) who spent nearly 20 years in the United States, where, besides doing much astronomical work, he edited, under the name of Dalloz, a Philadelphia journal, published in the interest of negro emancipation, called first *The Union* and later, *The Tribune.*

Halley Edmond Halley (1656–1742)
The son of a prosperous soap manufacturer, Halley was educated at St Paul's School and Queen's College, Oxford. He made observations of sunspots at Oxford in 1676 and in the same year obtained a grant to make an expedition to St Helena to compile a catalogue of southern hemisphere stars. This catalogue, containing the positions of 381 stars, was published in 1678.

In 1679 Halley joined Hevelius at Dantzig and in the following year worked with G. D. Cassini at Paris where he observed the Great Comet and first became interested in these objects. In 1705 he published a *Treatise of Cometary Astronomy* and calculated the parabolic orbits for 24 comets including the comet of 1681–82. Halley found that this comet had similar orbital elements to that observed by Kepler in 1607 and also by Apian in 1531. This led to his idea of a periodic comet with an elliptical orbit returning every 75 or 76 years and he made his famous prediction that it would return again at the end of 1758.

In the *Philosophical Transaction* for 1715 Halley published a memoir entitled *Of Nebulae or lucid Spots among the Fixt Stars*; this contained descriptions of six nebulous objects: M 11, M 13, M 22, M 31, M 42 and ω *Centauri*. Halley discovered M 13 by chance in 1714 and had seen ω *Centauri* while at St Helena in 1677; the other four had been found by other observers.

Halley, who became the second Astronomer Royal, made many major contributions to the science including the identification of proper motion of the stars and new orbital elements of the Moon. The prompt return of his comet in 1758 earned him an immense, if posthumous, fame.

Herschel   Caroline Lucretia Herschel (1750–1848)

She came to England to act as housekeeper for her brother William, and when he became engaged in astronomy acted also as his secretary and general assistant, often staying up all night to record her brother's observations.

She also became an observer in her own right and with a 27-inch focal length Newtonian 'sweeper' having a magnification of 20× she found many star clusters and nebulae, one of which was NGC 205, the second companion of M 31 which she observed in 1783. (Although Messier claimed, in about 1807, to have seen this object in 1773, the discovery of NGC 205 is usually credited to Caroline Herschel.)

She found 8 comets in a period of 11 years; notably those of 1786 and 1788. Jerome de Lalande was so impressed by these observational feats of Miss Herschel that he suggested to the French Academy of Sciences that she should be honoured with a suitable award. The learned members, however, resisted this on the grounds – uncharacteristic of the French – that they did not wish to be accused of an excess of gallantry!

After this it is pleasing to note that, for her published catalogue of 561 stars, compiled from Flamsteed's observations and a zone-by-zone catalogue of all the nebulae and clusters observed by her brother, she was awarded the Gold Medal of the Royal Astronomical Society in 1828 and made an honorary member of the Society in 1835. She retired eventually to her native Hanover, received a Gold Medal from the King of Prussia in 1846 and died two years later at the age of 98.

Herschel   Sir Frederick William Herschel (1738–1822)

Born in Hanover, the son of a musician in the Hanoverian Guard, he came to England in 1757. He taught music for a time and became an organist at Halifax in 1765 and organist and conductor at Bath in 1766 where he was known as a celebrated performer on many instruments and a composer of merit.

He became interested in astronomy in about 1767 and by 1774 he had learnt to make specula superior to any which had been made before. On March 13th, 1781, with his $6\frac{1}{2}$-inch, 7-foot F.L. reflector, he discovered at Bath, the planet which was later called Uranus and he became famous as the first discoverer of a new planet in historical times.

His energy and intellectual grasp were prodigious: he observed the heavens systematically on every clear night and wrote papers on almost every aspect of astronomy, showing phenomenal insight and breadth of vision. He became a Fellow of the Royal Society and in 1782 was appointed private astronomer to George III, moving his observatory to Datchet, near Windsor to be near the Royal household. There he built a 12-inch reflector of 20 foot F.L. and later, an 18.7-inch and started on his first series of observations of nebulae and star clusters. The first catalogue of 1000 new nebulae and clusters was later presented to the Royal Society in 1786.

His attempts to measure annual parallax from observations of close pairs of stars failed in its primary purpose but this led him to the important recognition of binary systems of stars which revolved in close gravitational contact. In 1782 he published a catalogue of 260 pairs and a further list of 284 new pairs in 1784. In 1802 he presented a paper on their binary relationship.

For many years he had to augment his meagre royal stipend by making and selling telescopes: these were eagerly sought after and he sold altogether about 76 of these fine instruments for a total of about £15 000. In 1788, however, he married the wealthy widow of a London brewer and was thus able to continue his observations and researches without the disturbance of having to earn a living. He moved his observatory to Slough, still near Windsor, and on Aug. 28th, 1789, finally set up his largest telescope of 48-inch aperture and 40 foot F.L. With this instrument he discovered a seventh satellite of Saturn, Enceladus on Aug. 28th, 1789 and rediscovered the sixth, Mimas, which he had earlier glimpsed in a smaller telescope.

Herschel corresponded with many other astronomers including Messier whom he first met when he visited Paris in 1802, and there is little doubt that it was the example of the French astronomer's *Catalogue of Nebulae* of 1781 which prompted Herschel to begin his own great survey.

Herschel made many other major contributions to astronomy: he established that the Sun was itself moving through space and he determined the solar apex to be near the star λ *Herculis*. He was also the first to conceive something of the overall size and shape of the Galaxy and the Sun's position within it. Although his estimates of the size of the galactic system were too small by a factor of about 10, his conception represented a revolutionary achievement for his time.

He was a Titan among astronomers; perhaps the greatest observer of all time and a pioneer in almost every branch of the science.

Herschel Sir John Frederick William Herschel, Bt (1792–1871)
The only son of Wm Herschel, he went to Eton and St John's College, Cambridge, graduating as senior Wrangler in 1813. He became one of the most eminent mathematicians of his day, especially in the field of analysis. After first reading for the Bar he turned to astronomy in 1816 and with the assistance of his father, built an 18-inch reflector of 20 foot F.L. With this telescope he extended his father's work on double stars with the collaboration of Sir James South. He also made further additions to his father's catalogue of star-clusters and nebulae, adding 525 new objects in a catalogue published in 1833.

On Nov. 13th, 1833, he and his family embarked for the Cape of Good

Hope and on March 4th, 1834, he began observing at Feldhausen near Capetown. During the next four years he made an immense number of observations of nebulae, clusters and double stars including a detailed study of both Magellanic Clouds.

The results of this concentrated and important work he later published in England in 1847 and in 1864 he produced a *General Catalogue* of his own and his father's nebulae which totalled some 5000 objects. He was made a baronet at Queen Victoria's Coronation in 1837 and was the recipient of many academic honours. He was an accomplished chemist and made important advances in early photography. He also published a great many papers and books, the best known of which is perhaps his *Outlines of Astronomy* published in 1849. His son, Sir Wm James Herschel was noted as the pioneer of the use of fingerprints in criminology.

Helvelius John Hevel or Hevelke (1611–87)

A wealthy brewer and city councillor of Dantzig, he was also a skilled engraver and an important early astronomer. He built a well-equipped observatory, known as the 'Sternenburg', covering the upper floors of four houses where he installed a 130-foot air telescope. He made observations of sunspots from 1642 to 1645 and obtained a reasonably accurate figure for the period of the Sun's rotation.

His *Selenographia*, published in 1647 contained detailed maps of the Moon based on 10 years of observation and included diagrams of the phase for each day of lunation. He also made some of the first estimates of the height of lunar mountains.

He began observations for a new star catalogue in 1657 and by 1679 had obtained positions for 1500 stars. This work was interrupted when his observatory was burnt down destroying most of his equipment. (Ironically, Hevelius, among his other offices, held the post of Dantzig's Chief of the Fire Guard!)

His second wife, Elizabeth Margarethe, often considered the first woman astronomer, assisted him in his observations and, after his death carried on his work. She published his catalogue of 1564 stars, *Prodromus Astronomiae* in 1690 together with an Atlas, *Uranographia* which contained 54 fine plates. These two works contained positions for 16 'nebulous stars' which were later looked for by many astronomers, notably Derham and Messier, but apart from M 31 and M 44 they are mostly asterisms or 'non-existent'. Hevelius used several telescopes for Moon and planet work but allowed nothing but plain sights on the instruments he employed for obtaining star positions.

Hodierna Giovanni Batista Hodierna (1597–1660)

Born and educated in Ragusa, Sicily, he became astronomer to the court of the Duke of Montechiaro, where he remained until his death. He was one of the first to grasp the significance of Galileo's astronomical discoveries, and corresponded with Huyghens and Hevelius. Little of his work was known until a booklet entitled *De Admirandis Coeli Caracteribus . . .* printed in Palermo in 1654, came to light in 1984.

This work was in part a discussion on comets but it also revealed that Hodierna had observed some forty-odd nebulae, including many original discoveries. He was the first such observer to attempt a classification of nebulae, dividing them into three categories, and his drawing of the Orion Nebula (M 42), showing three 'Trapezium' stars, seems to have been the first illustration of its kind.

His telescope was a simple Galilean refractor with a magnification of ×20 and narrow field of view. Nevertheless some nineteen of the objects he picked up have been identified with true nebulae or star clusters, and include an independent re-discovery of the Andromeda Nebula (M 31) which he described as 'unresolvable'. His true discoveries include six star clusters which later appeared in Messier's list; M 6, M 36, M 37, M 38, M 41 and M 47, together with four more, NGC 2362, NGC 6231, NGC 6530 and (possibly) NGC 2451, which were unknown to the later observer.

This is a remarkable achievement by any standards, and for his time, his work puts him in a class of his own as a pioneer observer of nebulae and star clusters.

Huggins   Sir William Huggins, KCB, OM (1824–1910)
An English amateur astronomer who built an observatory in 1856 at Tulse Hill in South London where he installed, in 1860, a fine 8-inch refractor with an objective made by the celebrated Alvan Clark.

He was a pioneer of the use of spectroscopy and presented a paper to the Royal Society in 1863 on *Lines of some of the Fixed Stars*. In 1864 he examined the planetary nebula NGC 6543 in Draco and found that it gave, not a continuous spectrum like the stars, but only one or two bright lines in the green and blue region, similar to the emission from a glowing gas. On May 18th, 1866, he made the first spectroscopic examination of a nova – nova Coronae – and found an emission of hydrogen. In 1868 he compared the spectrum of a comet with that of ethylene and found them similar: of this he wrote; 'The comet, though subtle as the Sphynx, has at last yielded up its secret.'

His wife, whom he married in 1875, was an active collaborator in his work and together they published in 1899, a fine *Atlas of Representative Stellar Spectra* for which they were awarded jointly the Actonian Prize of the Royal Institution. Huggins made many contributions to the *Philosophical Transactions* of the Royal Society and other journals and these were published as his collected *Scientific Papers* in 1909.

Huyghens   Christiaan Huyghens (1629–95)
A famous Dutch astronomer and horologist, he made great advances in the construction of pendulum clocks, inventing an efficient pallet escapement in 1656. He also constructed air telescopes of enormous size, one having a focal length of 250 feet! Despite the difficulties of observing with such unwieldy instruments, he made many important discoveries with them. He observed that Jupiter had an equatorial bulge and was the first to detect that the peculiar appearance of Saturn was due to its system of rings. He saw that Mars exhibited white polar caps and estimated the period of rotation of the

same planet. He also discovered Titan, the first satellite of Saturn on Mar. 25th, 1655, with a 12-foot telescope made by himself and his brother Constantyn: the $2\frac{1}{4}$-inch objective of this instrument is still in existence.

He came to Paris at the invitation of Louis XIV to establish, with G. D. Cassini, the Observatory of Paris and was also one of the founder members of the *Academie Royale des Sciences* in 1666.

He was the first to establish the idea of centrifugal force and developed the wave theory of light. He was also the inventor of the telescope eyepiece with two lenses of a combination which is still in use today. In 1656 he made detailed studies and sketches of the Orion Nebula, M 42, and in 1684 discovered the 'trapezium' of four stars in the centre of the nebula.

He visited London in 1665 and was made a member of the Royal Society. He left France in about 1686, fearing a persecution of Protestants after the Edict of Nantes was revoked by Louis XIV in 1685: he visited England again in 1689 and then retired to The Hague where he died in 1695.

Ihle M. J. Abraham Ihle (n.d.)

Little seems to be known about this German amateur astronomer. He was probably known to Hevelius who included in his catalogue *Prodromus Astronomiae* the 'nebulose' in Sagittarius which was later to be numbered by Messier as M 22.

Halley, in 1715 credited the discovery of this globular cluster to Ihle who is supposed to have seen it by chance when observing Saturn. Lacaille observed the same object at the Cape of Good Hope and included in it his first class of 'nebulae' as No. 12.

Kirch Gottfried Kirch (1639–1710)

A German astronomer who was attached to the Berlin Observatory, he was one of the early observers of variable stars. In 1686 he discovered that the star, χ *Cygni*, given by Bayer as 5th mag., was, in fact variable. He saw that it was about 5th mag. at maximum but at times declined to invisibility. He thought that this was due to its rotation and gave its period as $404\frac{1}{4}$ days. (The period is now 409 days.)

He discovered the fine galactic cluster in Scutum in 1681: Halley included this in his list of six 'nebulous stars' in 1715 and Derham observed it to be a 'cluster' in 1732. This is M 11.

In 1688 he charted a new constellation, 'The Brandenburg Sceptre', formed from some of the stars in Eridanus. Bode reviewed this in his *Uranographia* but it has now vanished from the charts.

Kirch observed a comet in Puppis on Oct. 26th, 1699: it was just visible to the naked eye and was moving toward the south. It was not seen again and it is not normally included in any list of known comets. However, the German astronomer, Schubart, has suggested that it may have been an apparition of the comet Tempel-Tuttle which occupies the same orbit as the Leonid meteors.

Kirch made many observations of the Great Comet of 1680 and these were used by Halley in his own comet investigations. In 1702 Kirch discovered the globular cluster in Serpens which Messier later included in his catalogue as No. 5.

Kirch's wife, Maria Magaretha Winckelmann, was also an astronomer: their son, Christfried Kirch (1694–1740) observed at Dantzig and Berlin and, with his three sisters, published an ephemeris in 1730.

Koehler or Köhler   Johann Gottfried Koehler (1745–1801)

He was born at Gauernik near Dresden on Dec. 15th, 1745. From 1771 to 1776 he was secretary of the Astronomical Society of Leipzig and in his final year was appointed Inspector or Superintendent of the combined Museums, Chamber of Arts and Salon of Mathematics in Dresden. He was a skilled maker and inventor of astronomical apparatus and demonstrated an improved astronomical pendulum clock at the astronomical congress at Seeburg. A folio of Moon-scape sketches, drawn by him, was preserved at Dresden until at least 1852.

Koehler was a very keen observer; he had an observatory at Dresden, equipped with two Dolland achromats of 6-foot and 10-foot F.L. and an 8-foot Gregorian reflector. He contributed various observations in the *Astronomisches Jahrbuch*, *Monatliche Correspondenz* and *Philosophical Transactions* of the Royal Society of London.

Starting in 1772 he compiled a list of some 20 nebulae and clusters which was published by J. E. Bode in the *Astronomisches Jahrbuch* for 1784. While following the comet of 1779, Koehler discovered M 59 and M 60 in Virgo on Apr. 11th–13th of that year. This was a few days before Messier found them and at about the same time as Oriani saw M 60 alone. Koehler also observed three more nebulae 'in the north shoulder of Virgo' on May 5th, 1779, but did not give precise positions for these.

He also seems to have been the first to have seen the fine galactic cluster of Cancer which Messier first recorded in 1780 as M 67. Koehler's observation, as reported by Bode, was made sometime before 1779.

He died at Dresden on Sept. 19th, 1801.

Lacaille   Abbé Nicholas-Louis de Lacaille (1713–62)

Born in Rumigny, a village in the Ardennes, his father died when he was young and the Duc de Bourbon obtained a bursary for him at the *Collège de Navarre*. He became a considerable classical scholar and his Latin prose was held to have great purity of style. He studied for the priesthood and after ordination became a professor at the *Collège de Quatre Nations*. Here he was introduced to Jacques Cassini who took a great liking to him and made him an assistant in his geodesic surveys.

He became a professor at the *Collège Mazarin* where he set up a small observatory in the attics and where, among his pupils he had Benoit, Lalande, Bailly and Turgot. He was made a member of the Academy of Sciences in 1741 and published textbooks on several science subjects.

In Nov. 1750 he left France on an expedition to the Cape of Good Hope and after being delayed at Rio de Janeiro, arrived at the Cape in March, 1751. Here he made more geodesic observations and also, in collaboration with his pupil Lalande, observing from Berlin on the same meridian, obtained a value of 57' for the lunar parallax, a result which is very close to the modern one.

While at the Cape he started work on his great star catalogue which was published after his death by J. D. Maraldi in 1763. This work, *Coelum Australe Stelliferum* gave details of more than 10 000 stars down to the 7th magnitude.

Lacaille also named many of the southern constellations many of which retain his nomenclature. However, the one he entitled 'Fleur de Lys' in honour of Louis XV has not been adopted.

While at the Cape he also observed a number of nebulae and a catalogue of these, entitled *Sur les étoiles nebuleuses du ciel Austral* was published in the *Memoirs* of the Academy of Sciences in 1755.

This catalogue contained three classes: class I, 'Nebulae'; class II, 'Nebulous Star Clusters'; class III, 'Stars accompanied by Nebulosity'. There were 14 objects in each class, making a total of 42. Lacaille had used only small telescopes and several of his clusters later turned out to be only asterisms, but eight others were later included in Messier's list, being visible from Paris. Among the remaining objects are the globular clusters 47 *Toucanae* (Lac. I. 1); ω *Centauri*, seen earlier by Halley at St Helena (Lac. I. 5); the 'Jewel Box' cluster around χ *Crucis* (Lac. II. 12) and the bright diffuse nebula around η *Carinae* (Lac. III. 5 and 6).

Lacaille was given a grant of 10 000 livres by the Treasury for his Cape expedition and he paid back 855 livres of this on his return as not having been expended. He was considered the foremost French astronomer of his day and his fruitful expedition to the Cape brought him well deserved recognition in many other countries.

Lacaille fell ill in Feb. 1762, crippled by rheumatism, due, it is said, to many long nights in cold observatories and he died on Mar. 21st, 1762. He is buried in the chapel of the *Collège de France*.

Lalande Joseph-Jerome le Francais de Lalande (1732–1807)
He first studied Law at the insistence of his parents but was soon attracted to astronomy and entered the *Collège de France* as a pupil of Delisle and Lacaille. He was a brilliant student and when Lacaille embarked for the Cape of Good Hope he was sent at the same time to Berlin to make measurements of the meridian and to co-operate in obtaining the lunar parallax. After this successful task he returned to France in 1752 – aged 20! – and was made a member of the Academy of Sciences.

In 1758 he was engaged with the mathematicians Clairaut and Mme. Lepaute in the laborious calculations necessary to obtain an accurate date for the return of Halley's comet and he afterwards maintained that this episode ruined his constitution for the rest of his life.

He became Professor of Astronomy at the *Collège de France* in 1762 where among his pupils he taught Méchain and Piazzi. In 1764 he published his *Traité de l'Astronomie* which was widely read and reprinted in 1771 and 1792. In 1793 and 1794 Lalande, despite his royalist sympathies, managed to avoid the worst excesses of the Terror and even helped some of his friends escape arrest by hiding them in the Paris Observatory.

In 1795 he became director of the observatory and in 1801 he published his *Histoire Céleste Français* which contained positions of nearly 50 000 stars to the 10th magnitude. Many of the observations were made by his nephew, Michel le Francais de Lalande and his wife. Lalande was a generous friend to many young astronomers including Messier and Méchain: he travelled to Switzerland, Italy, Holland and England and, at the age of 67 made a balloon ascent. He died in April 1807 and is buried in the cemetery of Père Lachaise.

Lassell   William Lassell (1799–1880)

An English amateur astronomer who, like Hevelius, made his fortune as a brewer and then devoted himself to astronomy. He had owned several small telescopes in his early years and in about 1844 he decided to build a 24-inch reflector. He visited the Rosse workshops at Birr Castle to get advice and made for himself an improved grinding machine. He was also friendly with James Nasmyth, the inventor of the steam hammer, and used his foundry to cast his speculum-metal mirrors.

With the 24-inch reflector he discovered Neptune's satellite, Triton, on Oct. 10th, 1846 and, in 1848, found, at the same time as G. P. Bond at Harvard, the eighth satellite of Saturn, Hyperion. In 1852 he shipped his 24-inch telescope to Malta to take advantage of the clearer skies and there he observed that Uranus had two inner satellites in addition to the two outer ones found by Wm Herschel.

Later, Lassell, like all telescope makers, decided to build bigger, and eventually he made a successful 48-inch equatorially mounted telescope: with this instrument he observed many nebulae and made detailed drawings of some of them, including M 27, M 57 and M 83.

This huge telescope – the movable parts alone weighed about 8 tons – was also taken to Malta. There, Lassell and his German assistant A. Marth made many observations of nebulae. They confirmed the spiral structure of many of those observed by Lord Rosse and also produced a catalogue of 600 previously unknown nebulae.

Lassell returned to England in 1865: the great 48-inch telescope, however, was left in its crates and shortly before his death, was broken up and sold for scrap.

Le Gentil   Guillaume-Joseph-Hyacinthe-Jean-Baptiste Le Gentil de la Galazière (1725–92)

After first intending to enter the Church, his lessons under Delisle at the *Collège de France* influenced him to take up astronomy. He later became an assistant to Jacques Cassini at the Paris Observatory and took part in some of the geodesic surveys under the same mentor.

In 1753 he became a member of the Academy of Sciences and in the *Memoirs* of the Academy for 1759 published an extensive memoir, *Remarks on the Nebulous Stars*. In this he discussed the suspected variability of the Andromeda Nebula and also reported observations made of the Orion Nebula, including drawings made by himself, Huyghens and Picard. In March 1760 he sailed from Brest on an expedition to observe a transit of Venus which was due in June 1761. Le Gentil's destination was the French colony at Pondicherry in India, but Britain then being at war with France, he was prevented from landing by the British blockade of the port. Le Gentil, unable to observe the transit with precision, did not return to France but decided to prolong his voyage for another eight years so that he could observe the next transit, due in 1769!

He visited many parts of the Pacific and Indian Oceans, including the Marianas, the Philippines and Madagascar. He intended to make his observations of the 1769 transit at Manila but was ordered to return, as before to Pondicherry. On the day of the transit, June 3rd, 1769, the sky clouded over just as the phenomenon was due to begin and cleared again after it was already over. His colleagues at Manila had an uninterrupted view.

Le Gentil returned to France in 1771 and eventually published the account of his travels, *A Voyage in the Indian Ocean*, in two volumes in 1779–81. He had discovered the companion to the Andromeda Nebula, M 32, in 1749 and also wrote a memoir to the Academy about it in 1749. He made many other observations of star clusters and nebulae and, besides M 32, was the first to discover the gaseous nebula in M 8. The two galactic clusters, M 36 and M 38 in Auriga, were found independently by him, but these had almost certainly been seen previously by Hodierna before 1654.

Lexell   Anders Lexell (1740–84)

Born at Abo (now Turku) in Finland, he was a mathematician and astronomer who worked for a time at the Observatory of St Petersburg. With the French mathematicians, Laplace and De Saron, he computed the path of the new object discovered by Wm Herschel in March 1781 and showed that it was a planet with an orbit outside that of Saturn.

Immediately following this discovery he predicted that another new planet might be found even further from the sun. (Neptune was not, in fact, discovered until 1846.)

Lexell was also very interested in comets and computed the path of the comet discovered by Messier in June 1770. He showed that the orbit was elliptical and computed its period which turned out to be the very short interval of 5.6 years.

It is thought that the comet had become short-period by being captured by Jupiter in 1767: the next return after Messier's discovery in 1770 was due in 1776 but this apparition occurred in daylight and the comet was not seen. In 1779 it was again perturbed by Jupiter and has not returned since.

It is generally thought to have been the comet of closest approach to the Earth, having come within less than $1\frac{1}{2}$ million miles in 1770. Although called 'Messier's comet' at one time, it is now usually referred to as 'Lexell's comet'.

Mairan   Jean-Jacques Dortous de Mairan (1678–1771)

Born at Beziers on Nov. 26th, 1678, into a family of the *petite noblesse*, he attended college at Toulouse and was, at first, interested in ancient languages and became a considerable Greek scholar.

He went to Paris in 1698 and took up the study of physics and mathematics: he was made a member of the Royal Academy of Sciences on Dec. 24th, 1718, as an Associate in the department of Geometry.

In 1721, he headed, jointly with Varignon, a commission to correct the then faulty and haphazard system of determining the draught and capacity of shipping. In 1723, at Beziers, he founded, under the patronage of Cardinal de Fleury, an academy to encourage the sciences in the Midi.

In 1740 he became permanent secretary of the Academy of Sciences and in 1743 was elected to the *Academie Française*. Soon after this honour, he was appointed editor of the important literary and scientific periodical, the famous *Journal des Sçavans*.

De Mairan's interests were wide; he made important contributions on the theory of heat, recorded observations of various meteorological phenomena and attempted a description of the orbit and rotation of the Moon.

His best known work was his *Traité physique et historique de l'Aurore Boreale*, published at Paris in 1733 and reprinted in the *Journal des Sçavans* in 1754 and 1755. In this work he mentioned the nebulosity around the star closely to the north of the Orion Nebula which Messier later catalogued separately as M 43.

De Mairan died of pneumonia in Paris on Feb. 20th, 1771.

Maraldi I   Jacques Philippe Maraldi (1665–1729)
A nephew of the great G. D. Cassini, he came to Paris in 1686 to assist his uncle in his geodesic work. He spent many years on measurements of the meridian between Dunkirk and Amiens. At the Paris Observatory he made a long series of meteorological observations and was also interested in variable stars. In 1704 and 1705 he attempted to obtain measurements of stellar parallax and also made observations for a catalogue of star positions.

Maraldi II   Jean-Dominique Maraldi (1709–88)
A nephew of Maraldi I, he came to Paris from Italy in 1727 and became a member of the Academy of Sciences in 1731. He was engaged on geodesic work with his uncle and he made successful use of the times of eclipses of Jupiter's satellites for longitude determination. He obtained a difference of longitude between the observatories of Greenwich and Paris of 9 m 23 s the modern value is 9 m 20.93 s.)

He made observations of transits of Mercury and Venus and also calculated the orbit of the comet of 1759. He assisted in the publication of 25 volumes of the *Connaissance des Temps* and in 1763 he published Lacaille's catalogue, *Coelum Australe Stelliferum*, for the author who had died in the previous year.

He discovered the globular cluster in Pegasus, M 15 in 1745 and on Sept. 7th, 1746, found another, M 2, in Aquarius.

He retired to Perinaldo in Italy in 1772 where he died in 1788.

Marius   Simon Marius or Mayer or Meyr (1570–1624)
Born at Guntzenhausen in Franconia he was first a musician and *kapellmeister* to the Margrave of Anspach. He later went to Italy, first to study medicine at Venice and Padua and then astronomy. He returned to Germany in 1604 and

became mathematician and astronomer in the household of the Elector of Brandenburg-Anspach.

In about 1614 he published at Nuremberg a booklet with a long Latin title now generally known as *Mundus Jovialis* . . . in which he claimed to have seen the four satellites of Jupiter with a Belgian telescope, on Dec. 28th, 1609. As Galileo first observed only three of these on Jan. 8th, 1610, and the fourth some time later, Marius maintained he had the prior claim and also gave the satellites their classical names of Io, Europa, Ganymede and Callisto.

There has been much dispute about this ever since and it is generally held that Marius' claim was false: the names he gave them are seldom used and the eight other satellites discovered since have never been given official names as it was thought that by doing so, it would condone the original fraud.

This accusation of fraud is quite unwarranted as Marius observed the four Jovian satellites before Galileo but did not realise what they were. By the time Galileo saw them, Marius had started keeping a record of their positions relative to the planet and this soon showed that they were bodies attendant upon Jupiter. While detracting nothing from Galileo's achievement, Marius must be given some credit for these observations.

There is no dispute, however, that he was the first to observe the Andromeda Nebula, M 31, with the telescope and his description of this object, included in the preface to his controversial *Mundus Jovialis* . . . where he compares it to the flame of a candle seen through a sheet of horn, has never been forgotten.

Maskelyne

Nevil Maskelyne (1732–1811)

He was educated at Westminster School and Trinity College, Cambridge where he graduated as seventh Wrangler in 1754. He was ordained in the following year but he had probably already decided to devote himself to astronomy rather than to the Church when he observed the famous annular eclipse of the Sun on July 25th, 1748, an event which influenced the young Charles Messier also.

In 1761 he undertook a journey to St Helena on behalf of the Royal Society, to observe the transit of Venus. During this voyage he tried out the method of determining longitude by means of lunar distances and later, introduced this method by publishing in 1763, *The British Mariner's Guide*. In 1765 he succeeded Bliss as Astronomer Royal and in 1766 published the first volume of the Nautical Almanac. He remained superintendent of the *Nautical Almanac* Office until his death.

He published many papers in scientific journals: in 1760, he recommended *A Proposal for discovering the Annual Parallax of Sirius*; *The Transit of Venus* in 1761 and 1769; the *Tides of St Helena* in 1762 and *Astronomical Phenomena in St Helena and Barbados* in 1764.

In 1772 he suggested to the Royal Society an idea for an experiment to obtain the Earth's density and, from Maskelyne's observations, Charles Hutton later deduced a specific gravity of 4.5 for the Earth.

Charles Messier in old age

Pierre Méchain, Messier's colleague and discoverer of many nebulae in Virgo and Coma Berenices (R.A.S.)

Admiral W. H. Smyth, author of A Cycle of Celestial Objects *which contained descriptions of many Messier Objects (B.A.A.)*

Barnabus Oriani, director of the Observatory of Brera, Milan and discoverer of M 61 (B.A.A.)

Méchain   Pierre François André Méchain (1744–1804)

His father was an architect and Méchain, with the intention of following an architectural career also, received a good grounding in physics and mathematics. Financial difficulties at home, however, forced him to leave his college and after working as a tutor for a while, he was helped by Jerome de Lalande who gave him some of the proofs of his book *L'astronomie* to correct and also gave him lessons in the same subject. Lalande also obtained for him the post of assistant hydrographer at the Depot of Maps and Charts of the Navy at Versailles in 1772. This was, at first, only a temporary position and for a time Méchain had to supplement his income by teaching mathematics.

In 1774 he obtained a more permanent post as a calculator with the Depot of the Navy and at about this time he made his first acquaintance with Messier who was with the same department but working at the small observatory in the *Hôtel de Cluny*.

Méchain was first engaged in making surveys of the French coastline but he made observations at Versailles also. In 1774 he observed an occultation of Aldebaran and later presented a memoir on his observation to the Academy of Sciences. Méchain, like Messier, spent a great deal of his spare time in looking for comets and in 1781 he discovered two: being better equipped mathematically than his colleague at Cluny, he was also able to calculate their orbits. In 1782 he was made a member of the Academy of Sciences for his work on comets.

During his searches for comets he came across many new nebulous objects especially in the Coma–Virgo region, and these observations he promptly passed on to Messier who included them in his catalogue. Although rivals in the field of comet discovery, the two Marine astronomers remained on the best of terms throughout their careers.

In 1786 Méchain became an associate editor of the *Connaissance des Temps* and in that year made the first observation of a comet which Caroline Herschel observed in 1792 and which was seen again by Pons in 1805. The German astronomer J. F. Encke was able to show that this comet was periodic, returning every $3\frac{1}{2}$ years and the comet now bears Encke's name.

In 1787 Méchain worked with J. D. Cassini and Legendre in verifying the accurate difference of longitude between Paris and Greenwich and in the same year, with the same two colleagues, paid a visit to Wm Herschel at his observatory in Slough near Windsor.

The comet which Méchain discovered in 1790 was observed again by Tuttle in 1858 who demonstrated that this, too, was a periodic comet with an orbital period of $13\frac{3}{4}$ years: it is now known as 'Tuttle's comet'.

In 1791 it was decided to undertake a new survey of the meridian from Dunkirk to Barcelona. The northern portion was alloted to Delambre while Méchain, with an assistant named Tranchot, undertook the southern half, beginning operations on June 25th, 1792. By this time the French Revolution was well under way and both parties, especially Delambre in the north, suffered considerable difficulty. Méchain and Tranchot were arrested by revolutionaries at Essone, their surveying instruments being considered as weapons of the counter-revolution, but eventually they were allowed to proceed.

335

While in Spain, Méchain had an accident which laid him up for several months and by the time he had recovered, war had broken out between France and Spain and he was interned. This did not prevent him from discovering another comet on Jan. 10th, 1793 at Barcelona. Méchain was away from Paris during the worst excesses of the Terror but his family suffered greatly and Méchain himself lost all his property in the upheaval.

Eventually he obtained permission to leave Spain for Italy and after remaining in Genoa for some time, he returned to Paris in 1795.

He became a member of the new Academy of Sciences and the *Bureau des Longitudes* and was, for a while, director of the Paris Observatory, where he discovered another comet on Dec. 26th, 1799 in Ophiuchus. This was his eighth comet discovery and Messier also took part in the observations necessary to obtain its orbit.

Méchain seems to have been acutely worried by some discrepancies in one of the latitude determinations during his survey and for a long time refused to publish the results. He became something of a recluse and then, obtaining permission from Napoleon to extend the Dunkirk–Barcelona survey as far as the Balearic Islands, he left Paris in 1803.

After completing part of his work he caught yellow fever and died at Castillion de la Plana in Spain on Sept. 20th, 1804.

Oriani   Barnabus Oriani (1752–1832)

An Italian astronomer who, although of humble origin, was given a good education by the Order of St Barnabus in Milan and after joining the staff of the observatory of Brera, eventually became its director.

While following the comet of 1779 – discovered by Bode on Jan. 6th and independently by Messier on Jan. 18th – he observed 'three nebulous stars' in Virgo. These were M 49, M 60 and M 61. Messier had first discovered M 49 in 1771 but Oriani appears to have seen M 60 on April 12th (three days earlier than Messier) and M 61 on May 5th (six days earlier). M 60, however, together with the neighbouring M 59, which Oriani seems to have missed, was observed by Koehler at Dresden on April 11th and 13th, 1779; also while following the comet.

Oriani played a major part in the determination of the orbit of Uranus soon after it was discovered by Wm Herschel in March 1781. After de Saron, Lexell and others had shown that the object could not be moving in a parabolic orbit and had obtained circular elements, Oriani improved on this by calculating an elliptical orbit in 1783. Later, in 1789, he was the first to obtain an even more accurate result allowing for the perturbations of Jupiter and Saturn.

He also became engaged in the great geodesic surveys in France and Italy and undertook, with Baron de Zach, the measurement of the meridian between Genoa and Milan in about 1810. He was a close friend of Piazzi and worked with him on the orbit of the minor planet Ceres.

Palitzch   Johann Georg Palitzch (1723–88)

A substantial farmer and amateur astronomer who lived at Prohlis, near Dresden in Saxony. He is known to have investigated the variability of Algol in 1744, although the first to detect that this star was regularly variable seems to have been Montanari in 1669.

Palitzch is remembered for his first sighting of the returning Halley's comet on the night of Dec. 25th–26th, 1758. He was very keen-sighted and it is not known with certainty whether he first saw the comet, which was very faint, with the naked eye or with a telescope. Smyth says that he used an 8-foot F.L. telescope. As an astronomer, he was known to Baron de Zach; but some writers, such as Delambre and Doublet, describe him merely as 'a Saxon peasant'.

Peiresc   Nicholas-Claude Fabri de Peiresc or Peirescus (1580–1637)
He came from an influential family of lawyers and magistrates and he himself studied Law at Aix and Avignon. He completed his studies of Padua and there met Galileo who was then under the restriction of the Holy Office. Peiresc attempted to intervene in Galileo's dispute and wrote to Cardinal Barberini, the nephew of Pope Urban VIII, to intercede for the astronomer.

Peiresc later became Councillor of the *Parlement de Provence* and was noted for his patronage of the sciences. His interest in astronomy had been aroused by his meetings with Galileo and he obtained a telescope from him with which he observed sunspots and the satellites of Jupiter. He is known to have observed the Orion Nebula, M42, with the telescope in 1610.

He made many geodesic measurements and realised that observations of the times of lunar and solar eclipses from different places could be used to determine longitude. From the times of the eclipse of the Moon on Aug. 27th, 1635, observed at Cairo and Aleppo as well as in Europe, he showed that the Mediterranean was nearly 600 miles shorter than had been supposed.

Peiresc turned part of his own house into an astronomy school where, among his pupils, he taught Pierre Gassendi, an astronomer who had great influence in the earlier part of the 17th century.

Piazzi   Giuseppe Piazzi (1746–1826)
The director of Palermo Observatory in Sicily and famous for his discovery of the first asteroid or minor planet, Ceres.

After first taking holy orders, Piazzi came to England to study astronomy where the rigours of his education, according to Caroline Herschel, involved 'getting broken shins by falling over the rack-bar' of her brother's giant 40-foot telescope at Slough.

Piazzi's patron was the Viceroy of Naples who allowed him to use the old Palace of Palermo as an observatory. Here, Piazzi installed a fine 5-foot meridian circle by Ramsden which was brought from England in 1789.

Piazzi was not a member of the conference at which Baron de Zach proposed a search for the suspected planet between Mars and Jupiter but he discovered Ceres on Jan. 1st, 1801, when examining a region in Taurus in connection with his star catalogue. He first detected it as an 8 mag. star but soon discovered its movement and subsequently followed it until Feb. 11th.

He maintained a long friendship with Admiral Smyth who called at Palermo in 1813 during his hydrographic survey of the Mediterranean.

Smyth so admired the Sicilian astronomer that he gave his son the name of Piazzi.

Apart from his discovery of Ceres, Piazzi's major contribution to astronomy was his star catalogue: this was an extension and revision of Flamsteed's earlier catalogue and the second part of this work, published in 1814, contained accurate positions for 7646 stars down to magnitude 10 between 90° N to 46° 21′ S declination.

After a long career as director at Palermo, Piazzi died at Naples on July 22nd, 1826.

Picard Abbé Jean Picard (1620–82)

He was born on July 21st, 1620, in the town of La Flèche, Anjou, and after a religious education and induction into the priesthood, later became the Prior of Rillé in Anjou. His early interest in astronomy, however, led him to Paris where he became, with G. D. Cassini and Christiaan Huyghens, a founder member of Louis XIV's newly created *Academie Royale des Sciences* in 1666.

Picard was one of the first to realise that the telescope could be applied to position-measuring instruments in such a way as to improve greatly the accuracy of observations. With his colleague, Adrien Auzout (1622–91), he constructed in 1667 or 1668 a simple telescopic sight and attached it to a 38-inch quadrant. With this instrument Picard made the first really accurate measurement of a degree of the meridian (between Malvoisine and Amiens), and in 1671, published a memoir entitled *Le Mesure de la Terre*.

In 1679 Picard founded the French ephemeris, *La Connaissance des Temps*, and continued to edit it until his death on July 12th, 1682.

He made some observations of the Orion Nebula, M 42, and, according to de Mairan, a manuscript drawing by Picard of March 20th, 1673, showed four stars near the centre of the nebula where Huyghens had seen only three. Huyghens' first observation of the fourth star of the 'trapezium' was in 1684.

Pingré Alexandre-Guy Pingré (1711–96)

He became interested in astronomy at the age of 14 and after an early academic career in Rouen where he started a school, he came to Paris in 1753. He suffered from poor sight and was not very strong and this limited his ability as an observer. He was, however, an outstanding mathematician and often collaborated with Messier in the reduction of his comet observations and in other work.

Despite his frail physique he travelled widely: in 1761 he went to the island of Rodrigues near Madagascar to observe the transit of Venus and in 1769 he travelled with the Chevalier de Fleurieu in the *Isis* on an Atlantic voyage to study the behaviour of marine chronometers.

In the previous year he had obtained Messier's assistance in rating the chronometers for a voyage to be undertaken in the Baltic for the same purpose; Messier made all the observations while Pingré did the necessary calculations. Pingré sailed again in 1771 with Borda and de la Crenne in the *Flore* for another navigation survey.

He also published from 1754 to 1757, a nautical almanac, *État du Ciel*, which contained tables of the Moon based on le Monnier's work and intended for nautical navigation.

Pingré was always interested in comets and he wrote a comprehensive work on this subject, entitled *Cometographie*, published in two volumes in 1783 and 1784. He lost all his resources during the Revolution but continued to work right up to his death in 1796.

Roberts Isaac Roberts (1829–1903)

A successful businessman of Liverpool and amateur astronomer, he became a pioneer of celestial photography. He built a 20-inch aperture silver-on-glass reflector of 8 foot focal length with the intention of preparing a series of photographic star charts. The first plates were taken in 1885 at Maghull in Lancashire and later, in 1886, he took his first, very successful photographs of M42 and M45.

He then moved to the south of England and built an observatory at Crowborough in Sussex and specialised in the photography of star clusters and nebulae. In 1888 he made a 3-hour exposure of the Andromeda Nebula, M31, which plainly showed its spiral structure. He was awarded the Gold Medal of the Royal Astronomical Society for his photographic work in 1894.

The first volume of his well known *Photographs of Stars, Star Clusters and Nebulae* was published in 1893, followed by a second in 1899. A third volume was prepared and issued by his widow in 1928.

Roberts' 20-inch reflector is now in the Science Museum in South Kensington, London.

Rosse William Parsons, 3rd Earl of Rosse (1800–67)

A wealthy amateur astronomer and telescope maker, he was educated at Trinity College, Dublin and Magdalen College, Oxford. As Lord Oxmantown, he became Member of Parliament in 1821 for King's County, a seat which he held until 1834. He was President of the Royal Society from 1849 to 1854 and was awarded the Royal Medal in 1851.

His chief claim to recognition is his construction and use of large reflecting telescopes. His largest was a 72-inch aperture speculum of 4 tons weight which, after 10 years of experiment and effort was cast and figured and set up at Birr Castle in Ireland. This mirror, which was the largest to be used for the next 70 years, was mounted in a wooden tube bound with iron measuring 7 feet diameter by 52 feet length. The tube itself was erected upon a huge scaffolding system between two high walls to shield it from the effects of the wind. The tube could be raised and lowered in altitude but could move only some 15° either side of the meridian.

Although the considerable light-gathering qualities of this huge instrument enabled many faint objects to be seen, and showed the spiral nature of many nebulae for the first time, the definition of the speculum was not particularly good. One German astronomer who visited the observatory said afterwards, 'They showed me something which they said was Saturn and I believed them.' Nevertheless, Lord Rosse can be credited with a lot of important pioneer work especially in the observation of nebulae; he made a special study of the Orion Nebula, M42, and his assistants produced many fine drawings of it.

Saron   Jean-Baptiste-Gaspard de Saron, generally known as Bochart de Saron (1730–94)

The last *Premier Président* of the *Parlement de Paris* who died on the guillotine on April 20th, 1794.

He was a man of many and varied talents; besides being a magistrate and an authority on jurisprudence, he was a skilled instrument maker and optical worker and also made a collection of fine astronomical instruments made by English and French craftsmen. He was also a mathematician of ability, being one of the first to realise, by calculating its orbit, that the object which Wm Herschel had discovered in 1781 was the planet Uranus.

His interest in comets and cometary orbits brought him into contact with Messier and a warm friendship grew up between them. His last calculations were done in prison while awaiting execution: these were of the comet which Messier had discovered on Sept. 27th, 1793, in Ophiuchus. It is said that Messier was able to recover the comet after perihelion from the results of Saron's predictions and was able to inform him secretly of this success just before de Saron was taken to the scaffold.

De Saron was noted for his generosity; he lent many of his instruments to other astronomers and is known to have assisted Laplace with the expense of publishing his *Théory du movement elliptique et de la figure des planetes*.

His interests also included chemistry, etching and printing. He and his wife set the type and printed secretly – for private printing presses were illegal in Paris at this time – an octavo volume of 266 pages of an unpublished manuscript by Counsellor d'Aguesseau, entitled *Discours sur le vie et la mort*. . . . Only 60 copies were run off and the work is now very rare and sought after by bibliophiles.

Smyth   Admiral William Henry Smyth, RN (1788–1865)

Smyth's father was an American who, as a loyalist emigré, came to England from New Jersey after the American Revolution. He entered the Royal Navy as a boy and was so exceptional a seaman that he obtained a commission from the lower deck. During the Napoleonic wars he served in the Mediterranean and commanded a brigantine. While undertaking a hydrographic survey in 1817 he put in at Palermo in Sicily and while there he met the astronomer Piazzi, already celebrated as the discoverer of Ceres. Smyth spent some time at Piazzi's observatory and this encounter fired an already latent interest in astronomy.

He retired from the Royal Navy in 1825 and settled near Bedford where he built an observatory. He was very interested in double stars, star clusters and nebulae and in 1844 first published his *Cycle of Celestial Objects*. This was issued in two volumes, the first being concerned with general astronomy and the second – known as the *Bedford Catalogue* – became widely and deservedly popular for its exhaustive information concerning nebulae, star clusters and double stars.

This work gained him the Gold Medal of the Royal Astronomical Society in 1845 and he also served as President of that society for two years. The *Cycle* was re-issued several times, the latest being published by George F. Chambers in 1881. Chambers was a barrister of the Inner Temple and a keen amateur astronomer and author of several books on astronomy including a comprehensive *Handbook of Descriptive Astronomy*. He acquired the copyright of Smyth's *Cycle* together with the plates and engravings and prepared a revised edition which was also extended to cover many objects in the southern hemisphere.

Admiral Smyth was an extremely popular and influential figure and was interested not only in astronomy but also in archaeology and numismatics. Smyth has often been criticised for his rather fanciful descriptions of the colours of his double stars, such as 'pale white' and 'dusky blue' but it must be remembered that the perception of colour is a very subjective matter and also that the acuteness of colour-perception differs greatly between individuals. Smyth may well have had an abnormally keen sense of colour and although his terms may be a little *outré*, this may be due to the difficulty of expressing in words such subtle nuances of sight.

Admiral Smyth's son, who was named Charles Piazzi Smyth in honour of the Sicilian astronomer, also became an astronomer. He was at one time attached to the observatory of the Cape of Good Hope and later, in 1845, he became Director of Edinburgh Observatory and Astronomer Royal of Scotland, a post he held until 1888.

He made an expedition to Tenerife in 1857 and later became influenced by Egyptology and especially by the significance of the measurements of the Great Pyramid.

Struve   Friedrich Georg Wilhelm Struve (1793–1864)
In 1815, at the age of 22, he became director of the Dorpat observatory in Estonia. His famous work on double stars began in 1824 when he obtained a fine $9\frac{1}{2}$-inch Fraunhofer refractor which was the first telescope to have a clock-driven equatorial mounting. Struve also made and used an accurate travelling-wire micrometer and in the years 1825 to 1827 made a complete survey of the sky down to declination 15° S. In a total catalogue of 120 000 stars he included about 2200 double stars. This was followed in 1837 by his *Mensurae Micrometricae* which gave details of 3112 double and multiple stars.

He also attempted measurements of stellar parallax between 1834 and 1837 and obtained a parallax for *Vega* of 0".26. This result actually preceded the successful measurement of the parallax of *61 Cygni* by Bessel in 1837 but the latter's results gained a wider acceptance.

In 1839 Struve was invited by Nicholas I of Russia to take up the directorship of the Pulkovo observatory near St Petersburg where he continued double star observations and other astronomical work with his son, Otto Struve (1819–1905), who succeeded him as director when he retired from the post in 1861.

Wilhelm Struve's double stars are labelled Σ in the catalogues and those found by his son Otto, $O\Sigma$.

Webb   Thomas William Webb (1807–85)
An English clergyman who was for many years Vicar of Hardwick in Herefordshire. Like many clergymen of his day, he was a keen amateur astronomer and was a close friend of G. H. With, the astronomical mirror maker.

341

Webb's first telescopes were a 4-inch fluid achromat of poor performance and a rather better 3.7-inch refractor by Tulley: later, his father, also a clergyman, gave him a 9.3-inch 'With' reflector and Webb helped to popularise the new silver-on-glass mirrors which gave a greatly superior performance to the older metal 'specula'.

Webb had read and used Smyth's *Cycle of Celestial Objects* and he decided to write a similar but more condensed guide for the amateur observer. This book, the renowned *Celestial Objects for Common Telescopes* was first published in 1859 and this was later re-issued in three revised editions. It proved to be an inestimable boon to amateur astronomers for very many years and was again issued in a revised 6th edition by T. E. Espin in 1917.

Although by this time Webb's *Celestial Objects* is somewhat out of date in much of its contents, it is still in demand and in 1962 an unrevised reprinting was issued by Dover Publications Inc. of New York in two volumes.

Besides his writing, and no doubt his care of souls also, Webb did a great deal of observing of his own: the long-period variable *S. Orionis* was discovered by him in 1869.

Zach Francis Xavier, Baron de Zach (1754–1832)
Born at Presburg in Austria–Hungary, he came from a military family; one of his brothers was a general in the Austrian army and was taken prisoner at the battle of Marengo.

De Zach also served in the army but later decided to take up astronomy and visited Paris and London where he made friendly contact with many astronomers. In 1786, Duke Ernest of Saxe-Coburg Gotha made him the director of the observatory at Seeburg where de Zach used a meridian telescope by Ramsden and two mural quadrants of 8-foot radius. In 1786, de Zach published a new star catalogue and in 1806 produced a revised and augmented version giving accurate positions for 1830 stars.

In 1800, de Zach, at a conference of astronomers at Gotha, revived a plan for a systematic search for the hypothetical planet between Mars and Jupiter which had been suggested earlier by several other astronomers including Lexell and Lalande. Before the scheme had really started, however, the first of the 'asteroids', as they were later to be called, was found by Piazzi on Jan. 1st, 1801, at Palermo by a chance observation.

After following the new planet, which was later called Ceres, Piazzi became ill, but from his observations, the mathematician Gauss was able to calculate its orbit. From these calculations the planet was recovered after perihelion by Olbers at Bremen and de Zach at Seeburg on Dec. 31st, 1801.

De Zach published many astronomical and geographical papers; he was a skilled observer and an accomplished linguist, speaking most European languages fluently.

After the death of Duke Ernest in 1806, de Zach left Seeburg and went to live in southern France and then went to Italy where he took part with Oriani in a survey of the meridian between Genoa and Milan.

He eventually retired to Paris where he died in 1832, one of the many victims of an epidemic of cholera at that time.

'THE INDEFATIGABLE OBSERVER'

A short biography of Charles Messier, 1730–1817

On Christmas night in the year 1758, a well-to-do farmer, living near Dresden in Saxony, left the convivial gathering in his farmhouse and took a turn around his paddock. He was a keen amateur astronomer and being blessed with acute eyesight, it was his habit to make a quick but thorough survey of the sky on every favourable occasion.

That night the stars were bright in the frosty air and as soon as his sight had become adapted to the darkness, he began to examine each sector of the sky carefully and systematically. Soon his attention was held by a faint, diffuse patch of light in the constellation Pisces not far above the SW horizon which, with his intimate knowledge of the heavens, he knew was something he had not seen there before. He turned away and then looked again: it was just about at the limit of visibility. From time to time it vanished but then re-appeared, sometimes shining quite strongly. There was no doubt that something new had appeared among the stars.

He hurried indoors and brought out his telescope: it was a cumbersome instrument with a tube about 8 feet long and it was awkward to set up. His fingers were cold too and his impatience to begin viewing made him even more clumsy. Finding the faint object was also difficult and frustrating, for the mounting did not turn very smoothly and the eyepiece gave only a small field of view. Sometimes he lost sight of the object altogether.

A freehand monogram inscribed by Charles Messier in a personal copy of his 1781 Catalogue

343

At last, however, he was able to focus his telescope upon it: now it appeared larger and brighter but still it was only a faint, diffuse glow. Carefully he noted the position of the patch relative to the few, faint stars in the field of view.

By 11.00 pm it had declined below 20° altitude and had become lost in the haze of the horizon: by that time, however, he was able to determine that the hazy object had perceptibly moved among the stars.

The farmer's name was Palitzsch and the object he had seen was Halley's comet, returning as Edmond Halley had predicted, 76 years after its last appearance in 1682.

This momentous event had been eagerly awaited by astronomers throughout the whole of Europe. If the comet were to return as Halley had forecast, it would show that the mysterious objects, previously thought to be merely chance visitors from space, were in fact moving in closed orbits and subject to the same laws of motion that governed the familiar and predictable planets.

Halley himself had died in 1742 but he had concluded as early as 1701 that the comet he had seen in 1682 was the same as those which had appeared in 1607, 1531 and 1456. He predicted that the same comet would return again in late 1758 or early 1759.

As the date of the comet's expected return grew close, the calculators of the day, such as Clairaut, Lalande and Madame Lepaute, laboriously checked their figures, taking into account the effect that the giant planets Saturn and Jupiter would have upon its orbit. They revised the date at which it would be at its closest to the Sun to April 13th, 1759.

More than a year before this date, however, the observers began their searches for the expected comet, each hoping to be the first to see it as it approached.

One of the keenest of these observers was a young man of 28 named Charles Messier. While farmer Palitzsch was making his historic sighting of Halley's comet, Messier was scanning the sky with the same purpose from the observatory on top of the tower of the old *Hôtel de Cluny* in Paris.

Charles Messier was not yet entitled to the style of 'astronomer': he was hardly more than a clerk-assistant to the titular Astronomer of the Navy, Nicholas Delisle, and his diligent searching of the heavens on every available night was only a semi-official duty. But it was a labour of love, for Messier was not only devoted to astronomical observing; he was also an ambitious young man.

He was of humble origin, having been born, the tenth of 12 children, in the small town of Badonviller, about 20 miles east of Luneville in Lorraine on June 26th, 1730. Before he was 11 years old his father had died and the young Charles Messier, finishing his schooling at an early age and finding little prospect of betterment at home, decided, at the age of 21, to seek his fortune in the capital.

When he arrived in Paris in Oct. 1751 he had, so Delambre tells us, 'hardly any other recommendation than a neat and legible handwriting and some little ability at draughtsmanship'.

Messier owed the first step in his astronomical career to Nicholas Delisle who, having spent 20 years in Russia where he had helped to establish the Observatory of St Petersburg, had returned to Paris in 1747. He set up the small observatory in the *Hôtel de Cluny*, a 15th-century building which had been built near the *Rue St Jacques* as a residence for the Abbots of Cluny, and enjoyed a salary of 3000 livres, a permanent residence in the *Hôtel* and the title of Astronomer of the Navy.

Delisle engaged the young Messier, no doubt because of his neat hand and careful drawing ability, to keep the register of observations at the observatory. One of his first tasks, however, had nothing to do with astronomy, for Delisle, whose passion was the collecting of old documents, gave him the task of copying a large map of the Great Wall of China and a plan of the Old City of Pekin.

It was the observatory, though, which attracted Messier, for he had been interested in astronomy from an early age. In his own memoirs he said that he always remembered the thrill with which he first observed, as a boy of 14, the comet of 1744. Four years later, the annular eclipse of the Sun which was visible in northern Europe on July 25th, 1748, made a profound impression on the young Messier – and, by many accounts also stimulated Jerome de Lalande, a contemporary and friend of Messier, and Nevil Maskelyne, later to become one of England's Astronomers Royal – to embark on their astronomical careers.

Messier was guided in his early apprenticeship at Cluny by Delisle's secretary, Libour, who gave him instruction in the keeping of the daily records of observations and also trained him to use the old-fashioned telescopes which were employed in the observatory. Delisle himself gave his assistant some lessons in elementary astronomy and, in particular, impressed upon him the necessity of obtaining exact positions during all his observations.

By 1754 Messier's position at Cluny had been regularised to that of clerk of the Depot of the Navy with the modest but useful salary of 500 francs per annum, together with his board and lodging. Some of his duties were still concerned with map-making for he worked for a while with the surveyor La Grive in preparing a plan of the city of Paris and part of a large-scale map of France.

It was the night's observing that was most to his liking; especially that of the observation of comets. Several of the Cluny telescopes, although somewhat crude and inefficient, even by the standards of their day, had a fair aperture and a reasonably wide field of view and were suited for the detection of faint, diffuse objects such as comets.

For the young Messier, too, comet hunting had more than a touch of glamour, for discoverers of new comets could expect to gain a certain amount of publicity and even fame. In his memoirs, Messier admitted that, at this time, he considered that he had found his *métier*. 'I was engaged', he wrote, 'in the work to which I was best suited and which gave me the most pleasure.'

Delisle had made his own calculations to facilitate the search for Halley's comet. These were based upon the earlier apparitions of 1531, 1607 and 1682 from which he concluded that the earliest date for its appearance would be 35

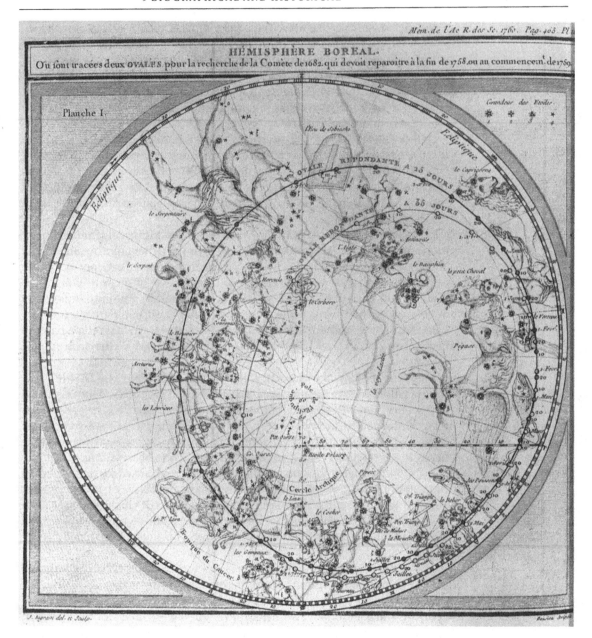

Messier's star chart, showing the two ovals calculated by Delisle to limit the area of search for Halley's comet in 1758. From Mémoirs de l'Academie Royale des Sciences *for 1760. (British Museum)*

days before perihelion and the latest, before it became lost in twilight, 25 days before perihelion. Accordingly he drew up a planisphere or star-chart with two ovals marked on it which corresponded with the probable positions of the comet for these two intervals.

Messier was given instructions by Delisle to search the areas marked on the chart and he actually began these observations at some time in 1757, nearly two years before the expected date of the comet's return. He used a 4½-foot F.L. Newtonian reflector for this work and during 1758 he observed another comet, which was not Halley's, following it from Aug. 15th until Nov. 2nd.* During the months of Nov. and Dec. 1758, the sky at Cluny was frequently clouded over and Messier was able to make only a few, hurried observations.

Through some hiatus of communication, the news of the sighting by Palitzch and the confirmation of the comet's identity by other German observers, did not reach Paris for several months. Messier continued his diligent but fruitless search into the new year and we can give an account in his own words of the events that followed.

'On Jan. 21st, 1759, the whole day was very fine and without cloud; in the evening I went over the sky with the telescope, keeping to the limits of the two ovals drawn on the celestial chart which was my guide. At about six o'clock I discovered a faint glow resembling that of the comet I had observed in the previous year: it was the Comet itself, appearing 52 days before perihelion!

'There is cause to presume that if M. Delisle had not made the limits of the two ovals so restricted, I would have discovered the comet much earlier, while it had a greater elongation from the Sun. It had been much nearer to the earth two months before Jan. 21st.

'This discovery I reported to M. Delisle for him to announce. This was one of the most important astronomical discoveries for it showed that comets could return. He went to the observatory to see the comet, recommended me to observe it assiduously – but not to reveal the discovery.

'I was a loyal servant of M. Delisle, I lived with him at his house and I conformed with his command.'

This narrative was written by Messier a long time after the event, appearing nearly 50 years later, in a memoir on the early observations at Cluny which he contributed to the *Connaissance des Temps* for 1810.

There is no doubt that Delisle's suppression of his early sighting of Halley's comet was one of the greatest disappointments in Messier's life. He had hoped to have been the first to observe this historical phenomenon and so to have made his reputation as an observer and thus advance his career. The bitter blow of Delisle's prohibition was never to be forgotten.

Whether Delisle's action was due to excessive caution or to some other,

---

* In the *Memoirs* of the Academy for 1759, Delisle wrote a long account of the observations of this comet and paid full tribute to 'the assiduous work of my assistant, Messier': he also mentioned that Messier had discovered a new nebula (M 1) while following the comet. An elaborate chart of the comet's track was prepared by Messier who then accompanied Delisle to present it to Louis XV on Oct. 31st, 1758. 'His Majesty had the grace to approve the work and listened with favour to the explanation which I had the honour to give to him.' wrote Delisle.

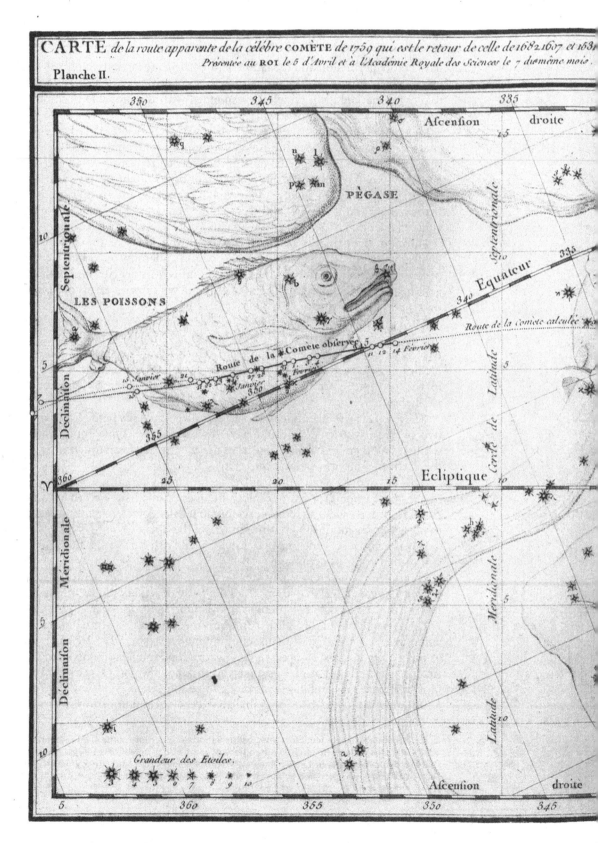

Messier's star chart showing portions of the actual track of Halley's comet between the head of Aquari
nebulae are also plotted on the chart; M 2 in *Royale des Sciences for 1760. (British Museum)*

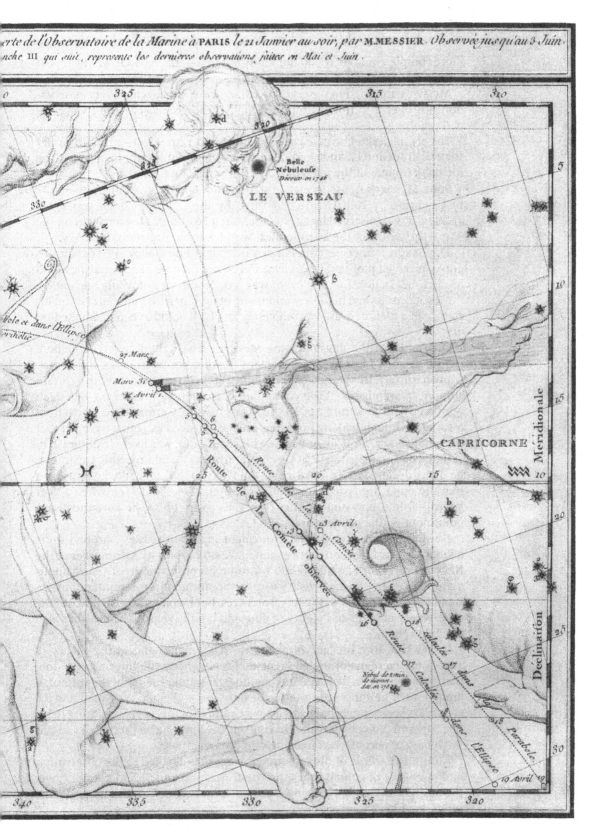

rte de l'Observatoire de la Marine à **PARIS** *le 21 Janvier au soir, par* **M. MESSIER** *Observée jusqu'au 3 Juin*

nche III qui suit, represente les dernieres observations faites en Mai et Juin.

Belle
Nébuleuse
Découv. en 1746

LE VERSEAU

CAPRICORNE

Méridionale

Déclination

Route de la Comète observée

Nebul. de 2 min.
de diametre
dec. en 1751

d M 30 in Capricornus. From Mémoirs de l'Académie Jan. 21st, and Apr. 16th, 1759. Two

more complex motive, we do not know but his later behaviour was to appear just as irrational.

Messier continued to observe the comet, carefully logging every observation, for over a fortnight until eventually it became lost in the twilight of perihelion.

The actual perihelion date occurred on March 12th, 1759 – a month earlier than Clairaut and Lalande had calculated – most of this error being due to an inexact estimate of the mass of the perturbing planet, Saturn.

Soon Messier's hopes received their second setback when the news of the first sighting by Palitzch reached Paris. By the time the comet was again visible – it was observed by La Nux on March 26 and by Messier again on the 31st – the whole of the astronomical world was buzzing with the news.

The Saxon farmer became famous overnight and Halley, 17 years in his grave, received more praise, at least in France, than he had ever achieved in his lifetime. In England, the event was recognised but not wildly applauded.

Messier continued his observations and maintained his imposed silence: his chance of a lifetime was lost completely and he would have to make the best of it.

On April 1st, the day following Messier's first sighting after perihelion, Delisle decided to publish his assistant's earlier observations of the comet. Quite naturally, this was received by other French astronomers with a certain amount of scepticism. If the comet had been seen by Messier on Jan. 21st, why keep it all dark until April? And in any case the Germans had seen it first! Messier, although making no claims on his own account, must have felt that the whole affair, far from enhancing his reputation, was likely to diminish it.

If he felt any resentment at this rather irrational and unfeeling treatment, he never declared it. He owed a debt of loyalty to the old man who had given him the opportunity to start upon his chosen career and even when a year later, Delisle suppressed another comet discovery of his, he bore the disappointment with a prudent and stoical silence.

However, soon after this latter incident in 1760, Delisle, who had some leanings toward monastic retirement, gave up his post as Astronomer of the Navy and Messier was allowed to continue his observations at Cluny on his own. He was not given Delisle's vacant appointment nor remuneration; he was still only the clerk of the naval archives, but he had all the facilities of the observatory at hand and plenty of time and limitless enthusiasm with which to exploit them.

It was true that his lack of ability in mathematics and in the theory of astronomy were major impediments to his advancement, but he knew, too, that, as an observer, he was acute, methodical and skilful. In this important, if limited, field he showed himself to be a tireless and gifted craftsman and his eventual success can be taken as a good example of the 'cobbler sticking to his last'. His ambition never wandered outside the limits he had accepted and this singleness of purpose was to bring him reward enough.

Almost the whole of his subsequent work was devoted to the observation and discovery of comets and soon the frustration and disappointment over the business of Halley's comet were, if not forgotten, at least submerged in the interests of his newer activities.

350

He observed the famous transit of Venus of 1761 and wrote a short memoir on his observations; he also observed Saturn and recorded his impressions of the appearance of its rings. But it was the comets which drew him irresistibly: he observed the comet of 1763 and made the discovery of a new one on Jan. 3rd, 1764. In 1766 he was even lucky enough to discover another new comet by a chance observation with the naked eye.

And while he sought for new comets, he also engaged in the pursuit of academic recognition. He corresponded with astronomers and other academicians in Britain, Germany and Russia where his name was already known for his comet discoveries and these were useful contacts. His correspondent in St Petersburg was the Swiss exile La Harpe, a member of the Russian Academy of Sciences and later to be tutor to the young Alexander I. Frederick La Harpe, at Messier's request, submitted his name to the Academy and, to his great satisfaction, he was accepted as a member.

On Dec. 6th, 1764, Messier was made a foreign member of the Royal Society of London. The admission to membership of this eminent society was not as exclusive then as it became in the 19th century and is today, but it was an important recognition by the scientific world of his contribution to astronomy.

The most important prize, however – the membership of the status-conscious *Academie Royale des Sciences* in Paris – eluded him for many years. He managed to get his name proposed on several occasions but the conservative Academicians could not bring themselves to accept the claims of a mere observer, however talented. And there was that 'doubtful business of Halley's comet in 1759 . . .'. In 1763 Messier got so far as to tie in the election with the mathematician Jeurat but he lost the place on the second ballot.

Writing in the *Connaissance des Temps* for 1809, Messier gives this account of the election in 1763.

'On Jan. 19th (1763) an associate membership became vacant in the Astronomy division of the Academy of Sciences which occurred when M. le Gentil became an ordinary member due to the death of the Abbé de Lacaille on Mar. 21st, 1762 at 6 o'clock in the morning, aged 48.

'I solicited this associate membership: I presented to the Academy my journals of observations since 1752 with an account of my principal observations which the Academy had printed in the 5th volume of *les Savans Étrangers*, page 611. The candidates for this place were presented in the following order: MM. Messier, Bailly, Jeurat and Thuiller: Bailly received 14 votes, Messier and Jeurat 13 each and Thuiller, 4. One paper was rejected in which Jeurat was the sole nominee; it was that of the Cardinal de Luynes, who had faithfully promised his votes to Jeurat and myself. If he had voted for me (as well) we would have had three of us with 14 votes; a very unusual event in these elections.

'A new ballot followed between Jeurat and myself: Jeurat had the majority: he was presented to the King with Bailly: the minister's letter of Jan. 29th announced the nomination of Bailly to the associate vacancy and Jeurat a supernumerary in the same class.'

Messier stuck to his observations: during the daytime he made meteorological records and also registered the movements of nearly 100 sunspots. At night, when the Moon made the sighting of faint comets impossible, he observed the occultations of stars: the clear and moonless nights were reserved exclusively for comets – or rather for comets and their imitators – the hazy, comet-like, nebulae and star clusters.

While he was observing the earlier comet of 1758, Messier had come upon the nebula in Taurus which had so much resembled a comet that he had made a note of its position to ensure that it would not be a source of confusion on future occasions. In 1760, again while looking for a comet, he had come upon another of these nebulous objects in the constellation Aquarius and he had noted this one also.

Early in May 1764 he evidently decided that it was worth making a list of as many of these objects as he could find; not for any great interest in the objects themselves, but only to facilitate his comet observing.

He treated this subsidiary task with his usual thoroughness: he first re-surveyed the 16 'nebulosae' which Hevelius had reported in his *Prodromus Astronomiae* in 1660. This was available to Messier through Maupertuis' *Figure des Astres* in which Derham's list was quoted, and it turned out to be a somewhat frustrating business. One of the 16 was the Andromeda Nebula which was already familiar, but most of the rest turned out to be either small groupings of stars which were in no way nebulous in the telescope, or else to be completely untraceable.

He then looked for the objects which had been reported by other observers such as Halley, Maraldi, de Chéseaux, Lacaille and Le Gentil. These were more fruitful sources and by the end of Oct. 1764 he had made a list of 40 objects, 22 of which had been originally found by others while the remainder were all newly discovered by Messier himself. Number 1 in his list was the nebula in Taurus he had seen first in 1758; number 2, the one in Aquarius of 1760, while the last, number 40, was an 'asterism' of only two stars he had encountered in Ursa Major while looking for one of Hevelius' non-existent nebulae. This last object hardly came into the class of comet-confusing bodies and possibly it was included merely to make the list up to the round figure of 40.

In Jan. 1765 he found another object, a small cluster of stars near Sirius, and this became object No. 41. Messier seems to have let the affair rest for a while for it was not until 1769 that he decided to prepare his list for publication. He wrote an explanatory preface and, besides giving the positions for each object, he also added a brief description to help in their identification.

At the last moment, before submitting his memoir, he possibly decided that the total of 41 objects seemed too untidy and on the night of March 4th, 1769, he observed and described four more objects to bring the number up to 45. These additions have all the appearance of 'make-weights'; Nos. 42 and 43 comprised the well-known Orion Nebula while Nos. 44 and 45 – the Praesepe and Pleiades clusters – were hardly to be mistaken for comets. However, they made the list look more impressive.

Messier was now busier than ever: some marine chronometers, constructed by Le Roy, were to be given their trials at sea and Messier had the task of making the observations necessary to regulate them. The results of the observations were calculated by his colleague Pingré who also took part in the Baltic trials in the frigate *L'Aurore*. Messier seems to have been a reluctant traveller but on this occasion he was at sea for nearly four months while Lalande took charge of the observing programme at Cluny.

In 1769 Messier discovered a new comet: the following is his account, written up and later included in the *Connaissance des Temps* for 1807.

'On Aug. 8th, 1769, while sweeping the sky for comets, I found one several degrees above the horizon: it appeared faintly in the telescope as a nebulosity several minutes in extent, one degree before the Ecliptic and between the stars 24, 29 and 31 Arietis, following Flamsteed's catalogue, 2nd edition.

'I made an announcement to the Academy on Sept. 16th and on the 28th, presented a map of its route to the King at M. Bouret's lodge at Croix-Fontaine. This comet became quite remarkable; it was one of the largest I had observed, its tail grew longer each day; on August. 31st it measured 34°; on Sept. 5th, 43°; Sept. 9th, 55°, and on the 10th, 60° with a dazzling light.

'M. Pingré, who was at sea between Teneriffe and Cadiz, found it 90° long on Sept. 11th, but so faint at its extremity that when Venus rose above the horizon, its light shortened the tail by several degrees.

'I ceased to view the comet on Sept. 16th when it entered the sun's rays. I found from these first observations that I should see it again in the evening of the second branch of its orbit.* I observed it again on Oct. 24th, to the left of Arcturus and continued observations until Dec. 1st, when it ceased to be visible, disappearing between the right arm of Ophiuchus and the tail of the Serpent.'

Messier prepared maps to show both branches of the comet's track and these were printed in the *Memoirs* of the Academy of Sciences for 1775 together with a plate showing five different aspects of its nucleus and tail. He also sent a map and a letter describing the discovery to the King of Prussia and was, no doubt, gratified to receive the following reply.

Kniegwitf in Silesia. Sept. 2nd, 1769

The letter which you wrote to me on the 15th Aug. last to inform me of the discovery you have made of a new comet in the constellation of Aries, greatly pleased me and I am very grateful to you. As I have passed on this information to the director of the department of mathematics of the Berlin Academy of Sciences, Sieur de la Grange, I would be delighted if you would correspond with him more fully on this subject; and in this, I pray that God will keep you in His holy care.

Frederick.

This royal recognition brought Messier his next foreign honour, the membership of the Berlin Academy of Sciences; the prize he coveted most, however, the membership of the all-important Paris Academy, still eluded him.

(This comet of 1769 was later to be dedicated by Messier to another monarch, although in a very different setting, in 1808.)

---

* The orbit and perihelion date were calculated by Bessel. Perihelion: 1769 Oct. 7 d 14 h.

In June 1770 Messier made another important comet discovery. Not only did the new comet become extraordinarily bright, but when its orbit was computed from Messier's observations by Lexell and Pingré, they found to their surprise that its path, like that of Halley's comet, was elliptical. More than that, the period of its orbit was extremely short, only about 5½ years compared with the 76 years of Halley's, and the news of this 'sprinter' of the comet family soon created a distinct stir.

'Messier's comet', as it was first called, was never seen again but the Academy of Sciences now came to the conclusion that Messier's claim to membership could no longer be refused and in the same year, 1770, he was elected an Academician. Messier's name was now so closely linked with the discovery of comets that Louis XV with typical Bourbon wit, dubbed him 'the ferret of comets', an epithet which has firmly stuck.*

Messier's catalogue of 45 nebulae and star clusters, too, was accepted for publication in the *Memoirs* of the Academy of Sciences and it appeared in 1771. In the same year he was made officially Astronomer of the Navy and his salary was increased to 1700 francs per annum. This was again increased to 2000 francs in 1774.

He must have been gratified, too, at the introduction to his catalogue which Jerome de Lalande contributed to the same 1771 volume of the *Memoirs*. This account is typical of Lalande's generosity towards his colleagues.

'ON THE NEBULAE'

'Many celebrated astronomers have observed the nebulae: because they often appear like the comets and, at first sight, may be confused with them, M. Messier has thought it incumbent upon him to prepare a catalogue. This work could only be undertaken by an indefatigable and experienced observer.

'The nebulae are more or less luminous patches, fixed, having a sensible diameter; some of them, seen in the telescope, are no more than a cluster of very small and very close stars; others retain their appearance in many instruments with which they are observed; but the strongest indication of analogy is that if one had even more perfect instruments, one might distinguish, as in the others, stars in the process of formation.

'M. Messier gives here a detailed description of the nebulae and determines their positions and indicates which stars are the closest.

'In reading the Memoir of M. Messier, one cannot help regretting that so precise and zealous an observer has not been situated in a pleasanter climate or under a purer sky, where he would not have to await the breaks in the cloud at long intervals, to uncover each point in the sky. The excellent instruments which he has used have been lent to him by a leading amateur astronomer, whose excess of modesty ishes to hide himself from the public eye, but which cannot escape the rightful esteem of the Savans.†

* In his eulogy of Messier, read before the Academy of Sciences on March 16th, 1818, Delambre said that the words used by Louis XV were 'the "bird-nester" of comets'.

† This, of course, was Messier's close friend, President Bochart de Saron.

TABLE des Nebuleufes, ainfi que des amas d'Étoiles, que l'on découvre parmi les Étoiles fixes fur l'horizon de Paris; obfervées à l'Obfervatoire de la Marine.

ANNÉES & JOURS.	ASCENSION droite.			DÉCLINAISON.			DIAM.		INDICATION DES NÉBULEUSES & amas d'Étoiles.
	D.	M.	S.	D.	M.	S.	D.	M.	
1758. Sept. 12	80.	0.	33	21.	45.	27. B.	....		nébuleufe placée au-deffus de la corne méridionale du Taureau.
1760. Sept. 11	320.	17.	0	1.	47.	0. A.	0.	4	nébuleufe fans étoile, dans la tête du Verfeau.
1764. Mai... 3	202.	51.	19	29.	32.	57. B.	0.	3	nébuleufe fans étoile, entre la queue & les pattes d'un des Chiens de chaffe d'Hévélius.
8	242.	16.	56	25.	55.	40. A.	0.	2½	amas de très - petites étoiles, près d'*Antarès* & fur fon parallèle.
23	226.	39.	4	2.	57.	16. B.	0.	3	belle nébuleufe fans étoile, entre le Serpent & la Balance, près de l'étoile de 6.ᵉ grandeur, cinquième du Serpent, fuivant le catalogue de Flamfteed.

The beginning of Messier's first catalogue, listing M 1 to M 5
(From *Mémoirs de l'Académie des Sciences*, Paris 1771)

'To the catalogue of nebulae which he has observed, M. Messier has added a table of nebulae which other astronomers have described, and for which he has looked in vain. It is very probable that these supposed nebulae might be far-distant comets and have been visible only in instruments and in a very small part of their orbit.'

Messier now had every reason to be proud of his achievement: it had taken him 20 years of patient effort but at last, after all his early frustrations and disappointments, he had made good. The young provincial with only a 'neat hand and some skill at drawing' was now Astronomer to the Navy, Director of the Observatory of Cluny, a member of the Royal Academy of Sciences, a notable, even famous 'ferret of comets', one of which bore his name, and as a little extra, the author of a memoir upon some *Nebulae and Star Clusters*. He cannot possibly have conceived that it would be only this last distinction which would illuminate his name for posterity.

In this year, 1771, soon after his list was published, Messier observed on the night of Feb. 19th, four more objects that could be added to his collection. Two of these, Nos. 47 and 48, were not checked with his usual care and the positions he gave for them, being in error, were to cause later observers much speculation and fruitless search. Only one more, No. 50, was added in 1772, another two in 1774 and after that no more were discovered until No. 53 was found in 1777.

In 1777 an old speculation was revived on the possible existence of another planet whose orbit should lie inside that of Mercury. Messier contributed to this affair by reporting on June 17th that he saw 'a number of small bodies very rapidly cross the Sun's disc in parallel directions. These', he added, 'may have been hailstones or seeds in the air but more probably small meteorites.'

In 1778 he observed two more nebulae and in 1779 and early 1780 he had accumulated enough new objects to justify publishing an augmented list of 68. This, with its accompanying notes was printed in the French almanac *Connaissance des Temps* for 1783, published three years in advance in 1780.

In 1779 Messier discovered another comet: he first observed it on Jan. 18th but it turned out that J. E. Bode, at Berlin, had seen it first, 12 days earlier on Jan. 6th. While observing this comet, Messier encountered three new nebulae in Virgo on April 15th: these were M 58, M 59 and M 60. Unknown to Messier, one of them (M 60) had been seen by Oriani at Milan just three days earlier, also while observing the same comet. This near-coincidence was repeated during the next month; Oriani finding M 61 on May 5th, 1779, and Messier, this time six days later, on May 11th. Messier, in fact, *saw* M 61 on May 5th and 6th but mistook it for the comet; it was only on the 11th that he realised his mistake.

To add to this conjunction of events, Bode reported in 1780 that Koehler, at Dresden, while following the same comet, also observed M 59 and M 60 on April 11th and 13th, 1779. Koehler's description of these two objects was 'Two very small nebulae, hardly visible in a 3-foot telescope, NW in a triangle above rho and 34 Virginis; one above the other.' In the case of M 60 at least, the only fair result would be to declare a triple dead heat.

Starting again at the end of Aug. 1780, Messier pursued his search for new nebulae and clusters with considerable vigour and, by the middle of April 1781 he had observed another 32 new objects which brought the number to the tidy total of exactly 100.

In this latter part of the catalogue he was given much help by his colleague, Pierre Méchain. Méchain, who was some 14 years his junior, was a calculator in the Department of the Navy at Versailles. He was a protégé of the leading French astronomer, Jerome de Lalande, who had introduced him to Messier in 1774. Méchain became a rival of his elder colleague in comet finding – he discovered no less than eight between 1781 and 1799 – but the two astronomers always remained the best of friends.

Messier, having little mathematical ability, never attempted to calculate the orbits of his comets: he had many friends and colleagues such as Pingré and the eminent lawyer and amateur mathematician, de Saron, who were well able and always willing to do this. Méchain, too, often performed this service for his friend and in addition to this he also looked out for nebulae while observing comets of his own.

It was Méchain who discovered the rich field of faint nebulae in the constellations Virgo and Coma Berenices and some 20 or so of the objects in Messier's final catalogue were first discovered by Méchain. Messier was always careful to make acknowledgement of Méchain's contributions although he naturally made a point of checking their positions before entering them in his list.

Messier plotted the positions of the nebulae on various sheets of the maps he used for marking the tracks of the comets he observed. For this purpose he made meticulous tracings of plates from both Flamsteed's and Bevis' star atlases.

He also made sketches of some of the clusters and nebulae and at the end of his descriptions of the objects in his first list of 1771, he hinted that he was preparing a planisphere on which to show them all. As far as is known, he never completed this task, but on the chart of the comet of 1779, published in the *Memoirs* of the academy for 1779, page 371, Plate XIV, he plotted the positions of M 3, M 13, M 27, M 29, M 49, M 51, M 53, M 56, M 57, M 58, M 59, M 60, M 61, M 63, M 64, M 84, M 85, M 86, M 87, M 88, M 89, M 90 (M 91), M 98, M 99, M 100.

Messier also claimed to have made drawings of each of the objects as he observed them but none of these seem to have been published except for a drawing of the Andromeda Nebulae, M 31, with its two satellite elliptical nebulae, and a fine drawing of the Orion Nebulae, M 42.

This latter drawing was included in the 1771 *Memoirs* of the Academy and was accompanied by Messier's explanatory note which is worth quoting in full as it shows how thoroughly and carefully he observed and recorded.

ADDITION TO THIS MEMOIR
ORION NEBULA

'The drawing of the Orion Nebula which I present to the Academy has been executed with the greatest possible care. The nebula is represented here just as I have seen it many times with an excellent achromatic telescope of $3\frac{1}{2}$ feet F.L., having a triple objective of 40 lignes ($3\frac{1}{2}$ inches) aperture and which magnifies 68 times. This telescope, made in London by Dollond, belongs to *M. le President de S* . . .

'I have examined this nebula with the greatest attention, in an entirely serene sky as follows:

25th and 26th February 1773. Orion in the meridian.
19th March between 8 and 9 in the evening.
23rd March between 7 and 8.
25th and 26th of the same month at the same time.

'These observations, and the drawings combined and compared with each other, I have tried to render with equal care and precision with regard to both shape and appearance.

'This drawing will serve to show, later on, if this nebula is subject to any change. There may be, even now, cause to presume this; for if the drawing is compared with those given by MM. Huyghens, Picard, Mairan and le Gentil, one finds there such a change that one would have difficulty to make out that it was the same. I will repeat these observations later with the same telescope and same magnification.

'In the figure which I give, the circle represents the field of the telescope at its true aperture; it contains the Nebula and thirty stars of various magnitudes. The figure is inverted as shown in the instrument: the extent and limits of this nebula and the appreciable difference between its clearest or most distinct light and that which merges gradually with the background of the sky, can clearly be detected.

'The jet of light, directed from star No. 8 to star No. 9, passing by the side of a small star of tenth magnitude, becomes extremely diffuse, as well as the light directed towards star No. 10 and that which is opposed to it, where are the eight stars contained in the nebula: among these stars there is one of eighth magnitude, six of the tenth; the eighth being of the eleventh magnitude.

'M. de Mairan, in his treatise on the Aurora Borealis, speaks of star No. 7. I give it in my drawing just as it is at present, that is to say, surrounded by a thin nebulosity.

'On the night of the 14th and 15th October 1764, in a serene sky, I determined, relative to theta in the nebula, the positions in R.A. and Dec., of the more conspicuous stars by means of a micrometer adapted to a New-tonian telescope of $4\frac{1}{2}$ feet F.L. These stars, numbered up to ten, are reported on the drawing, contained in the field of the telescope; together with an eleventh which is outside the circle.

'The stars in this table can be easily recognised in the drawing, either by the numbers given to each or by the differences in R.A. and Dec. relative to theta, the drawing being divided by a 5-minute graticule.

'The positions of the stars which have no numbers have been fixed by estimation of their relative alignments. The magnitudes of the stars can easily be obtained by the scale included in the drawing.

'Those of the tenth and eleventh magnitudes are absolutely telescopic and are very difficult to find.'

	No. of Star	Diff. R.A. from theta		Diff. Dec. from theta		
		m	s	m	s	
	Theta	0	0	0	0	Theta Orionis
To the	1	7	$7\frac{1}{2}$	11	23	above theta
left	2	5	$22\frac{1}{2}$	7	0	above theta
of theta	3	3	$22\frac{1}{2}$	11	23	below theta
	4	1	$37\frac{1}{2}$	4	50	above theta
	5	1	$22\frac{1}{2}$	1	43	above theta
	6	2	$22\frac{1}{2}$	1	43	above theta
To the	7	3	$22\frac{1}{2}$	7	29	below theta, nebulous
right	8	3	$37\frac{1}{2}$	1	45	above theta
of theta	9	9	$37\frac{1}{2}$	13	46	above theta
	10	9	$52\frac{1}{2}$	1	36	below theta
	11	14	45	14	42	above theta

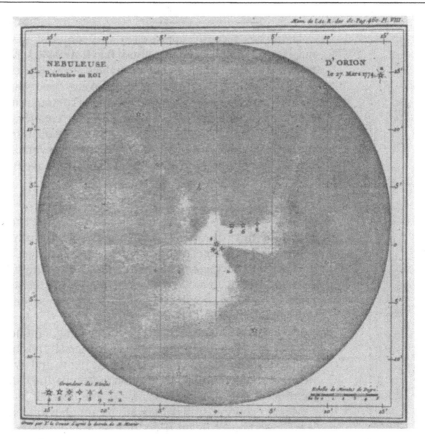

Messier's drawing of the Orion Nebula which was included in the Mémoirs
de l'Académie. Royale des Sciences. 1771 (*British Museum*)

Just before Messier submitted his final list of 100 objects for printing, he
received a note from Méchain giving particulars of three new objects he had
just seen. Messier had no time to check these but they were added to the list to
swell the new total to 103 and this was published in the *Connaissance des
Temps* for 1784; printed in 1781. The second of these three additional objects,
No. 102, was later admitted by Méchain to be a duplicate observation of No.
101 and this, too, has been the subject of controversy ever since.

This was Messier's final catalogue and although he had ideas of extending it
at a later date, apart from making notes of a few additional objects, he never
got as far as putting them into print.

On March 13th of this same year, a new and epoch-making discovery was
made – this time in England – and Messier, always quick to appraise any new
feat of observation, was soon involved.

On April 14th, 1781 he received the following letter from Britain's Astronomer Royal.

Greenwich. April 5th 1781

There has recently appeared a very unusual new star, of 5th or 6th magnitude: on the last two nights it had R.A. 84° 23′ 35″ and Dec. 23° 34′ 14″ North: it is not nebulous but resembles a tiny planet of 4″ to 5″ diameter. When magnified 100 times it looks like a fixed star almost resembling a planet. It appears to me to be a new kind of comet: its movement is very slow, of 2′ a day to the east of R.A., parallel to the equator.

In the telescope, its light is as white and brilliant as that of Jupiter. It is with the respect which I owe you, Sir, who have discovered so many comets, that I inform you of this by first post as perhaps you have not yet seen it.

Nevil Maskelyne

Messier wasted no time: here is his account of the events which followed.

'On the next day, the 15th, in a perfectly clear sky, I looked for his new comet with my large achromatic telescope of 3½ feet F.L. I found only stars and nothing having any different character: of those I saw I obtained their R.A. and Dec. with a micrometer adapted to my telescope. As I had no other means of finding the object, I left it until the next day to verify those I had determined the night before.

'On the 16th, about 8 o'clock in the evening, I found the object among the stars which I had determined: it had changed its position and exactly resembled a star of 6th mag., of the same light and without atmosphere or nebulosity. I determined its position by comparing it directly with the star H. Geminorum: its R.A. was 84° 50′ 56″ and Dec. 23° 34′ 52″ North.'

Messier continued his observations and, having learnt that its discoverer was a certain William Herschel, a well-known musician, but hardly recognised as an astronomer, wrote to him as follows:

Paris. April 29th, 1781

*To Monsieur Herschel at Bath**

Having learnt from M. Maskelyne, Astronomer Royal of England, in a letter which he has done me the honour to write to me, that he was observing a new comet of very singular character, having no atmosphere, no coma nor tail, resembling a little planet with a diameter of 4″–5″, a whitish light like that of Jupiter and having the appearance when seen in the telescope of a star of the sixth magnitude, I searched for it and as it resembles the stars I had some little difficulty in recognising it and I found it had the character which M. Maskelyne noticed.

I am constantly astonished at this comet which has none of the distinctive characteristics of comets as it does not resemble any of those I have observed, whose number is 18. I will add here the observations I have made . . .

I have since learnt by a letter from London that it is to you, Sir, that we owe this discovery. It does you the more honour as nothing could be more difficult

---

* Messier spelt the name 'Hertsthel'.

than to recognise it and I cannot conceive how you were able to return several times to this star or comet as it was necessary to observe it several days in succession to perceive that it had motion since it had none of the usual characters of a comet.

For the rest, this discovery does you much honour and allow me to compliment you for it. I should be very curious, Sir, to learn the details of this discovery and you will oblige me, Sir, if you will be so good as to inform me of them.*

With equal consideration and respect,

Messier,

'Astronomer of the Navy of France and of the Academy of Sciences, Paris.'

Messier's account of this affair continues:

'My first observations were communicated to M. de Saron in order to obtain its orbital elements. He employed the usual methods for comet calculations and spent a lot of time at this work but was unable to justify four observations; the last being too far from the three others, implying that the object was very distant.

'At last, on May 8th, 1781, he succeeded in proving that its (perihelion) distance was too great to be a comet and that it was a new planet. He made its distance 12 times the distance of the Sun. It was only by using this great distance that the observations would tally.'

Other calculators such as Boscovich, Lexell, Lalande and Méchain also engaged themselves with the same problem and the two first-mentioned came to similar conclusions – that the path was not parabolic but circular – or nearly so – and with its orbit considerably larger than that of Saturn.

The first recognition of the object as a new major planet, however, appears by a short head, to belong to de Saron, and Messier is entitled to some credit for the timely and precise observations on which this conclusion was based.

Wm Herschel, thus the first discoverer of a planet in historical times, gained an immediate and widespread fame: he was awarded the Copley medal of the Royal Society together with a fee-exempt Fellowship and numerous foreign honours soon followed.

Herschel named the new planet Georgium Sidus in honour of George III of England but Lalande, in Paris, thought it should be called Herschel's Planet after its discoverer. Other names, including Hypercronius (by Bernoulli), Minerva, Neptune (by Lexell) and Cybele were put forward at various times: the name it now bears – Uranus – was suggested by the Berlin astronomer J. E. Bode in 1782.

Messier made many more observations of the new planet but, later in the year, his work was unfortunately interrupted by a serious accident. On Nov. 6th, Messier went with his close friend, Bochart de Saron and several others, including de Saron's children, to visit the *Parc Monceau*. This was part of the

---

* Herschel, whose telescope was considerably superior to Messier's, replied later, explaining that he was able to recognise that it had a well-defined disc which immediately distinguished it from a star.

old *Plaine de Monceau* belonging to the Orleans branch of the royal family and here, the Duke of Orleans, later '*Philippe Égalité*' had commissioned the landscapist, Carmontelle, to build a 'Chinese Garden', something after the style of those created in England by William Chambers. This delightful extravagance included among its paths and groves, many novelties such as pagodas, pyramids, castles, tombs and windmills.

While walking in this profusion of 'follies', and apparently separated from de Saron, Messier noticed a little door which he imagined led to some sort of grotto. The door was open and unguarded and he stepped inside. The grotto was, in fact, an underground ice-store: the interior was dark and Messier fell some 25 feet on to the ice in the cellar below.

He suffered severe injuries, breaking an arm, thigh, wrist and several ribs, besides losing a lot of blood through an injury to his head. Fortunately someone noticed the open door and the injured Messier was rescued by workmen with ropes and ladders.

Messier was given the best attention by the eminent doctors and surgeons of the Academy but he was incapacitated for a long time and when the surgeon, Valdajou, had to re-break his leg after finding it improperly set, his complaints were loud and long. His wrist was never properly reset and his right thigh, shortened by more than an inch, left him with a permanent limp. Bochart de Saron, who, besides being a mathematician was also a presiding judge of the *Parlement de Paris*, managed to obtain for his friend several grants during his convalescence. One of these was for 1200 livres and another of 2400 livres and this was also accompanied by a pension of 1000 livres.

On Nov. 9th, 1782, just over a year after his accident, Messier revisited his observatory in the tower of Cluny to prepare for the observation of a transit of Mercury: this event he was able to observe three days later on Nov. 12th.

Soon he was observing again as assiduously as ever with comets still as his main interest. On May 14th, 1784, Messier received a letter from a Chevalier d'Angos who claimed to have observed at Malta, a faint new comet in the constellation of Vulpecula and gave two positions for it. Messier, always keen to see a new comet, spent several nights looking for the elusive object but was unable to find it. As no other observer had seen it either, it was assumed that the comet must have been receding from the Earth and had now disappeared.

Some 25 years later it was found that the Chevalier d'Angos, who claimed to have 'discovered' other comets too, had deliberately invented his observations. J. F. Encke, who finally unravelled the whole wretched affair, wrote, 'D'Angos had the audacity to forge observations he had never made, of a comet he had never seen, based on an orbit he had gratuitously invented, all to give himself the glory of having discovered a comet.'

Jerome de Lalande, whose orbital elements for a real comet d'Angos had used in his forgeries, was angrier still and his comments, according to his friends, were too outspoken to be repeated.

In 1785, Messier was appointed an associate editor of the *Connaissance des Temps* in succession to Jeurat who had beaten him in the ballot for membership of the Academy in 1763. Messier served a five-year term in this important post until 1790.

These were now years of great upheaval in France; the storming of the

Bastille occurred in July 1789; Louis XVI was led to the guillotine on Jan. 21st, 1793, and in July of that year, with the assassination of Marat, the 'Year of the Terror' began. Messier kept his eyes prudently on the heavens.

As if to lend credence to the old superstition that comets were harbingers of war, pestilence and famine, four new ones were reported in 1793; one of them, that of May 17th was seen only to the unhappy d'Angos at Tarbes and must be discounted. The comet of Jan. 10th was found by Méchain at Barcelona and on Sept. 24th, Perny found a small comet in Cassiopeia which Messier observed two days later.

On Sept. 27th, 1793 Messier discovered a comet in Ophiuchus and continued to observe it until October 11th, 1793.

As he had done so often before, Messier sent the positions he had determined for the comet to his friend de Saron in order to have the elements of its orbit calculated. The circumstances on this occasion, however, were sadly altered. Bochart de Saron as an ex-President of the *Parlement de Paris* (it was abolished by the Constituent Assembly in 1791) was thus labelled as one of those 'merciless enemies of reform' and was in prison.

During his imprisonment, de Saron, probably grateful for any activity to take his mind off his approaching fate, managed to complete the necessary calculations and these were given to Messier in good time. Before de Saron died on the guillotine on April 20th, 1794, we are told that Messier was able to inform his old friend, by a note, secretly concealed in a newspaper, that his calculations had proved correct and that he had recovered the comet in the predicted part of the sky.

Messier must have been very distressed at the cruel fate which overtook his wealthy and aristocratic friend, for the two had been on intimate terms for many years. Writing after the Revolution J. D. Cassini (Cassini IV) commented at length upon their happy relationship in his *Eulogy on President de Saron*, published in 1810.

'Whenever a new comet appeared, it was soon picked up by that vigilant Argus, M. Messier, who hastened to inform President de Saron, bringing the first observations. While one calculated the orbital elements, the other followed the object, not letting it out of his sight until its remoteness rendered it invisible. Until that time, no matter what deviation occurred in its course, the calculations of de Saron immediately put M. Messier on its track again and the comet was soon retrieved.

'This happy association between one of our most celebrated observers and one of our most adept calculators has considerably augmented the list of comets whose orbits have been calculated. This amicable partnership during thirty years forged the bonds of a steadfast and rewarding friendship between two scholars whose conformity of taste, ability and character made them naturally attracted to each other.'

Robespierre's 'Reign of the Guillotine' ended with his own overthrow on July 27th, 1794 (9th Thermidor) and the people of France now began to count the cost. Many had suffered during the famine of 1793; Messier lost his pension and salary from the Navy* and, having little by the way of savings,

---

* In 1796 he complained, somewhat bitterly, '. . . during those years I was much distracted by scarcity of provisions; I lacked, in my observatory, oil and candles and was not able to pro-

was given, with many others, generous help in the necessities of life by Jerome de Lalande.

Besides de Saron, three other prominent French scientists lost their lives during the Revolution.

J. S. Bailly, the astronomer and mathematician, who, like de Saron was interested in comet orbits, became famous as the President of the Third Estate, opposing Louis at the historic 'Tennis Court' proceedings. He was Mayor of Paris from July 1789 to Nov. 16th, 1791, and incurred the enmity of the revolutionaries by his order to disperse the insurrection in the *Champ de Mars*. He was guillotined on Nov. 12th, 1793.

A. L. Lavoisier, the eminent chemist and Director of the Academy of Sciences, became involved in politics through his work on revenue and taxation – unpopular activities even in times of peace: he was attacked by Marat, arrested in 1792, and after a long imprisonment, summarily tried and executed on May 8th, 1794.

The Marquis de Condorcet, another mathematician who did pioneer work on probability theory, became perpetual secretary of the Academy of Sciences and inspector-general of the Mint. Although himself a revolutionary, he was later outlawed for his independent attitude and refusal to vote for the death penalty against Louis XVI. He was found dead in prison – it is thought that he was poisoned – immediately after his arrest on April 7th, 1794.

Lalande, who had strong Royalist sympathies must have been very adroit to escape the attentions of the extremists. His well-known generosity to his friends extended to hiding those who were hunted, such as the economist Dupont de Lemours and several priests, in hideaways in the Observatory of Paris.

After the Revolution, Messier was made a member of the newly-constituted Academy of Sciences and he also joined the important Bureau of Longitude. He continued to make observations at Cluny and was as occupied in comet hunting as ever but his contributions to the *Connaissance des Temps* were now signed, not 'Charles', but 'Citoyen' Messier.

On April 12th, 1798 he found a new comet near the Pleiades; it was small and had no tail and though bright, did not become visible to the naked eye. This, according to Lalande was the twenty-first comet discovered by Messier and brought the number he had observed to forty-one.

Although by modern methods of accounting, Messier's total of new discoveries would probably be reduced to about 13 or 14 it is still a very creditable achievement. It was not to endure very long as a record, however, for Jean Louis Pons (1761–1831) at Marseilles accumulated a 'bag' of no less than thirty-seven new comets during his lifetime.

In 1798 Messier's wife died. Messier was apparently very proud of the fact that she came from a very good family of Lorraine and they had always been a very devoted couple. La Harpe, who had helped to obtain for Messier his membership of the Russian Academy of Sciences, tells a tale of Messier and his wife that has often been quoted.

cure them, through the suppression of the salary which, as the fruit of 40 years' work, I used to enjoy . . .'

'Messier is, at all events, a very good man', says La Harpe, 'and as simple as a child. He lost his wife some years ago and his attendance upon her death bed prevented him being the discoverer of a comet for which he had been lying in wait and which was snatched from him by Montaigne de Limoges. This made him desperate. A visitor began to offer him condolences for his recent bereavement, when Messier, thinking only of the comet, answered, "I had discovered twelve; alas, to be robbed of the thirteenth by that Montaigne!" and his eyes filled with tears. Then, recollecting that it was necessary to deplore the loss of his wife, exclaimed; "Ah! that poor woman!" – and wept again for his comet.'

It is a faintly amusing, if slightly malicious anecdote – and rather too well told. The comet which Montaigne, an apothecary and amateur astronomer of Limoges, had discovered, appeared in 1772 (it is now known as 'Biela's comet') and, in 1798, when his wife died, Messier had found, if Lalande is correct, not 12 comets but 21. La Harpe's tale is probably apocryphal, but it shows how others viewed his comet obsession and in this sense it is true enough.

Messier and his wife had no children and after her death he lived alone for some time until his sister and later, a brother, came to live with him. To his great sorrow, he soon lost these members of his family also and for the remainder of his life he was well looked after by a widowed niece, a Mme Bertrand.

In 1799 he observed a comet which Méchain had discovered in Ophiuchus on Dec. 26th. In 1801, on March 30th and May 24th, he took part in some important observations of the occultation by the Moon of the first magnitude star, *Spica*. This is a comparatively rare event and as it could be easily observed and timed with great precision, it provided a means of obtaining the lunar parallax. Messier's times for the second of these events were: 14 h 7 min $50\frac{1}{2}$ s for immersion and 15 h 17 min $7\frac{1}{2}$ s for emersion.

About this time he began to think about making further additions to his catalogue of nebulae and star clusters. He also felt some need to explain how he had come to compile it in the first place. Messier wrote this memoir in the *Connaissance des Temps* for 1801.

'What caused me to undertake the catalogue was the nebula I discovered above the southern horn of Taurus on Sept. 12th, 1758, whilst observing the comet of that year. This nebula had such a resemblance to a comet in its form and brightness that I endeavoured to find others so that astronomers would not confuse these same nebulae with comets just beginning to appear. I observed further with suitable refractors for the discovery of comets and this is the purpose I had in mind in forming the catalogue.

'After me the celebrated Herschel published a catalogue of 2000 which he has observed. This unveiling of the heavens, made with instruments of great aperture, does not help in the perusal of the sky for faint comets. Thus my object is different from his and I need only nebulae visible in a telescope of two feet. Since the publication of my catalogue I have observed still others: I will publish them in the future in the order of right ascension for the purpose of making them more easy to recognise and for those searching for comets to have less uncertainty.'

We know that Messier did not carry out his intention of republishing his catalogue with the objects in order of R.A. nor did he publish an additional list. His explanation that the object of his catalogue was different from that of Herschel's is perfectly valid but the statement that large apertures were not helpful in looking for comets does not really hold water. Messier probably realised that he had no hope of competing with Herschel whose powers so greatly outstripped his own, and he was no doubt content to rest upon the fact that his own modest list of 103 objects at least had seniority.

It is very probable that Herschel had in fact been stimulated by Messier's observations to start his own great survey of nebulae. Alexander Aubert had sent Herschel a copy of the *Connaissance des Temps* for 1784, containing Messier's list, soon after its publication.

Herschel began his own 'sweeps' for nebulae in Oct. 1783 and in his paper, *Account of some Observations tending to investigate the Construction of the Heavens*, read to the Royal Society on June 17th, 1784, he referred to Messier's catalogue as follows: 'The excellent collection of nebulae and clusters of stars which has lately been given in the *Connaissance des Temps* for 1783 and 1784, leads me next to a subject which, indeed, must open a new view of the heavens. As soon as the first of these volumes came to my hands, I applied my former 20-feet reflector of 12 inches aperture to them; and saw, with the greatest pleasure, that most of the nebulae, which I had an opportunity of examining in proper situations, yielded to the force of my light and power, and were resolved into stars.' Herschel was careful, when compiling his own catalogue, not to include any of Messier's objects in it, nor to give them numbers in any of his eight classes. (The 'missing' objects, M 47 and M 48 and the additional objects from M 104 on, were not thought to be members of Messier's original list and Herschel, who found them for himself, gave his own designation to them.)

Méchain, with three of his colleagues, had been to England in 1787 to pay a visit to Herschel's observatory at Slough. Messier did not go with them: apart from a journey to his native Lorraine in the autumn of 1760 and one or two excursions into Champagne with de Saron, he seemed quite content to stay in Paris. Herschel, however, made a return visit to Paris in 1802 and met Messier on at least two occasions.

Herschel, in his diary, gives this short account of one of these meetings.

'I saw M. Messier at his lodgings; he complained of having suffered much from his accident of falling into an ice-cellar. He was still very assiduous of observing and regretted that he had not enough interest to get the windows mended in the tower of the Abbey of Cluny where his instruments are but he keeps his spirits. He appeared to be a very sensible man in conversation: merit is not always rewarded as it ought to be.'

Herschel's remarks raise a host of questions: Messier's accident at Monceau had occurred almost 20 years previously: was he still not recovered from his injuries? Perhaps they were an old man's grumbles. He was now 71 and the stairs up to his observatory in the tower were long and steep.

The last phrase, 'merit is not always rewarded as it ought to be', is very enigmatic and the biographer can only regret that Herschel was not a little more explicit. No doubt he thought that the French astronomer's capabilities deserved a better equipped observatory than the old and ricketty structure on the tower of Cluny, and Messier's out-of-date and inefficient telescopes would have astonished the builder of the great 40-foot reflector if he had seen them.

Herschel was not the only one to consider that Messier was ill-rewarded: the French mathematician and astronomer, D'Alembert, observed, 'One can only regret that so careful and zealous an observer was not more happily situated.' We find, in E. Doublet's *Histoire de L'Astronomie*, another, half-revealing remark: 'Messier had failed, for several years before the Revolution, to be commissioned with important work, because of his opinions.'

We do not know what these opinions were: Messier does not seem the sort of man who spoke out too often or too freely and his opinions, either astronomical or political, never seem to have appeared in writing.

He was not to be left without honours in his own country however. The old religious and military orders were abolished after the Revolution but Napoleon instituted the new Legion of Honour in 1802. In 1806, Messier received the Cross of this Order from the hands of the Emperor himself. Many others of the Academy were accorded the same distinction but one feels that very few would have felt the same pride and gratification as Charles Messier, Astronomer of the Navy.

He later expressed his gratitude for this honour in a manner which, by its very oddity, seems to reveal much of his character. This took the form of an elaborate memoir entitled, *Sur la Grande Comete qui a paru a la Naissance de Napoleon le Grand*. Napoleon was born in 1769: the comet referred to was the famous one which Messier had himself discovered in that year about which he had written to the King of Prussia.

This unscientific and flattering monograph, although possibly well received by Napoleon, was not at all appreciated by Messier's colleagues of the Academy. However, in view of that fact that he was now 78 years of age, it was probably humoured as evidence of a senile eccentricity. Admiral Smyth, commenting on the memoir some 35 years later, described it as, 'the last comet astrologically put before the public by an orthodox astronomer'.

Messier now went more or less into retirement, for although he continued to make observations from time to time, the observatory grew more and more in need of repair and funds for this were not forthcoming. The long climb up the tower stairs, too, became more arduous with the years.

In 1815 he suffered a stroke which left him partially paralysed. Although he recovered enough to get about with some assistance and attended one or two meetings of the Academy, even this became increasingly difficult. In 1817 he contracted dropsy and after being confined to bed for only a few days, died on the night of the 11th–12th April in his eighty-seventh year.

Messier's eulogy was read before the members of the Acadamy of Sciences and tributes were paid to his accomplishments as an observer. No comet, however, attended the passing of this most devoted of comet discoverers and he has never been considered one of the 'great' astronomers. He had only a single talent but this he exploited to the full. His record of comet discovery made him a well known figure during his lifetime and his catalogue of nebulae and star clusters has preserved at least his name for posterity.

Although Messier received many rebuffs from the more conservative of the

Academicians, he was well liked by all his close colleagues. Jerome de Lalande, who was always generous in his praise, had proposed in 1775 that a constellation should be named after him. This was to be formed out of stars on the borders of Cepheus, Cassiopeia and Camelopardus and was actually printed on the new celestial globe drawn by Lalande and was later included in the French edition of Flamsteed's atlas and Bode's *Uranographia*.

The constellation was called 'Custos Messium' – The Guardian of the Crops – Lalande suggesting that 'the name will recall to the memory and recognition of astronomers yet to come, the courage and zeal of our most indefatigable observer, Messier, who, since 1757, seems to be in charge of guarding the sky and discovering the comets'. This well-meant tribute,* however, has not withstood the erosion of time and Messier's constellation was only short-lived. A later, though less exclusive attribute of a Moon crater on the Mare Foecunditatis is happily preserved.

Many astronomers since Messier's time have paid tribute to his work – not for his comet discoveries, but for his catalogue. Admiral Smyth, in his *Cycle of Celestial Objects*, wrote, 'It is a pity that this active and assiduous astronomer could not have been furnished with one of the giant telescopes of the present day. Had he possessed efficient means, his useful and, in its day, unique catalogue, for which sidereal astronomers must ever remain indebted to him, would, no doubt, have been greatly augmented . . . One is only surprised that with his methods and means, so much was accomplished.'

Although Messier seems to have been quite content with the instruments he had and claimed to have little use for 'great apertures', it must be admitted that his telescopes were hardly efficient. He used several types; the early ones having speculum mirrors of poor definition. One of them had a 7½-inch aperture with a focal length of 32 inches and a magnification of about 100×. Later, in about 1774 he began to use one of the achromatic refractors which began to be manufactured by Dollond in about 1758. Messier's achromat had a 3½-inch aperture and a focal length of 42 inches, giving a magnification of 120×. Neither of these instruments had a performance anywhere near that of amateur telescopes of a similar specification today.†

The positions of Messier's nebulae, as he found them, were plotted on his own maps which he had drawn for the purpose of recording his comet locations. We know that he used Flamsteed's catalogue for star positions and he probably used Flamsteed's popular star atlas also. Lalande, however, said that Messier also possessed one of the few copies of John Bevis' Atlas *Uranographia Britannica* of 1750 and Messier, in his catalogue, refers to

---

* The compliment was not quite as exceptional as it seems for Lalande was much prone to forming new constellations. One of his propositions was 'Felis' (the Cat) which he formed out of stars between Antlia and Hydra with the following citation:

'I am very fond of cats. I will let this figure scratch on the chart. The starry sky has worried me quite enough in my life, so that now I can have my joke with it.'

Another was 'Globus Aerostaticus' (the Balloon), formed to commemorate his own balloon ascent in 1799. All these new creations of Lalande have vanished from the charts.

It is not to be assumed that the formation of 'Custos Messium' and its dedication to Messier was merely a joke, however. Lalande must certainly have meant it quite sincerely.

† A more complete list of Messier's telescopes is given in Table 7 on page 371.

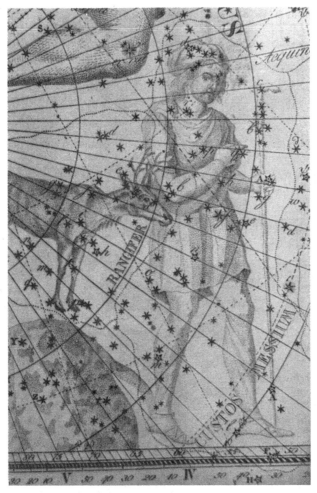

*The Constellation 'Custos Messium' proposed by Lalande
in honour of Messier*

'L'atlas Celeste Anglais' in connection with M 1, M 11, M 13, M 22, M 31, M 35 and M 42 as if these objects were already included there.

A copy of this rare atlas, reprinted from the original plates in 1818, is in the Adams collection at Cambridge University Library and, apart from the southerly nebulae, ω *Centauri* and the Magellanic Clouds, the only other nebulous objects included in its 51 beautifully engraved copperplates are the ones mentioned by Messier above. This evidence would seem to show that Messier must, at least, have had access to *Uranographia Britannica*.

The *Hôtel de Cluny*, where Messier spent most of his working life is now the *Musée de Cluny*: it stands near the intersection of the *Boulevard St Germain* and the *Boulevard Michelin* and quite close to the *Sorbonne*. As a museum, it contains one of the finest collections of medieval arts and crafts in the world and

Table 6. *List of comets discovered by Messier*

Perihelion Date	Date of Discovery	Discoverer	Calculator	Remarks
1759 March 12th	1758 Dec. 25th	Palitzch	Lalande Rosenberger	First predicted return of Halley's comet. Messier discovered independently, Jan. 21st 1759
1759 Nov. 27th	1760 Jan. 25th	Messier	Lacaille	Vis. naked eye. Tail 5° long
1763 Nov. 1st	1763 Sept. 28th	Messier	Burckhardt	Vis. for 8 weeks
1764 Feb. 12th	1764 Jan. 3rd	Messier	Pingré	Vis. naked eye. Tail 2½° long
1766 Feb. 17th	1766 Mar. 8th	Messier	Pingré	Vis. for 9 weeks
1766 April 26th	1766 April 1st	Helfenzrieda	Burckhardt	Disc. indep. Messier, April 8th 1766. Vis. naked eye
1769 Oct. 7th	1769 Aug. 8th	Messier	Bessel	Vis. naked eye. tail 60°–80° long. ('Napoleon's Comet')
1770 Aug. 13th	1770 June 14th	Messier	Lexell Leverrier	Known as Lexell's comet
1771 April 19th	1771 April 1st	Messier	Encke	Tail about 2° long
1773 Sept. 5th	1773 Oct. 12th	Messier	Burckhardt	Just vis. naked eye
1779 Jan. 4th	1779 Jan. 6th	Bode	von Zach	Disc. independ, by Messier Jan. 18th 1779
1780 Sept. 30th	1780 Oct. 27th	Messier	Clüver	Long-period comet
1785 Jan. 27th	1785 Jan. 7th	Messier	Méchain	Vis. for 5 weeks
1788 Nov. 10th	1788 Nov. 25th	Messier	Méchain	Vis. naked eye. Tail 2½° long
1793 Nov. 4th	1793 Sept. 27th	Messier	de Saron	Vis. for 15 weeks
1798 April 4th	1798 April 12th	Messier	Burckhardt	Found near Pleiades. Vis. for 6 weeks

it is famous for its fine tapestries, including the series of the 'Lady and the Unicorn', housed in the Rotunda.

The octagonal tower from which Messier observed, still stands but the observatory itself,* consisting of a pyramidal framework with opening windows which was built by Delisle, has now disappeared. The Observatory of Paris, designed for Louis XIV by Claude Perrault and directed in Messier's day by J. D. Cassini (Cassini IV) and later by Lalande, is only a mile away down the nearby *Rue de St Jacques*.

---

* Messier gave its position as Lat. 48° 51′ 14″ N, long. 1″.8 E (in time) of the Paris Royal Observatory.

There are two portraits of Messier in existence, both executed in about 1770–71, soon after his election to the Academy. One is an oil painting by Desportes and is included as a *frontispiece* to this book. On the back of this portrait there is a long notice in Messier's handwriting, enumerating all his titles: Member of the Royal Academy of Sciences, Astronomer of the Navy, Member of the Royal Society of London, the Royal Society of Science and Belles-Lettres of Prussia, the Academy of Sciences of Sweden, the Imperial Academy of St. Petersburg, the Institute of Bologna, the Society of Sciences of Holland, the Academy of Arts established in England, The Academy of Stockholm and the Academy of Uppsala – a revealing glimpse of his simple, and surely pardonable, vanity.

The other; a pastel by Ansiaume, also painted when Messier was about 40, shows him wearing conspicuously his Cross of the Legion of Honour – a strange anachronism as Messier did not receive this order until 1806.

Messier's true memorial, however, is his *Catalogue of Nebulae and Star Clusters* and we can take leave of him with the assessment of an eminent astronomer of our own time. As Harlow Shapley wrote in his own work on star clusters in 1930: 'For open and globular clusters as well as for bright nebulae of all kinds, the systematic listing by Messier in 1784 marked an epoch in the recording of observations . . . He is remembered for his catalogue; forgotten as the applause-seeking discoverer of comets.'

Montaigne de Limoges, who 'robbed' Messier of his 'thirteenth' comet, discovered three new comets. His first was on March 8th, 1772, and this appears to be the first recorded apparition of Biela's comet. The other two were found on Aug. 11th, 1774 and Oct. 18th, 1780, the last of these being discovered independently by Olbers on the same day.

Pierre Méchain is credited with having discovered eight new comets on the following dates: June 28th, 1781; Oct. 9th, 1781; March 11th, 1785; Jan. 17th, 1786 (the first recorded sighting of Encke's comet); April 10th, 1787; Jan. 9th, 1790 (Tuttle's comet); Aug. 7th, 1799; Dec. 26th, 1799.

Table 7. *List of some of the telescopes used by Messier*
In the *Connaissance des Temps* for 1807, Messier listed the refractors (*lunettes*) and reflectors (*télescopes*) which he used during the years 1765–69.

Instrument	Owner	Mag.
1. Ordinary refractor of 25 foot F.L.		138×
2. Achromatic refractor, 10½ foot F.L.	M. de Courtanvaux	120×
3. Achromatic refractor, 3¼ foot F.L. (Dollond)	Duc de Chaulnes	120×
4. Ordinary refractor of 23 foot F.L.		102×
5. Ordinary refractor of 30 foot F.L.	M. Baudouin	117×
6. Campani refractor	M. Maraldi	64×
7. Gregorian reflector ('Short') 6 feet F.L.	M. Lemonnier	110×
8. Gregorian reflector, 30 inches F.L. 6-inch aperture		104×
9. Newtonian reflector, 4½ foot F.L.		60×
10. Reflector, 1 foot F.L. 3-inch aperture	M. de Saron	44×
11. Refractor, 19 foot F.L. of Paris Observatory		76×

The Hôtel de Cluny as it appeared in the 18th century. The Observatory of the Marine can be seen as the wood and glass structure on top of the tower (Photo: Dr Owen Gingerich)

The Musée de Cluny as it appears today: the observatory was removed from the octagonal tower in the early part of the 19th century (Photo: Dr Owen Gingerich)

THE DISCOVERERS OF THE MESSIER OBJECTS

Many of the objects which Messier included in his first catalogue of 45 nebulae and star clusters, published in 1771, had previously been discovered by other observers – in fact, only about 17 or 18 of them can be established as having been first seen by him. In some cases, however, Messier made completely independent rediscoveries, and for all the objects, he endeavoured to obtain their accurate positions.

An example of Messier's rediscovery is that of M 1 which prompted him to start compiling his list. Messier came upon this object while looking for the comet of 1758, but it had been first seen by Dr John Bevis in 1731 who had marked it, with a few other nebulae, on his atlas, *Uranographia Britannica*, printed, but not published, in 1750. When Bevis recognised this entry in Messier's catalogue, he immediately wrote, on June 10th, 1771, to the French astronomer to inform him of his previous observation and Messier acknowledged this in the final edition of the catalogue, published in the *Connaissance des Temps* for 1784. During this correspondence, Bevis probably sent Messier a copy of his own atlas for Messier reveals, in his 1781 catalogue, that he used *Uranographia Britannica* during his observations of new nebulae and clusters.

In the latter half of Messier's catalogue, many of the objects were first found by Méchain at Versailles who passed on the information to his colleague: Messier then re-observed them and established their precise positions.

The table which follows gives the Messier Objects in order of the date of their discovery, together with the identity of their probable first discoverer. Not all of this information can be established with certainty but most of it has been obtained from original sources.

Table 8. *The Messier Objects in order of date of discovery*

M. No.	Date	Discoverer	Remarks
45	—	—	Known from earliest times. Mentioned by Hesiod about 1000 BC
44	—	—	Known to Hipparchus in 130 BC. Galileo saw it as a cluster, 1609
7	About AD 138	Ptolemy	(Pt. No. 567) 'Nebula following sting of Scorpius'
31	Before AD 905	—	Known to Al-Sufi as 'A little cloud'. S. Marius, telescope description, 1612
42	1610	N. Peiresc	Cysatus observed in 1618. Huyghens rediscovered in 1656
6	Before 1654	G. B. Hodierna	(Ha II. 4) De Chéseaux (No. 1) rediscovered in 1745–46
36	Before 1654	,,	(Ha II. 7) Le Gentil rediscovered in 1749
37	Before 1654	,,	(Ha II. 7) Messier rediscovered , Sept. 2nd 1764
38	Before 1654	,,	(Ha II. 7) Le Gentil rediscovered in 1749
41	Before 1654	,,	(Ha IV. 2) Flamsteed (Flam. No. 965) rediscovered Feb. 16th 1702

M. No.	Date	Discoverer	Remarks
47	Before 1654	G. B. Hodierna	(Ha IV. 1) Messier rediscovered Feb. 19th 1771. 'Missing' Object, Identified by Oswald Thomas, 1934
22	Aug. 26th 1665	A. Ihle	De Chéseaux (No. 17) 1745–46. Le Gentil observed in 1747. Lacaille observed 1751–2 (Lac. I. 12)
8	About 1680	Flamsteed	(Flam. No. 2446) (Cluster NGC 6530 only. Le Gentil disc. neb. 6523 in 1749)
11	1681	G. Kirch	First resolved by W. Derham, *c.* 1733
5	May 5th 1702	G. Kirch	
13	1714	Halley	
43	Before 1731	De Mairan	Part of Orion Nebula
1	1731	J. Bevis	Messier rediscovered, Sept. 12th 1758
16	1745–46	De Chéseaux	(De Ch. No. 4) Messier, June 3rd 1764
25	1745–46	,,	(De Ch. No. 5) Messier, June 20th 1764
35	1745–46	,,	(De Ch. No. 12) On Bevis' Atlas of 1750
71	1745–46?	,,	(De Ch. No. 13) Koehler, 1772–78
4	1745–46	,,	(De Ch. No. 19) (Lac. I. 9)
17	1745–46	,,	(De Ch. No. 20) Messier, June 3rd 1764
15	Sept. 7th 1746	J.-D. Maraldi	Messier, June 3rd 1764
2	Sept. 11th 1746	,,	Messier, Sept. 11th 1760
32	Oct. 29th 1749	Le Gentil	Messier, Aug. 3rd 1764
55	1751–52	Lacaille	(Lac. I. 14) Messier, July 24th 1778
69	1751–52	,,	(Lac. I. 11) Messier, Aug. 31st 1780
83	1751–52	,,	(Lac. I. 6) Messier, Mar. 18th 1781
3	May 3rd 1764	Messier	
9	May 28th 1764	,,	
10	May 29th 1764	,,	Bode (No. 33) Aug. 14th 1774
12	May 30th 1764	,,	Bode (No. 32) Aug. 14th 1774
14	June 1st 1764	,,	
18	June 3rd 1764	,,	
19	June 5th 1764	,,	
20	June 5th 1764	,,	
21	June 5th 1764	,,	
23	June 20th 1764	,,	
24	June 20th 1764	,,	Part of M.W. $1\frac{1}{2}°$ diam. (Not 6603)
26	June 20th 1764	,,	
27	July 12th 1764	,,	
28	July 27th 1764	,,	
29	July 29th 1764	,,	Bode (No. 69) Dec. 5th 1774
30	Aug. 3rd 1764	,,	
33	Aug. 25th 1764	,,	
34	Aug. 25th 1764	,,	Bode (No. 7) Sept. 2nd 1774
39	Oct. 27th 1764	,,	
46	Feb. 19th 1771	,,	
48	Feb. 19th 1771	,,	'Missing' Object. Identified by T. F. Morris in 1959
49	Feb. 19th 1771	,,	Obs. B. Oriani, April 22nd 1779
62	June 7th 1771	,,	

M. No.	Date	Discoverer	Remarks
50	April 5th 1772	Messier	Bode (No. 16) Dec. 2nd 1774. Possibly discovered by G. D. Cassini before 1771
110	Aug. 10th 1773	,,	Rediscovered by Caroline Herschel, Aug. 27th 1783. 'Additional' Object included by Glyn Jones, 1966
51	Oct. 13th 1773	,,	Main portion, NGC 5194, only. Companion neb. disc. Méchain in 1781
52	Sept. 7th 1774	,,	
81	Dec. 31st 1774	Bode	Bode (No. 17) Koehler (No. 8A) 1772–78
82	Dec. 31st 1774	,,	Bode (No. 18) Koehler (No. 8B) 1772–78
53	Feb. 3rd 1775	,,	Bode (No. 26) Messier, Feb. 26th 1777
92	Dec. 27th 1777	,,	Bode (No. 76) Messier, Mar. 18th 1781
54	July 24th 1778	Messier	
67	Before 1779	Koehler	(K. No. 19) Messier, April 6th 1780
57	Jan. 1779	Darquier	Messier, Jan. 31st 1779
56	Jan. 19th 1779	Messier	
64	April 4th 1779	Bode	Bode (No. 77) Messier, March 1st 1780
59	April 11th 1779	Koehler	Messier, April 15th 1779
60	April 11th 1779	,,	Oriani, April 12th 1779. Messier, April 15th 1779
58	April 15th 1779	Messier	
61	May 5th 1779	Oriani	
63	June 14th 1779	Méchain	
78	Beginning 1780	,,	
65	March 1st 1780	,,	
66	March 1st 1780	,,	
68	April 9th 1780	,,	
75	Aug. 27th 1780	,,	Messier, Oct. 5th 1780
72	Aug. 30th 1780	,,	Messier, Oct. 4th 1780
70	Aug. 31st 1780	Messier	
76	Sept. 5th 1780	Méchain	Messier, Oct. 21st 1780
74	End Sept. 1780	,,	Messier, Oct. 18th 1780
73	Oct. 4th 1780	Messier	Asterism of 4 stars
79	Oct. 26th 1780	Méchain	Messier, Dec. 17th 1780
77	Oct. 29th 1780	,,	Messier, Dec. 17th 1780
80	Jan. 4th 1781	Messier	Méchain, Jan. 27th 1781
97	Feb. 16th 1781	Méchain	Messier, March 24th 1781
85	March 4th 1781	,,	Messier, March 18th 1781
98	March 15th 1781	,,	Messier, April 13th 1781
99	March 15th 1781	,,	Messier, April 13th 1781
100	March 15th 1781	,,	Messier, April 13th 1781
84	March 18th 1781	Messier	
86	March 18th 1781	,,	
87	March 18th 1781	,,	
88	March 18th 1781	,,	
89	March 18th 1781	,,	
90	March 18th 1781	,,	

M. No.	Date	Discoverer	Remarks
91	March 18th 1781	Messier	'Missing' Object, identified as NGC 4548 by W. C. Williams, 1969
93	March 20th 1781	,,	
95	March 20th 1781	Méchain	Messier, March 24th 1781
96	March 20th 1781	,,	Messier, March 24th 1781
94	March 22nd 1781	,,	Messier, March 24th 1781
105	March 24th 1781	,,	'Additional' Object included by H. S. Hogg, 1947
101	March 27th 1781	,,	Messier, March 27th 1781
103	1781	,,	
104	May 11th 1781	,,	Messier, May 11th 1781. 'Additional' Object included by Flammarion, 1921
106	July 1781	,,	'Additional' Object included by H. S. Hogg, 1947
107	April 1782	,,	'Additional' Object included by H. S. Hogg, 1947
108	1781 or 1782	,,	'Additional' Object included by O. Gingerich, 1960
109	1781 or 1782	,,	'Additional' Object included by O. Gingerich, 1960

Note: If M 44 and M 45 are excluded as having no known discoverers, and M 40 and M 102 disregarded as 'non objects', then the total list comprises 106 objects. On the basis of this table the 'score' is as follows: Messier 41, Méchain 27, Hodierna 6, De Chéseaux 6, Bode 5, Le Gentil 3, Koehler 3, Kirsch 2, Maraldi 2, and 1 each to Ptolemy, Al-Sufi, Peiresc, Flamsteed, Ihle, Le Gentil, Halley, Bevis, De Mairan, Darquier and Oriani.

Appendix 1

Table 9. *Positions of the Messier Objects for epoch 2000.0*

M. No.	R.A. h	R.A. m	Dec. °	Dec. ′	M. No.	R.A. h	R.A. m	Dec. °	Dec. ′	M. No.	R.A. h	R.A. m	Dec. °	Dec. ′
1	5	34.5	+22	01	36	5	36.3	+34	08	71	19	53.7	+18	47
2	21	33.5	−00	49	37	5	52.2	+32	33	72	20	53.5	−12	32
3	13	42.2	+28	23	38	5	28.7	+35	50	73	20	59.0	−12	38
4	16	23.6	−26	31	39	21	32.3	+48	26	74	1	36.7	+15	47
5	15	18.5	+02	05	40	12	22.2	+58	05	75	20	06.1	−21	55
6	17	40.0	−32	12	41	6	47.0	−20	46	76	1	42.2	+51	34
7	17	54.0	−34	49	42	5	35.3	−05	27	77	2	42.7	−00	01
8	18	03.7	−24	23	43	5	35.5	−05	16	78	5	46.7	+00	04
9	17	19.2	−18	31	44	8	40.0	+20	00	79	5	24.2	−24	31
10	16	57.2	−04	06	45	3	47.0	+24	07	80	16	17.0	−22	59
11	18	51.1	−06	16	46	7	41.8	−14	49	81	9	55.8	+69	04
12	16	47.2	−01	57	47	7	36.6	−14	29	82	9	55.8	+69	42
13	16	41.7	+36	28	48	8	13.8	−05	48	83	13	37.0	−29	52
14	17	37.6	−03	15	49	12	29.8	+08	00	84	12	25.1	+12	53
15	21	30.0	+12	10	50	7	03.0	−08	21	85	12	25.4	+18	11
16	18	18.9	−13	47	51	13	29.9	+47	12	86	12	26.2	+12	57
17	18	20.8	−16	10	52	23	24.2	+61	36	87	12	30.8	+12	23
18	18	19.9	−17	08	53	13	12.9	+18	10	88	12	32.0	+14	25
19	17	02.6	−26	16	54	18	55.1	−30	28	89	12	35.7	+12	33
20	18	02.6	−23	02	55	19	40.0	−30	57	90	12	36.8	+13	10
21	18	04.7	−22	30	56	19	16.6	+30	11	91	12	35.4	+14	30
22	18	36.4	−23	54	57	18	53.6	+33	02	92	17	17.1	+43	08
23	17	56.9	−19	01	58	12	37.7	+11	49	93	7	44.6	−23	53
24	18	18.4	−18	25	59	12	42.0	+11	39	94	12	50.9	+41	07
25	18	31.7	−19	14	60	12	43.7	+11	33	95	10	44.0	+11	42
26	18	45.2	−09	24	61	12	21.9	+04	28	96	10	46.8	+11	49
27	19	59.6	+22	43	62	17	01.2	−30	07	97	11	14.9	+55	01
28	18	24.6	−24	52	63	13	15.8	+42	02	98	12	13.8	+14	54
29	20	24.0	+38	31	64	12	56.7	+21	41	99	12	18.8	+14	25
30	21	40.4	−23	11	65	11	18.9	+13	06	100	12	22.9	+15	49
31	0	42.7	+41	16	66	11	20.3	+13	00	101	14	03.2	+54	21
32	0	42.7	+40	52	67	08	51.4	+11	48	102	—	—	—	—
33	1	33.8	+30	39	68	12	39.5	−26	45	103	1	33.1	+60	42
34	2	42.0	+42	47	69	18	31.4	−32	21	104	12	40.0	−11	37
35	6	08.8	+24	20	70	18	43.2	−32	17	105	10	47.9	+12	35

M. No.	R.A.		Dec.	
	h	m	°	'
106	12	19.0	+47	18
107	16	32.5	−13	03
108	11	11.6	+55	40
109	11	57.7	+53	22
110	0	40.3	+41	41

YEAR MAPS OF THE MESSIER OBJECTS

Key to symbols
(Maps 1 to 6)

M OBJECTS

- Gaseous Nebulae
- Planetary Nebulae
- Globular Clusters
- Galactic Clusters
- M 1 (Crab Nebula)
- Elliptical Galaxies
- Normal Spiral Galaxies
- Barred Spiral Galaxies
- Irregular Galaxies

STARS

- Magnitude 1
- Magnitude 2
- Magnitude 3
- Magnitude 4

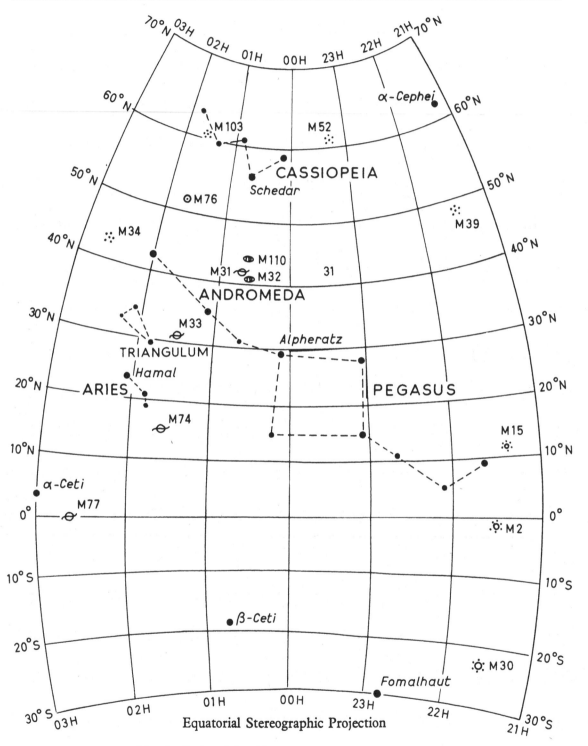

Equatorial Stereographic Projection

Map 1   Sept. — Oct. — Nov.

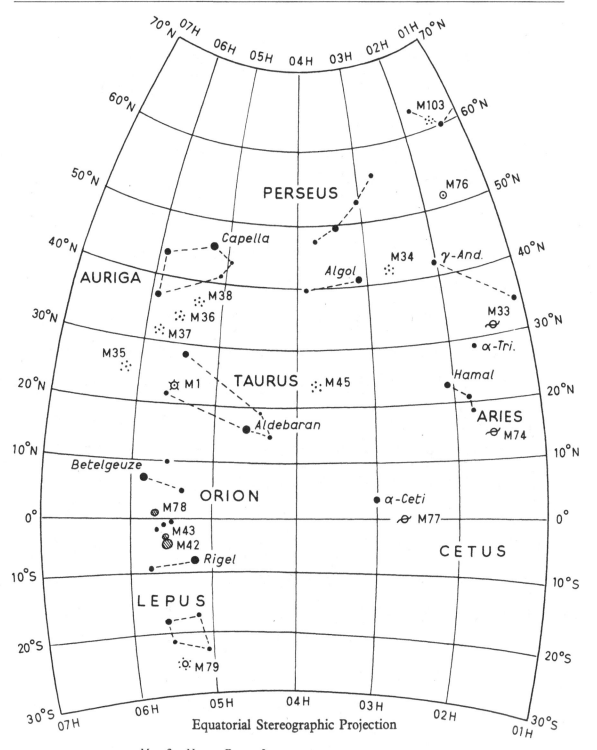

Map 2   Nov. – Dec. – Jan.

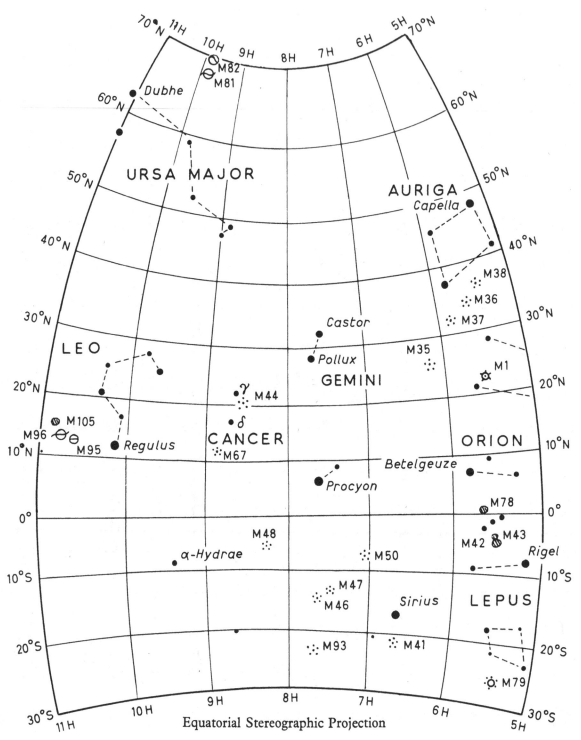

Equatorial Stereographic Projection

Map 3   Jan. – Feb. – Mar.

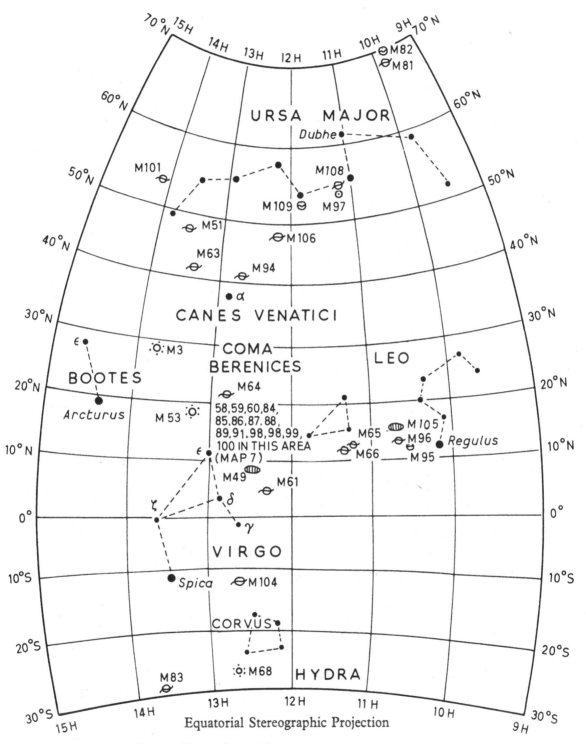

Map 4   Mar. – Apr. – May.

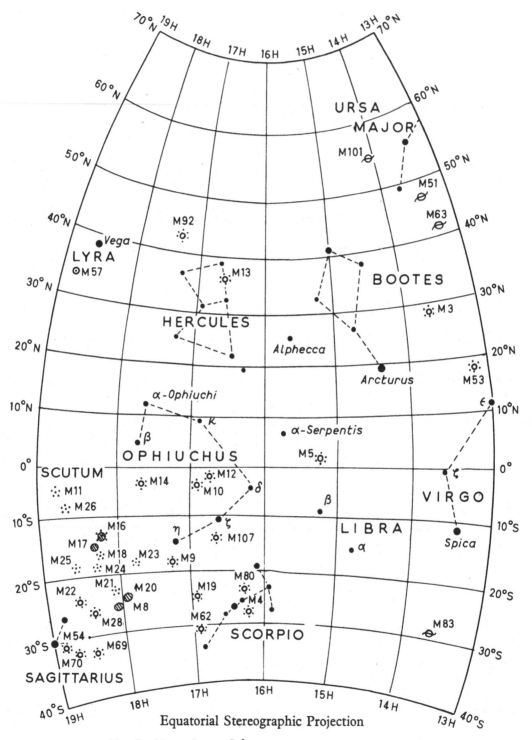

Equatorial Stereographic Projection

Map 5   May – June – July

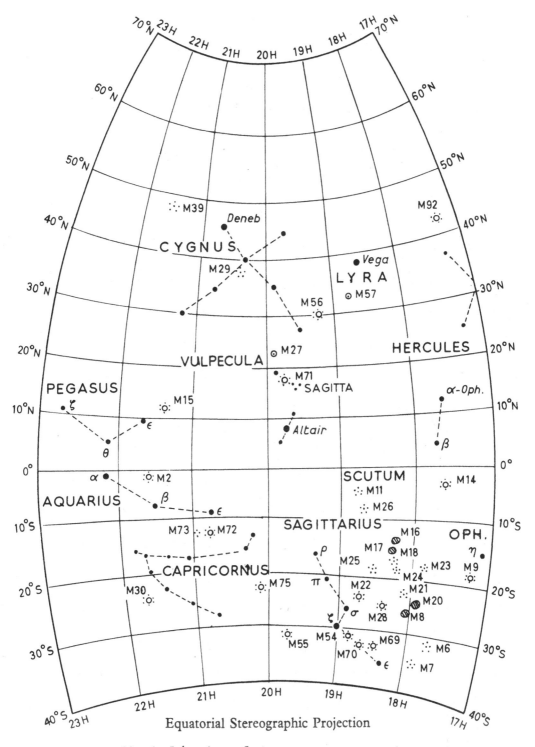

Equatorial Stereographic Projection

Map 6   July – Aug. – Sept.

Appendix 2 *A guide to the Messier Objects in the Virgo Group*

(To be used with Map 7)

The Virgo Group of galaxies contains at least 3000 members. Most of these are detectable only by photography but Wm Herschel found more than 300 nebulae in this area and some 30 or so of these may be seen in an 8-inch reflector in good conditions. Messier and Méchain between them discovered 17 members of this group, and probably would have found several more had they continued their searches.

As this book deals exclusively with the Messier Objects, no attempt will be made to explore the area fully but where other bright galaxies appear in the same (one degree) field, an identification and brief description of each is given in Chapter 4.

Even for exploring the Messier galaxies alone, a systematic approach is necessary if the observer is to avoid getting 'lost' and an outline scheme of search is given below. As many of the objects are fairly faint and diffuse, a dark sky and good seeing conditions are essential to obtain good results in this rich and rewarding field.

The natural line of attack would seem to be that of starting from *Denebola (β Leonis)*, moving east to *6 Comae Berenices*. This, however, means that the first galaxies encountered would be M 98 and M 99 which are about the faintest and most difficult to see of all the Messier galaxies in the group.

The safer course is to start from the east, at 3rd mag. ε *Virginis (Vindemiatrix)*. From this bright star move 5° W to pick up 5th mag. ϱ *Virginis* with 6th mag. *27 Virginis* close to the NW. These two stars provide an easily identifiable starting point.

1. From ϱ *Virginis* move 1½° N to pick up M 59 and M 60 which should appear in the same L.P. field, lying nearly E–W.
2. From M 59 move 1° W to M 58 which is the brightest of the Virgo galaxies. (If any difficulty is found in seeing these three galaxies it means that conditions are not clear enough. It will then be unlikely that any of the fainter objects will be found and the search may well be called off to await a more favourable night.)
3. From M 58 move slightly W and then 1° N to pick up M 89 and M 90 lying almost N–S and visible in the same L.P. field.
4. From the spindle-shaped M 90 now move 1° W and 1½° N to M 88 which has two small stars close to the south and is quite distinctive. The once 'missing' M 91 can now be found by moving 1° E and slightly N from M 88. This object is rather faint and its location should be checked by depressing the telescope a little more than 1° S to reveal M 90 again. Return to M 88.
5. From M 88 move back to M 89 and M 90 and from the more southerly M 89 move slightly S and 1¼° W to M 87 which is round and bright.
6. From M 87 move another 1¼° W and ½° N to M 84 and M 86 which appear in the same L.P. field oriented approximately E–W.
7. The next step is a longer one. From M 84 move 20′ W and then 3° N to M 100 which is rather more easily seen than its published magnitude of 10.6 would imply.

8. From M 100 move $\frac{1}{2}°$ W and 2° N to pick up the 5th mag. star, *11 Comae Berenices*. From this star move $\frac{1}{2}°$ N and just over 1° E to M 85 which, being small, may require higher magnification to see well.

9. From M 85 which is the most northerly of the Messier galaxies in the Virgo Group, return to the star *11 Comae Berenices* and then move 3° S and $1\frac{1}{4}°$ W to the 5th mag. star, *6 Comae Berenices*.

10. From *6 Comae Berenices*, M 98 is a little less than 1° due W and M 99 a little less than 1° SE. The former is very pale and may need averted vision to see at all: M 99 is a little brighter but also pale.

This excursion covers 14 of the 16 accepted Messier Objects in this region: to pick up the remaining two it is necessary to return to the starting point of the two stars, ϱ and *27 Virginis*.

11. From ϱ *Virginis*, move $2\frac{1}{4}°$ S and 3° W to M 49 which is bright and 'pearly' and very easy to see.

12. The last member* is M 61 which can be found from M 49 by moving $3\frac{1}{2}°$ S and 2° W. It is a little faint and pale and if the object proves elusive it may be located by examining the area midway between the two stars *16 Virginis* (5th mag.) and *17 Virginis* (6th mag.).

   If the observer loses his bearings at this point, a fresh start may be made from *Vindemiatrix*. From this bright star move $7\frac{1}{2}°$ S and $1\frac{1}{2}°$ W to $3\frac{1}{2}$ mag. δ *Virginis* and then sweep a further 9° W to 5th mag. *16 Virginis* again. M 61 is just over 1° N and $\frac{1}{2}°$ E of this star.

The observer with a small telescope, making his first attempt at exploring the Virgo Group should exert every effort to identify each object positively before moving on to the next. The telescope drawings and detailed descriptions on the pages of the individual Messier Objects should be used for this purpose.

The 'route' provided in this brief guide provides steps of more or less increasing difficulty and having completed the 'course' the observer should then go over it once, or several times, again. The experience thus gained will be invaluable and if he subsequently possesses a 6-inch or larger telescope, he can use the now familiar Messier Objects as starting points to explore the many fainter galaxies in this wonderful region.

---

* M 104 is also a true member of the Virgo Group but so far south of the others that it is dealt with separately (see M 104). M 86 is not a true member and M 98 probably a non-member (see M 86 and M 98).

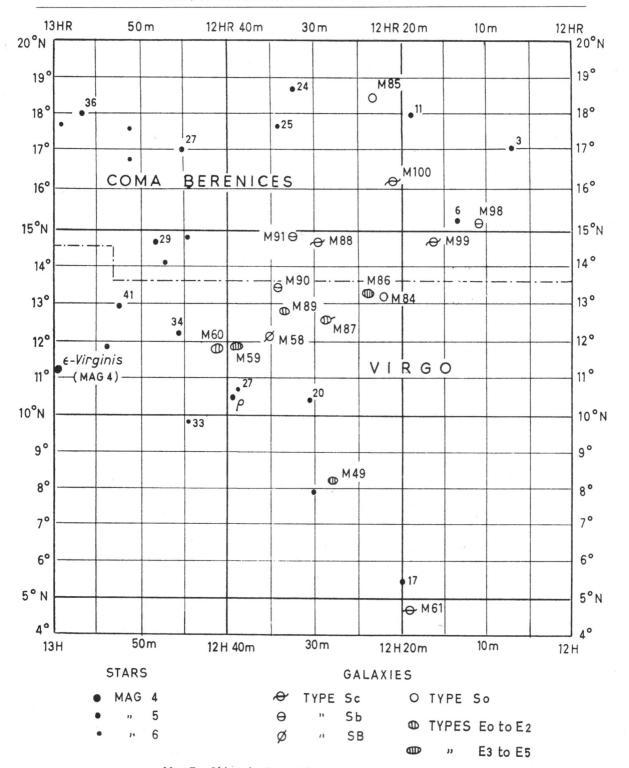

Map 7   Objects in Coma–Virgo

MAP 8

Appendix 3 Messier Objects 16 to 20 hrs. R.A. Map 8 10° to 40° S. Dec

(*In Sagittarius: Ophiuchus: Scorpius: Serpens*)

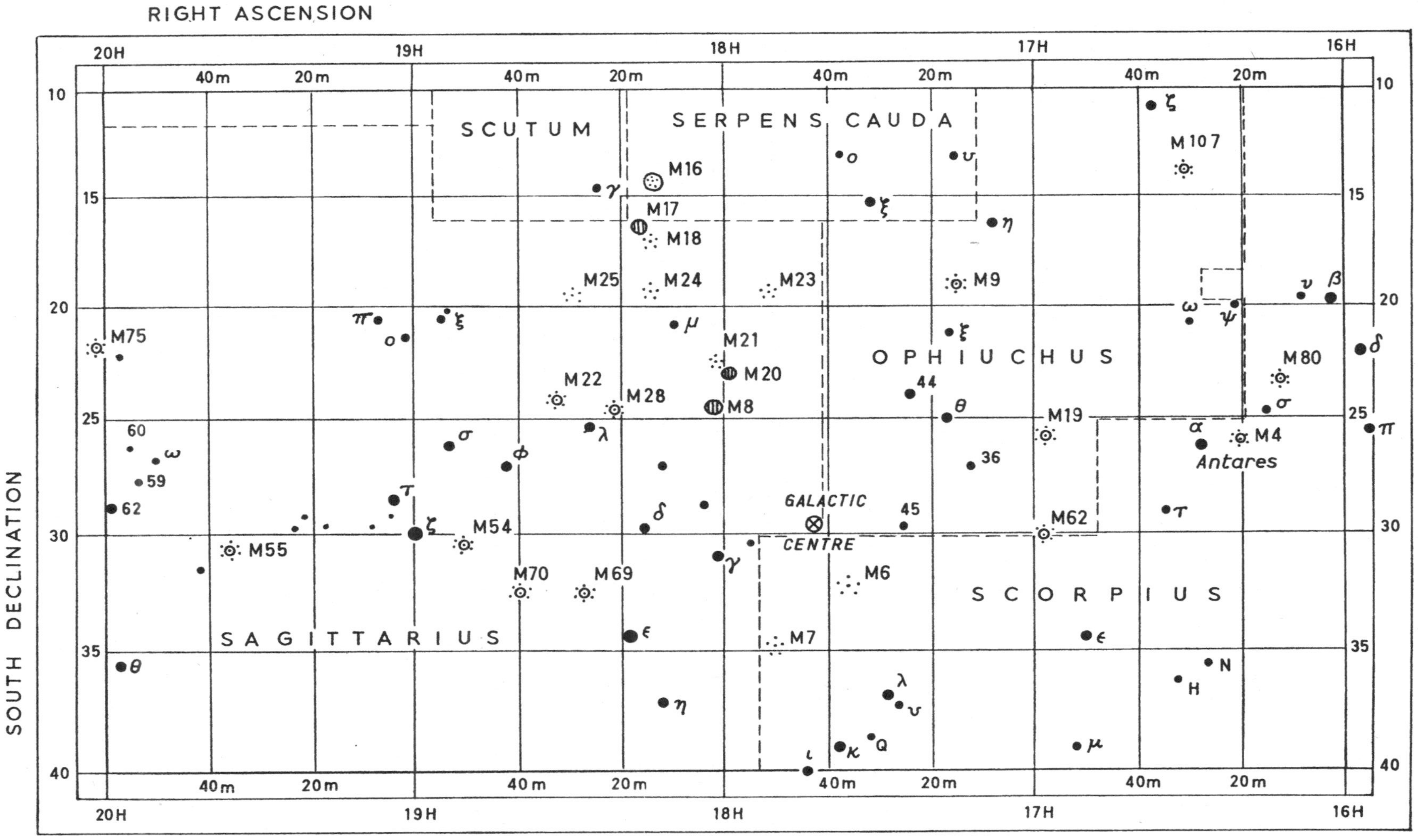

RIGHT ASCENSION

SOUTH DECLINATION

SCUTUM

SERPENS CAUDA

OPHIUCHUS

SCORPIUS

SAGITTARIUS

GALACTIC CENTRE

Antares

M107   M80   M4   M19   M62   M9   M6   M7   M23   M16   M17   M18   M24   M25   M20   M8   M21   M28   M22   M54   M55   M69   M70   M75

(For Key to Symbols see p. 379)

Appendix 4 Catalogue of stars in the Pleiades

Table 10. *The Pleiades*

H.D. Cat.	BOSS.G.C.	RA 1950 h. m. s.	Dec. ° ′ ″	Vis. mag.		
23630	4541	3 44 30.4	N. 23 57 08	2.96	*Alcyone*	η *Tauri*
23850	4586	3 45 11.0	23 54 07	3.80	*Atlas*	*27 Tauri*
23302	4477	3 41 54.1	23 57 28	3.81	*Electra*	*17 Tauri*
23408	4500	3 42 50.8	24 12 47	4.02	*Maia*	*20 Tauri*
23480	4512	3 43 21.2	23 47 39	4.25	*Merope*	*23 Tauri*
23338	4486	3 42 13.6	24 18 43	4.37	*Taygeta*	*19 Tauri*
23862	4587	3 46 12.4	23 59 07	5.00 var.	*Pleione*	*28 Tauri*
23288	4475	3 41 49.5	24 08 01	5.43	*Celaeno*	*16 Tauri*
23753	4564	3 45 22.9	23 16 09	5.51		*HR 1172*
23324	4485	3 42 10.4	24 41 02	5.63		*18 Tauri*
23432	4502	3 42 55.4	24 24 00	5.85	*Asterope*	*21 Tauri*
23923	4603	3 46 45.2	23 33 40	6.11		*HR 1183*
23629		3 44 22.4	23 57 47	6.29		*24 Tauri*
23441		3 43 03.9	24 22 24	6.43		*22 Tauri*
23712*		3 45 07.0	24 50 09	6.46	x	
23873		3 46 22.6	24 13 47	6.60		
23964		3 46 59.6	23 41 53	6.74	A.D.S. 2795	
23642		3 44 30.6	24 08 08	6.81		
23568		3 44 00.3	24 22 00	6.82		
23410		3 42 51.5	22 59 32	6.85	A.D.S. 2748	
24076		3 47 53.7	23 48 44	6.93		
23763		3 45 31.1	24 11 36	6.95		
23632		3 44 22.7	23 29 02	6.99		
23387		3 42 39.0	24 10 51	7.18		
23631		3 44 25.9	23 45 42	7.26	A.D.S. 2767	
23489		3 43 28.5	24 06 04	7.35		
24013*		3 47 28.8	24 20 42	7.42	x	
23155		3 40 43.6	24 55 28	7.51		
23872		3 46 17.6	24 14 43	7.52		
23948		3 46 57.5	24 11 54	7.54		

* Are not members of Pleiades cluster.

    A.D.S. 2748 has 11 mag. comp. P.A. 339°, dist. 3″.7

    A.D.S. 2767 has 10 mag. comp. P.A. 266°, dist. 5″.6

    A.D.S. 2767 has 10 mag. comp. P.A. 239°, dist. 3″.1 Q Triple

    A.D.S. 2795 has 11 mag. comp. P.A. 236°, dist. 10″.0 Q

(Adapted from a table in the *Handbook* of the British Astronomical Association, 1962, p. 52.)

Map 9   **The Pleiades**

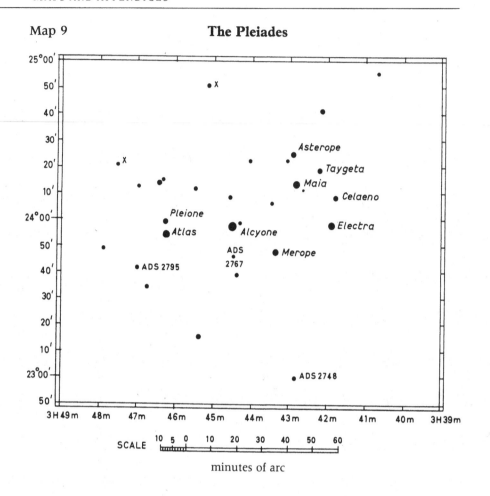

SCALE

minutes of arc

Appendix 5 Photographs of Messier Objects

INTRODUCTION

The photographs on the following pages include all the objects in Messier's catalogue except M 40 and M 102. No attempt has been made to keep to any sort of uniform scale: the object, in fact, has been to include a wide variety, from the first, historic photographs taken by Isaac Roberts, to those obtained by the 200-inch Mt Palomar telescope. Many of the photographs were taken by amateur astronomers of outstanding skill in this field.

Dr Hans Vehrenberg, well known for his fine *Photographic Star Atlas* and *Atlas of Deep-Sky Splendours*, has kindly made available a number of photographs taken with the Schmidt camera at his private observatory at Falkau. The main mirror of this fine instrument has a diameter of 45 cm, the corrector plate, 30 cm diameter and the telescope's focal length is 101 cm. The photo of M 25 was taken with a different instrument, a 62.5 cm F.L. Schmidt camera at Harvard's Boyden Station in South Africa.

Mr Christos Papadopoulos of Westcliff, Johannesburg, South Africa, has generously contributed many photographs, especially of the more southerly objects. Most of these were taken with a 1000 mm Zeiss Mirotar Lens and Contarex camera or at the 192-inch focus of a Tinsley 12-inch Cassegrain reflector. The film used was Spectro 103 a-F which has an equal sensitivity over the visual range. (All photographs are orientated with North down and East to the right.)

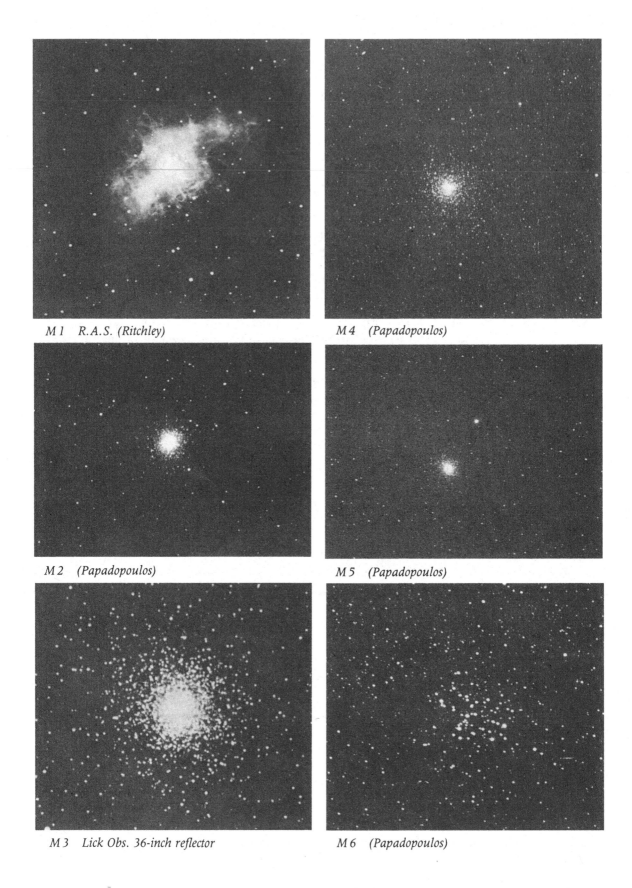

M 1   R.A.S. (Ritchley)

M 4   (Papadopoulos)

M 2   (Papadopoulos)

M 5   (Papadopoulos)

M 3   Lick Obs. 36-inch reflector

M 6   (Papadopoulos)

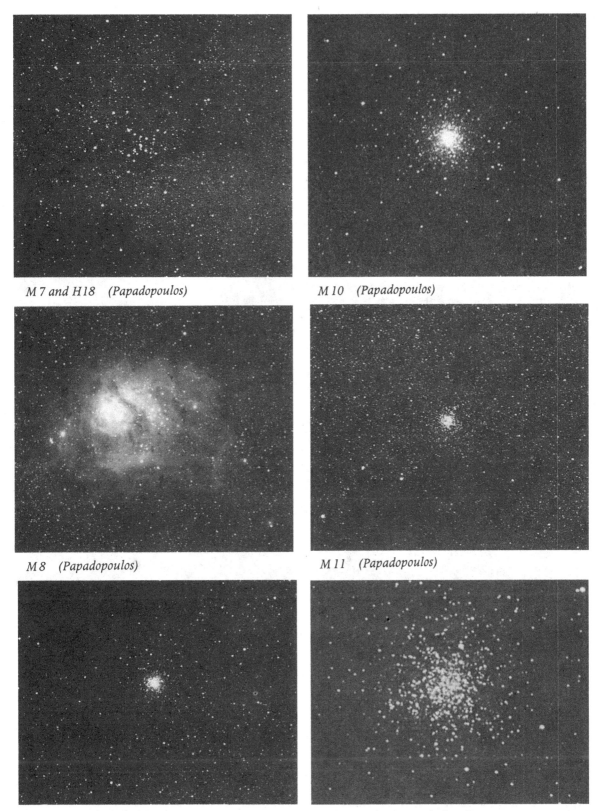

M 7 and H18   (Papadopoulos)

M 10   (Papadopoulos)

M 8   (Papadopoulos)

M 11   (Papadopoulos)

M 9   (Papadopoulos)

M 12   Lick Obs. 36-inch reflector

M 13   Mt. Wilson and Palomar Obs.

M 16   (Papadopoulos)

M 17   (Papadopoulos)

M 14   David Dunlap Obs. (Dr H. S. Hogg)

M 15   Yerkes Obs. (R.A.S.)

M 17 and M 18   (Papadopoulos)

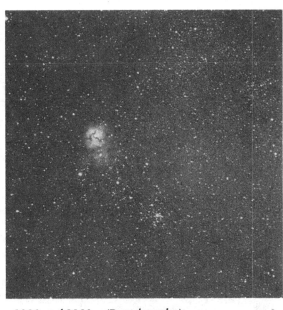

M 20 and M 21   (Papadopoulos)

M 19   (Papadopoulos)

M 22   (Papadopoulos)

M 20   (Papadopoulos)

M 23   (Papadopoulos)

M 24 with NGC 6603   (Papadopoulos)

M 26   (Papadopoulos)

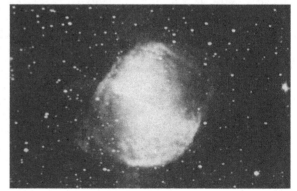

M 25   (Vehrenberg)

M 27   Lick Obs. 36-inch reflector

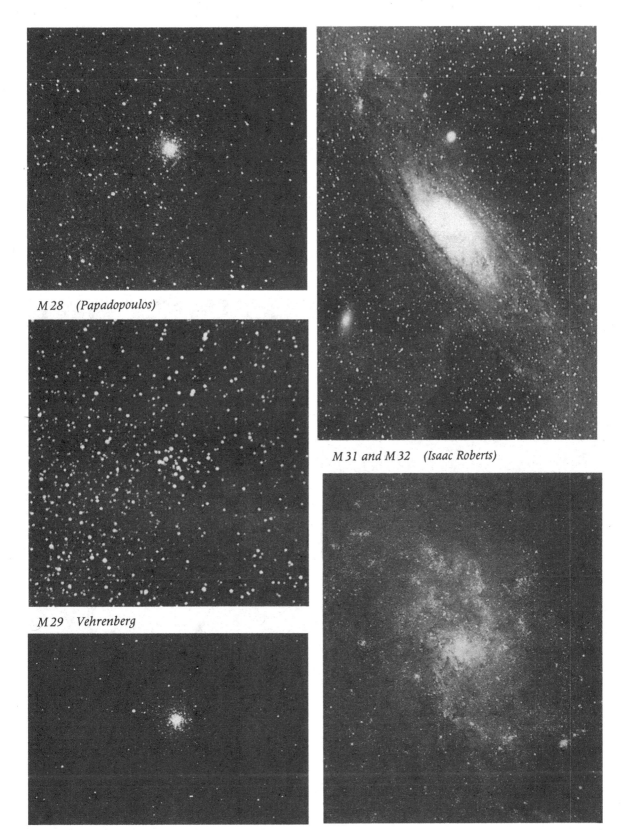

M 28   (Papadopoulos)

M 29   Vehrenberg

M 30   (Papadopoulos)

M 31 and M 32   (Isaac Roberts)

M 33   Mt. Wilson and Palomar Obs.

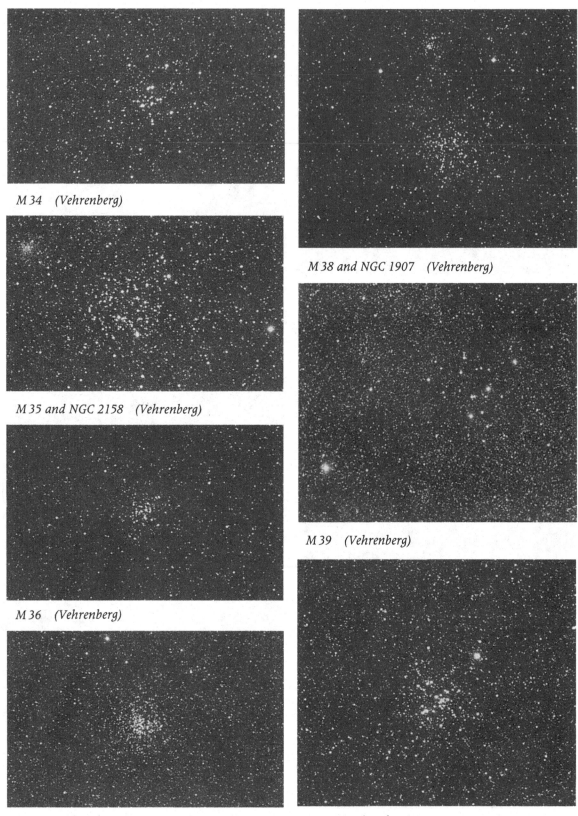

M 34   *(Vehrenberg)*

M 35 and NGC 2158   *(Vehrenberg)*

M 36   *(Vehrenberg)*

M 37   *(Vehrenberg)*

M 38 and NGC 1907   *(Vehrenberg)*

M 39   *(Vehrenberg)*

M 41   *(Vehrenberg)*

M 42 and M 43   (Papadopoulos)

M 45   Lick Obs. (Babcock)

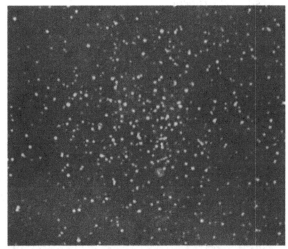

M 46 with NGC 2438   (Isaac Roberts)

M 44   (Vehrenberg)

M 47   Mt. Wilson and Palomar Obs.

M 48   (Vehrenberg)

M 51   Lick Obs. 36-inch reflector

M 49   (Papadopoulos)

M 52   (Vehrenberg)

M 50   (Vehrenberg)

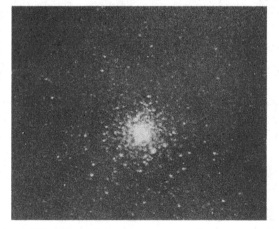

M 53   (Isaac Roberts)

M 54   (Papadopoulos)

M 55   (Papadopoulos)

M 56   (Isaac Roberts)

M 57   R.A.S. (J. S. Plaskett)

M 58   Harvard Obs. (Sky and Telescope)

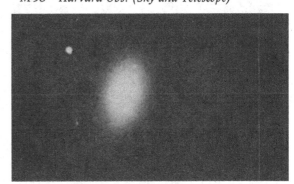

M 59   Mt. Wilson and Palomar Obs.

M 60 and NGC 4647   Mt. Wilson Obs.

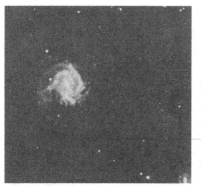

M 61   Licks Obs. 36-inch reflector   M 62   (Papadopoulos)   M 63   (De Kerolyr) (R.A.S.)

M 64   Lick Obs. 36-inch reflector   M 65 and M 66   (De Kerolyr) (R.A.S.)

M 67   Mt. Wilson and Palomar Obs.

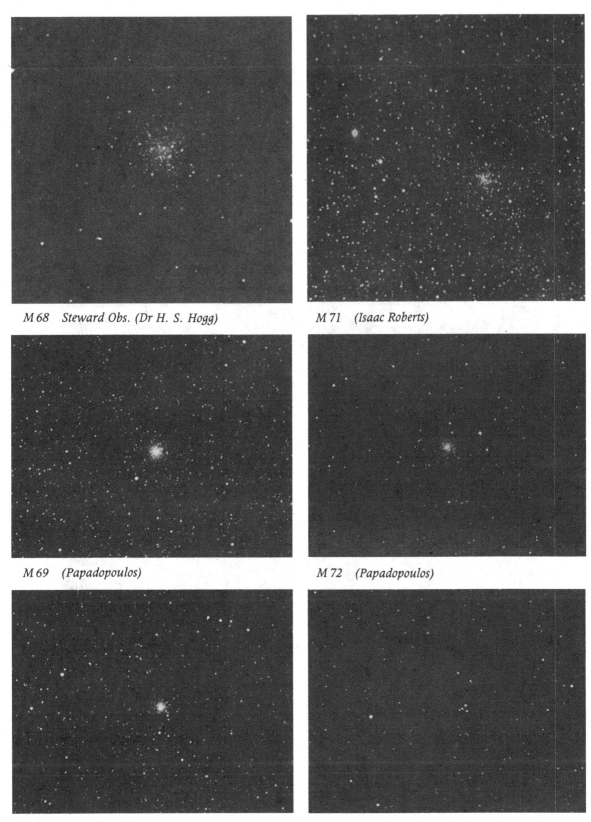

M 68   Steward Obs. (Dr H. S. Hogg)

M 71   (Isaac Roberts)

M 69   (Papadopoulos)

M 72   (Papadopoulos)

M 70   (Papadopoulos)

M 73   (Papadopoulos)

M 74   Lick Obs. 36-inch reflector

M 77   Lick Obs. 36-inch reflector

M 75   (Papadopoulos)

M 78   Lick Obs. 36-inch reflector

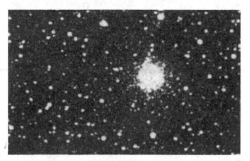

M 79   (Vehrenberg)

M 76   Lick Obs. 36-inch reflector

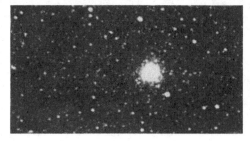

M 80   (Vehrenberg)

M 81 and M 82, with NGC 3077 and NGC 2976
Mt. Wilson and Palomar Obs.

M 83   Mt. Wilson Obs., 100-inch reflector

M 86   Harvard Obs. (Sky and Telescope)

M 84 and M 86   Harvard Obs. (Sky and Telescope)

M 87   Mt. Wilson and Palomar Obs.
200-inch reflector

M 88   Lick Obs. 36-inch reflector

M 85 and NGC 4394   (Papadopoulos)

M 89   Harvard Obs. (Sky and Telescope)

M 90   Mt. Wilson and Palomar Obs.
200-inch reflector

M 92   (Vehrenberg)

M 93   (Vehrenberg)

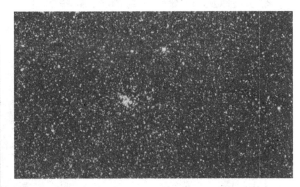

M 94   Lick Obs. 36-inch reflector

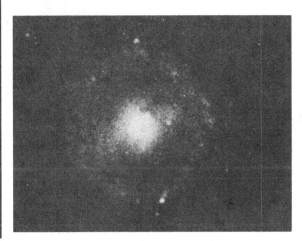

M 95   Lick Obs. 36-inch reflector

M 91

M 96   Mt. Wilson and Palomar Obs.
200-inch reflector

M 99   Lick Obs. 36-inch reflector

M 97   Mt. Wilson Obs. 100-inch reflector

M 100   Lick Obs. 36-inch reflector

M 98   Lick Obs. 36-inch reflector

M 101   Mt. Wilson and Palomar Obs.

M 103   (Vehrenberg)

M 106

M 104   Mt. Wilson and Palomar Obs.
200-inch reflector

M 107   (NGC 6171) Globular cluster

M 105

M 108   Lick Obs. 36-inch reflector

M 109

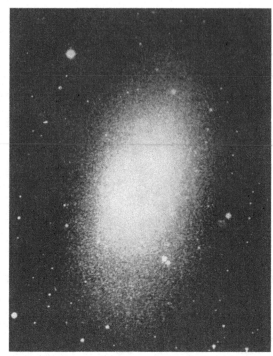

M 110   Mt Wilson and Palomar Obs.
200-inch reflector

Appendix 6 Bibliography

The following is a short list of the main books and publications consulted; the list is arranged in relation to the chapters of the book but several sources have provided material which was used in more than one chapter.

In each section the arrangement is in approximate chronological order of date of publication. Some abbreviations used are:

Phil. Trans. – Philosophical Transactions of the Royal Society.
J.B.A.A. – Journal of the British Astronomical Association.
W.S.Q.J. – Quarterly Journal of the Webb Society.
R.A.S.Q.J. – Quarterly Journal of the Royal Astronomical Society.

Introduction   *Description des Etoiles Fixes . . . par Al-Sufi*. H. C. F. C. Schjellerup. St Petersburg, 1874.
The Sidereal Messenger of Galileo Galilei. Trans. E. S. Carlos. London, 1880.
Ptolemy's Catalogue of Stars. C. H. F. Peters and E. B. Knobel. Carnegie Institution of Washington, 1915.
'Messier's Catalogue'. Harlow Shapley and Helen Davis. *Proceedings* of the Astronomical Society of the Pacific, 1917. (Also printed in *The Observatory*, Vol. 41.)
'The Additional Messier Objects'. Helen Sawyer Hogg. *Journal* of the R.A.S. of Canada. Sept–Oct., 1947.
'Charles Messier and his Catalogue'. Owen Gingerich. *Sky & Telescope*, Vol. 12, 1953; Vol. 13, 1954.
'The Missing Messier Objects'. Owen Gingerich. *Sky & Telescope*, Vol. 20, 1960.

Chapter 1   *Splendour of the Heavens*. T. E. R. Phillips and W. H. Steavenson. London, 1923.
Stars in the Making. Cecilia Payne-Gapochkin. London, 1953.
Discovery of the Universe. Gerard de Vaucouleurs. London, 1957.
The Structure and Evolution of the Universe. G. J. Whitrow. London, 1959.
Evolution of Stars and Galaxies. Harvard Lectures, 1958. Walter Baade, Harvard University Press, 1963.
'The Search for M 91.' W. C. Williams, *W.S.Q.J.* No. 11, April, 1972.

Chapter 2   *Star Clusters*. Harvard Observatory Monographs, No. 2. Harlow Shapley. New York, 1930.
Galaxies. Harlow Shapley. Harvard books on Astronomy, 1943.
The Size of the Universe. F. J. Hargreaves. Pelican Books, 1948.
The Modern Universe. Raymond A. Lyttleton. London, 1956.
Gaseous Nebulae. L. H. Aller. London, 1956.
The Inner Metagalaxy. Harlow Shapley. London, 1957.
Nebulae and Galaxies. G. Abetti and M. Hack. Trans. V. Barocas. London, 1963.
'The Size of the Galaxy.' R.v.d.R. Woolley. *J.B.A.A.*, Vol. 73, No. 4, 1963.

Chapter 3   'Mounting a Barlow Lens'. F. H. Thornton. *J.B.A.A.*, Vol. 60, No. 3, 1950.
'Constructing an Observing Chair'. F. J. Sellers. *J.B.A.A.*, Vol. 63, No. 1, 1952.

'The Night Sky', M. J. Hendrie. *J.B.A.A.*, Vol. 74. No. 4. 1964.

'Observatories for Amateurs'. M. J. Hendrie. *J.B.A.A.*, Vol. 74, No. 5, 1964.

Amateur Astronomer's Handbook. J. B. Sidgwick. London, 1963.

Astronomy for Amateurs. James Muirden. London, 1969.

Astronomy with Binoculars. James Muirden. London, 1976.

Chapter 4   (Historical)

'Of Nebulae or Lucid Spots among the Fixt Stars'. Edmond Halley. *Phil. Trans.* No. 347, 1715.

'Observations of . . . Nebulous Stars'. Wm Derham. *Phil. Trans.* No. 428, 1730.

'Observations of the Comet of 1746'. M. Maraldi. *Mem.* Roy. Acad. of Sciences. Paris, 1746.

'Nebulae of the Southern Sky'. N. L. de la Caille. *Mem.* Roy. Acad. of Sciences. Paris, 1755.

'Upon some Newly-discovered Nebulous Stars . . .'. J. E. Bode. *Astronomisches Jahrbuch* for 1779. Berlin, 1777.

'Catalogue of Nebulae and Star Clusters'. Charles Messier. *Connaissance des Temps* for 1784. Paris, 1781.

'Positions of Three new Nebulous Stars . . . in Virgo'. Oriani-Bode. *Astronomisches Jahrbuch* for 1784. Berlin, 1782.

Siderum Nebulosorum. Observationes Havnienses. H. L. D'Arrest. Copenhagen, 1867.

Handbook of Descriptive Astronomy. Geo. F. Chambers. Oxford, 1877.

'Observations of Nebulae and Star Clusters'. M. G. Bigourdan. *Paris Annales (Observationes) 1884*, Paris 1891. (Contains de Chéseaux' List.)

A Cycle of Celestial Objects. W. H. Smyth (Revised by G. F. Chambers). Vol. II., Oxford, 1898.

History of Astronomy during the Nineteenth Century. A. M. Clerke. 3rd Edn. London, 1897.

Celestial Objects for Common Telescopes. T. W. Webb. (Revised T. E. Espin) Vol. 2 (6th Edn.). London, 1898.

Star Names and their Meanings. R. H. Allen. 1899. (Reprinted New York, 1963.)

'Variable Stars in Star Clusters'. *Harvard Annals*, No. 38, 1902.

Studies in Astronomy. J. E. Gore. London, 1904. (Contains a survey of most of the Messier Objects. Previously printed in *The Observatory*, Vol. 25, 1902.)

Problems in Astrophysics. A. M. Clerke. London, 1903.

The System of the Stars. A. M. Clerke. (2nd Edn.). London, 1905.

Astronomical Essays. J. E. Gore. London, 1907.

Studies of the Nebulae. H. D. Curtis. Lick Obs. Pub. XIII. 1918.

'The Catalogue of Charles Messier'. C. Flammarion. *Bulletin of the Astr. Soc. of France*, Vols. 31–35, 1917–21.

The Search for the Nebulae. K. Glyn Jones. Chalfont St Giles, 1975.

Chapter 4   (Modern)

'A Half-Century of Globular Clusters'. Harlow Shapley. *Popular Astronomy*, May, 1949.

'Distances of Globular Clusters'. T. D. Kinman. *Mon. Not, Astr. Soc. Southern Africa.* Vol. XVII, No. 3, 1958.

'Star Clusters'. H. S. Hogg, *Handbuch der Physik.* Vol. 53, 1959.

'Properties of Galactic Clusters'. A. Wallenquist. *Annals of Uppsala Astronomical Observatory*, Band 4, No. 6, 1959.

Catalogue of External Galaxies. E. Holmberg. Stockholm, 1964.

Principles of Astronomy. S. P. Wyatt. Boston, 1964.

Catalogue of Galactic Planetary Nebulae. L. Perek and L. Kohoutek. Prague, 1967.

General Catalogue of Globular Star Clusters of our Galaxy. B.V. Kukarkin. Moscow, 1974.

Second Reference Catalogue of Bright Galaxies. de Vaucouleurs, de Vaucouleurs and Corwin. Austin, Texas, 1976.

The Messier Album. J. H. Mallas and E. Kramer. Cambridge, 1978.

Burnham's Celestial Handbook. Robert Burnham Jr., New York, 1978.

The Crab Nebula. Simon Mitton. London, 1979.

Webb Society Deep-Sky Observer's Handbook. Ed. K. Glyn Jones.

    Vol. 1 (Double Stars) 2nd Edn. Hillside, N.J. 1986.

    Vol. 2 (Planetary & Gaseous Nebulae). Hillside N.J. 1979.

    Vol. 3 (Open & Globular Clusters). Hillside N.J. 1980.

    Vol. 4 (Galaxies). Hillside N.J. 1981.

    Vol. 5 (Clusters of Galaxies). Hillside N.J. 1982.

Amateur Astronomer's Catalogue, Vol. 1. R. J. Morales. Tucson, 1986.

Chapter 5

Élements d'Astronomie. Jacques Cassini. Paris, 1740.

'Catalogues des Nebuleuses et des Amas d'Étoiles'. Charles Messier, *Mémoires de l'Academie Royale des Sciences.* Paris, 1771.

'Observations at Cluny Observatory, 1752 to 1760'. Charles Messier. *Connaissance des Temps* for 1810. Paris, 1807.

Mémoires pour servir a l'histoire des Sciences. J. D. Cassini. Paris, 1810.

Histoire de l'Astronomie au 18me Siecle. J. B. J. Delambre. Paris, 1827.

Histoire Abregée de l'Astronomie. E. Lebon. Paris, 1899.

Histoire de l'Astronomie. E. Doublet. Paris, 1922.

Histoire de l'Astronomie. F. Boquet. Paris, 1925.

Hevelius, Flamsteed and Halley. E. F. MacPike. London, 1937.

William Herschel. J. B. Sidgwick. London, 1953.

The French Revolution. M. J. Sydenham. London, 1959.

William Herschel and the Construction of the Heavens. M. A. Hoskin. London, 1963.

The Astronomers. Colin Ronan. London, 1964.

The Planet Uranus. A. F. O'D. Alexander. London, 1965.

'G. B. Hodierna's Observations of Nebulae . . .'. G. F. Serio, L. Indorato and P. Nastasi. *Journal for the History of Astronomy* No. 45, Feb. 1985.

'Some Notes on Hodierna's Nebulae'. K. Glyn Jones. *Journal for the History of Astronomy* No. 50, Aug. 1986.

'The First Deep-Sky Observer?' K. Glyn Jones. *W.S.Q.J.* No. 63, Jan. 1986.

General Catalogues, Atlases, etc

Atlas Coelestis. J. Flamsteed. 1753.

Uranographia Britannica. John Bevis. 1750. (Reprinted in 1818).

Uranographia. J. E. Bode. Berlin, 1801.

'General Catalogue of Nebulae and Clusters of Stars.' J. F. W. Herschel. *Phil. Trans.,* 1864.

Photographs of Stars, Star Clusters and Nebulae. Isaac Roberts. Vol. I, 1893; Vol. II, 1899.

New General Catalogue of Nebulae and Clusters of Stars. J. L. E. Dreyer. 1888. *Index Catalogue,* 1895. *Second Index Catalogue,* 1908. (Issued together as Royal Astronomical Society Reprint, 1962.)

Photographs of the Nebulae. Lick Obs. Pub. No. VIII, 1908.

Atlas Coeli 1950.0. Antonin Becvar, Prague, 1958.

Atlas Coeli II. Katalog 1950.0. Antonin Becvar, Prague, 1964.

Atlas of Deep-Sky Splendours. H. Vehrenberg, Dusseldorf, 1967.

Norton's Star Atlas and Reference Handbook, Edinburgh, 1968.

Catalogue of the Universe. P. Murdin. D. Allen and D. Malin. Cambridge Univ. Press, 1979.

Revised New General Catalogue . . . Sulentic and Tifft. Tucson, 1980.

Atlas of Galactic Nebulae. Neckel and Vehrenberg. Dusseldorf. Vol. I, 1986; Vol. II, 1987.

Sky Atlas 2000.0. Wil Tirion. Cambridge Univ. Press, 1988.

Miscellaneous *Handbook of Descriptive and Practical Astronomy.* G. F. Chambers, 3rd Edn. Oxford, 1877.

Spherical Astronomy. W. M. Smart. 4th Edn. Cambridge Univ. Press, 1949.

History of the Telescope. H. C. King. London, 1955.

Handbook of Aviation Meteorology. H.M.S.O. London, 1961.

'The Messier Objects'. R. H. Garstang. B.A.A. *Handbook,* 1964.

Positional Astronomy. D. McNally. London, 1974.

Astronomical Scrapbook. J. Ashbrook. Cambridge Univ. Press, 1984.

Index

I. Name index

References in **bold** type apply to biographical notes in Chapter 5. *Italics* refer to illustrations or their captions.

Abell, G. O. 24
Adams, Penn & Seaton 92
Allen, D. A. 229
Allen, R. H. 167
Aller, L. H. 120
Al-Sufi (903–986) 1, 127, 168
Apian 322
Aratus (*c.* 315–245 BC) 165
Arp, H. C. 59, 62, 67, 87, 130, 249
Aubert, A. 366
Auwers, G. F. A. (1838–1915) 11, 169, 244, 313
Auzout, A. (1622–91) 338
Awadalla, N. S. & Budding, E. 166

Baade, W. (1893–1960) 10, 29, 54, Ch. 4 *passim*
Baade & Spitzer 28
Bailey, S. I. (1854–1931) 107, 206, 311
Bailly, J. S. (1736–93) 328, 351, 464
Balanowsky, I. 258
Baldwin, J. E. & Smith, F. G. 258
Barnard, E. E. (1857–1923) 93, 113, 159, 197, 240, **312–13**
Bayer, J. (1572–1625) 22, 155, 327
Becvar, A. Ch. 4. *passim*
Bennett, J. C. 251
Benoit 360
Bernouilli, J. (1744–1807) 14, 240, 293, 361
Bertrand, Mme 365
Bessel, F. W. (1784–1846) **313**, 341, 353 f/n, 370
Bevis, J. (1695–1771) 3, 53, **313–14**, 368, 373
Bigourdan, M. G. (1851–1932) 102, 116, 149 and f/n, 163, 307, 318
Bion 318
Blaauw & Morgan 161
Bliss, N. (1700–64) 333
Bode, J. E. (1747–1826) 5, 51, Ch. 4. *passim*, 310, **314–15**, 328, 330, 356, 361, 370
Boggis, N. W. 246
Bok, B. J. 74, 93
Bond, G. P. (1826–65) 50, 130, 159, **315**, 330
Bond, W. C. (1789–1859) **315**
Borda, 338

Boscovich, Abbe R. G. (1711–87) 361
Boulliau, I. (Bullialdus) (1605–94) **127**, 129
Bowen, I. S. 18
Bradley, J. (1693–1762) 313
Braham, P. *160*
Brahe, Tycho (1546–1601) 1, 155, 165
Brodie, F. (1823–96) 78, 81, 82, 240, 287
Burbridge, E. M. & G. R. 208, 272
Burbridge, G. R. & Hoyle, F. 248
Burbridge, E. M. 213
Burckhardt, J. C. (1773–1825) 370
Burnham, S. W. (1838–1921) 169
Burton 146

Cacciatore 59
Calver, G. 311
Campbell & Moore 160
Carrington, R. C. (1826–75) 168
Cassini (I), G. D. (1625–1712) 180, 181, 310, **316**, 319, 327, 338
Cassini (II) J. (1677–1756) 181, 310, **316**, 318–19, 330
Cassini (III) C. F. (1714–84) **316**
Cassini (IV) J. D. (1748–1825) **316–17**, 335, 363, 370
Catchpole, Feast & Menzies 221
Chacornac, J. (1823–73) 245
Chambers, G. F. 52, *95*, 96, 97, 340–1
Clairaut, A. C. (1713–65) 329, 344, 350
Clark, Alvan (1832–97) 159, 313, 326
Clerke, A. M. (1842–1907) 111, 212, 214, 282
Clüver, 370
Cocke, Disney & Taylor 55
Common, A. A. (1841–1903) 67, 160, 311
Condorcet, Marquis de (1743–94) 364
Coolidge, S. 315
Courtes, G. 161
Crossley 311
Cuffey, J. 116, 172, 188, 225

Curtis, H. D. (1872–1942) 52, 285, 305, 311
Cysatus (1588–1675) 155

D'Aguesseau 340
D'Alembert, J. (1717–83) 367
D'Angos, Chevalier J. A. 70, **317**, 362–3
Danber 208
Darquier, A. (1718–1802) 25, 196, 307, **317**
D'Arrest, H. L. (1822–75) 16 f/n, Ch. 4. *passim*, 311, 312, **317–18**
De Chéseaux, P. L. (1718–51) 3, 51, Ch. 4. *passim*, 307, **318–19**, 352
De Crouzas 307
De Fleurieu 338
De la Crenne 338
Delambre, J. B. J. (1749–1822) 335, 337, 344, 354 f/n
Delisle, Claude 319
Deslsle, Guillaume (1675–1726) 319
Delisle, N. L. (1688–1768) 4, 318, **319**, 329, 330, 344, *346*, 347–50, 370
Delisle, Simon 319
De Mairan, J.-J. D. (1678–1771) *156* and f/n, 157, 163, 309, **331–2**, 338, 357–8
Denning, W. F. 168
Derham, Rev. Wm (1657–1735) 3, 71, 80, 129, 151, 307, 309, 318, **320**, 327, 352
De Rheita, A. M. S. (1597–1660) 169
Dewhirst, D. W. 57
Dollond, J. (1706–61) 357
Dollond, P. (1730–1820) 371
Doublet, E. 367
Draper, H. (1837–82) 160
Draper, J. W. (1811–82) 315
Dreyer, J. L. E. (1852–1926) 5, 12 f/n, 14, Ch. 4 *passim*, 311, 318
Duncan J. C. (1882–1967) 96, 198
Dunlop, J. 23

Encke, J. F. (1791–1865) 315, 317, 335, 362, 370
Espin, Rev. T. E. 52, 146, 370

Fath, 238

II. *Object index*

Under the entry 'Messier Objects', references in **bold** type apply in each case to the principal description in Chapter 4, the remainder cover additional references elsewhere. All other entries concern individual objects not included in Messier's catalogue.

III. *Subject index*

Numbers in *italics* refer to illustrations or their captions.